Flow Visualization: Experiments and Techniques

Flow Visualization: Experiments and Techniques

Guest Editors

Mingming Ge
Guangjian Zhang
Xin-Lei Zhang

MDPI

Basel • Beijing • Wuhan • Barcelona • Belgrade • Novi Sad • Cluj • Manchester

Guest Editors

Mingming Ge
Guangdong Provincial Key Laboratory IRADS and FST
Beijing Normal-Hong Kong Baptist University
Zhuhai
China

Guangjian Zhang
Research Center of Fluid Machinery Engineering and Technology
Jiangsu University
Zhenjiang
China

Xin-Lei Zhang
The State Key Laboratory of Nonlinear Mechanics
Chinese Academy of Sciences
Beijing
China

Editorial Office

MDPI AG
Grosspeteranlage 5
4052 Basel, Switzerland

This is a reprint of the Special Issue, published open access by the journal *Fluids* (ISSN 2311-5521), freely accessible at: https://www.mdpi.com/journal/fluids/special_issues/H56R7MTMK5.

For citation purposes, cite each article independently as indicated on the article page online and as indicated below:

Lastname, A.A.; Lastname, B.B. Article Title. *Journal Name* **Year**, *Volume Number*, Page Range.

ISBN 978-3-7258-3536-2 (Hbk)
ISBN 978-3-7258-3535-5 (PDF)
https://doi.org/10.3390/books978-3-7258-3535-5

Contents

About the Editors

Mingming Ge

Mingming Ge is Assistant Professor at UIC, Visiting Fellow at Macao University of Science and Technology. Research interests include green technology and sustainable energy, experimental fluid mechanics, microfluidics, and cavitation.

Guangjian Zhang

Guangjian Zhang is Associate Professor at the Research Center of Fluid Machinery Engineering and Technology, Jiangsu University. Research interests include experimental fluid mechanics and cavitation.

Xin-Lei Zhang

Xinlei Zhang is Researcher at The State Key Laboratory of Nonlinear Mechanics, Institute of Mechanics, Chinese Academy of Sciences. Research interests include fluid mechanics and AI-driven modeling.

Preface

As we strive to develop more efficient engines, design better medical devices, and understand natural phenomena such as weather patterns, the need for accurate and detailed flow visualization becomes ever more pressing. This motivated this Special Issue, which is addressed to a diverse audience. For researchers, it serves as a valuable resource for staying updated on the latest research findings and techniques. Engineers can benefit from the practical applications of flow visualization presented here, which can further aid in the design and optimization of their products. Students, on the other hand, can use this as a learning tool to gain a deeper understanding of fluid dynamics and experimental techniques.

This Special Issue features contributions from an international cohort of authors in the field of flow visualization. These experts have dedicated their time and expertise to sharing their latest research findings, and we are truly grateful for their contributions. I would also like to express my heartfelt gratitude to my wife, Katherine. On countless evenings and weekends, as I buried myself in research papers and editorial tasks, she provided understanding, encouragement, and unwavering support. Her love and patience have been a constant source of strength, allowing me to focus on bringing this Special Issue to fruition.

Finally, I sincerely thank the peer reviewers for their constructive feedback, the institutions that facilitated this collaboration, and the editorial team, especially Cori Jia, for her professional guidance during the entire editorial process, making the publication of this Special Issue in book format a reality. Without her, this endeavor would have been far more challenging.

Mingming Ge, Guangjian Zhang, and Xin-Lei Zhang
Guest Editors

fluids

MDPI

Editorial

Recent Developments and Future Directions in Flow Visualization: Experiments and Techniques

Mingming Ge [1,*], Guangjian Zhang [2,*] and Xinlei Zhang [3]

1 Faculty of Science and Technology, Beijing Normal University-Hong Kong Baptist University United International College, Zhuhai 519087, China
2 Research Center of Fluid Machinery Engineering and Technology, Jiangsu University, Zhenjiang 212013, China
3 The State Key Laboratory of Nonlinear Mechanics, Institute of Mechanics, Chinese Academy of Sciences, Beijing 100190, China; zhangxinlei@imech.ac.cn
* Correspondence: mingmingge@uic.edu.cn (M.G.); eddyzhang@ujs.edu.cn (G.Z.)

Flow visualization has long been a critical tool for understanding complex fluid dynamics in both natural and engineered systems [1–3]. Over the past few decades, advancements in experimental techniques, imaging technologies, and computational methods have significantly enhanced our ability to observe, quantify, and analyze fluid flows. From classic methods such as dye injection and particle image velocimetry (PIV) to more cutting-edge approaches like high-speed imaging, laser-induced fluorescence, and digital holography, the field has evolved to meet the demands of increasingly complex research questions [4–12].

Despite these impressive advancements, there remain significant gaps in knowledge, particularly regarding the visualization of turbulent flows, multiphase systems, and the interaction of flows with deformable surfaces such as flexible wings and aquatic vegetation [13,14]. While progress has been made in capturing instantaneous flow fields and high-resolution images, challenges persist in achieving real-time and three-dimensional visualization under challenging environmental conditions (e.g., underwater flows, extreme turbulence, or highly unsteady flows) [15,16]. Moreover, there is a need for improved techniques that can seamlessly combine visual data with quantitative analysis to bridge the gap between theory and experiment.

This Special Issue on *Flow Visualization: Experiments and Techniques* serves to address some of these knowledge gaps by presenting a collection of papers that highlight the latest experimental advancements in flow visualization. The articles within this Special Issue explore novel imaging systems, innovative experimental setups, and advanced data processing techniques that enable a more accurate and detailed visualization of fluid phenomena. By focusing on areas such as turbulent boundary layers, vortex dynamics, and the flow–structure interactions in natural environments, this Special Issue brings together a diverse set of studies that push the boundaries of current flow visualization techniques [17–19].

The field of flow visualization is rapidly evolving, with new techniques and applications emerging that promise to deepen our understanding of fluid dynamics [20–24]. The advancements presented in recent research underscore significant contributions across diverse applications and methodologies, including PIV, Laser Doppler Velocimetry (LDV), and Particle-Tracking Velocimetry (PTV) [25–36]. This Editorial aims to provide a brief overview of these developments, identify the gaps in the knowledge, and discuss how the current Special Issue addresses those gaps, with a focus on future research directions. Nobes et al. [25] propose a novel combined vortex detection algorithm (CVD) designed

Received: 30 December 2024
Revised: 15 January 2025
Accepted: 20 January 2025
Published: 22 January 2025

Citation: Ge, M.; Zhang, G.; Zhang, X. Recent Developments and Future Directions in Flow Visualization: Experiments and Techniques. *Fluids* **2025**, *10*, 23. https://doi.org/10.3390/fluids10020023

to enhance the accuracy and reliability of vortex identification and analysis in oscillating airfoil wakes. The CVD method improves upon traditional techniques, such as the Q-criterion, by offering a more robust framework for detecting and quantifying vortex structures, particularly in complex, time-varying flows. The algorithm integrates multiple flow-field metrics to identify vortex boundaries and evaluate key flow parameters with greater precision. This advancement is particularly beneficial in experimental fluid dynamics, where accurate vortex characterization is crucial for understanding wake dynamics and optimizing aerodynamic performance. The study highlights the effectiveness of CVD in providing clearer insights into vortex behaviors and offers a promising tool for future research in vortex dynamics and turbulence.

Elaswad et al. [26] highlight key advancements in visualizing flow behaviors in complex geometries, such as toroidal systems, through methods like PIV and LDV. The paper discusses the influence of parameters like Reynolds and Dean numbers on secondary flow phenomena, particularly in curved or rotating conduits, and how these can be optimized to enhance the performance of fluid dynamic systems like the PIVG (Particle Image Velocimetry Gyroscope). Through both numerical simulations and experimental validations, the authors examine the impact of angular acceleration on fluid behavior, revealing insights into the development of primary and secondary flow components. They further emphasize the importance of precise pressure and velocity field measurements for improving the accuracy and reliability of flow measurements, which are critical for applications in fluid mechanics, engineering, and sensor technologies. The study provides a valuable framework for optimizing fluid dynamics in rotational systems, contributing to a better design and understanding of experimental fluid flow investigations.

Shirinzad et al. [27] present a comprehensive study on the enhancement of PIV software, focusing on an improved algorithm optimized for Central Processing Units (CPUs) to facilitate accessible and efficient flow analysis. By leveraging Python's versatility, the authors developed an algorithm that precisely captures time-averaged flow, velocity fields, and vortices, offering an alternative to GPU-optimized PIV software. The algorithm was validated through rigorous testing on various platforms, including supercomputing clusters and Google Colaboratory, demonstrating its robustness in experimental flow studies. The primary contribution of this work lies in providing an open-source, CPU-based solution for real-time and offline PIV analysis, expanding its applicability to a wider range of research environments, particularly for those without access to specialized GPU hardware.

Colli et al. [28] investigate the effects of parallel blade–vortex interactions (BVIs) on the aerodynamic performance of an airfoil, with a particular focus on its relationship to blade stall. Their study utilizes wind tunnel experiments to reproduce parallel BVI on an NACA 23012 blade model at a Reynolds number of 300,000. The vortex was generated by impulsively pitching a second airfoil upstream, and the aerodynamic loads acting on the blade were measured using unsteady Kulite pressure transducers. Notably, the authors employed PIV techniques to visualize and analyze the flow field over the blade model. The paper contributes significantly to the field of flow visualization, particularly by applying PIV to study flow dynamics in the context of parallel BVI, with additional novelty in the investigation of oscillating sinusoidal motion of the blade. This work exemplifies how advanced experimental techniques like PIV can offer deep insights into complex aerodynamic phenomena.

Hassan et al. [29] investigate the interaction between flow dynamics and acoustic phenomena in rectangular deep cavities, focusing on passive control strategies. Using advanced flow visualization techniques, the authors analyze the coupling between aerodynamic forces and the resulting acoustic fields, particularly in the context of cavities, which are known to generate strong noise due to vortex shedding and flow instabilities. Their experiments utilize PIV and other visual techniques to map flow structures and identify the dynamics that contribute to aeroacoustic noise. By applying passive flow control methods, the authors demonstrate how altering flow characteristics can reduce unwanted sound emissions, offering valuable insights into noise mitigation strategies in engineering applications such as aerospace and automotive design. This study significantly contributes to the field of flow visualization by showing how experimental techniques can help visualize and control complex flow–acoustic interactions.

Mehta et al. [30] present a detailed experimental and computational study focused on the flow characteristics and acoustic behaviors of supersonic rectangular impinging jets. Their research specifically examines how the orientation of the jet—whether aligned along the major or minor axis—affects both the flow dynamics and the noise produced. Through different flow visualization techniques, including PIV and Schlieren, the authors capture the complex flow structures that arise in each configuration. Their results highlight how these different orientations lead to variations in pressure fields, shear layers, and jet impingement patterns, directly influencing noise emissions. This work contributes to the broader field of flow visualization by using cutting-edge experimental methods to explore the relationship between flow configurations and acoustics, providing valuable insights into optimizing jet design for noise control in high-speed fluid dynamics, especially in aerospace engineering.

Xi et al. [31] focus on the experimental visualization and analysis of airflow patterns around different types of face coverings. The authors utilize a variety of visualization methods, including the Schlieren optical system, laser/LED particle imaging system, thermal imaging camera, and vapor-SarGel system, to study how mask leakage and the resulting flow patterns affect the interpersonal transmission of airborne particles. Through their experiments, they examine various face masks, quantifying the leakage flows and their potential implications for reducing disease transmission. The findings highlight the critical role of mask design and fit in minimizing leakage, contributing valuable insights to public health and safety practices, particularly in the context of respiratory disease prevention.

Prisăcariu et al. [32] present a novel application of the quantitative color Schlieren technique to analyze the gasodynamic parameters of an H_2O_2 exhaust jet in air. By leveraging a calibrated color filter within a Z-type Schlieren setup, the study achieves the extraction of density and temperature gradients of the turbulent jet, produced by a micro-thruster designed for small satellites. The authors compare their experimental results with CFD simulations to validate the Schlieren method's measurement accuracy. Despite challenges such as calibration errors and the reduced accuracy in 3D flows compared to 2D cases, this work advances the Schlieren technique's capability to provide quantitative insights into complex jet dynamics, offering potential applications in aerospace engineering and combustion analysis.

Fan et al. [33] conduct an experimental study to investigate the flow dynamics of oil–water mixtures downstream of a restriction in a horizontal pipe. They employ two advanced techniques for flow visualization—a high-speed camera and an Electrical Capacitance Volume Tomography (ECVT) system, the latter of which is a non-intrusive tool for measuring the volumetric phase distribution at pipe cross-sections. The study examines how varying valve openings, flow rates, and water cuts affect the flow pattern and pressure drop. The results show a significant correlation between the oil–water flow pattern and the pressure gradient, with variations depending on the valve openings and water cuts. In particular, smaller valve openings lead to more complex flow behaviors, including oil-in-water dispersions. The findings provide valuable insights into how flow conditions influence both flow patterns and pressure drops in oil–water mixtures, with implications for optimizing flow management in industrial applications.

Liu et al. [34] explore the turbulent flow dynamics within a gas-stirred cylindrical water tank, with a particular focus on ladle metallurgy, which is critical in steelmaking. The study investigates how turbulence affects key steelmaking processes, such as the mixing and distribution of additives and the transport of inclusions. By employing an advanced PTV system, specifically the "Shake-the-Box" method, the authors simulate the stirred flow field in a water ladle model, using compressed air injections at the tank's bottom to actively stir the flow. This method aims to optimize ladle design and improve the precision of process control strategies in steelmaking, thus enhancing overall efficiency and steel quality. Additionally, the paper addresses the challenge of mitigating distortion in particle images caused by the cylindrical plexiglass walls of the model, thereby improving the accuracy of flow field measurements.

Takeyama et al. [35] introduce a novel technique aimed at enhancing the acquisition of velocity vectors in fluid dynamics experiments using 3D3C Rainbow PTV. By integrating an innovative in-picture tracking method, the authors significantly improve the number of velocity vectors that can be accurately captured in complex, three-dimensional flows. This advancement addresses a key challenge in flow visualization, particularly in high-dimensional flow fields where traditional PTV methods struggle with limitations in data acquisition and resolution. The paper demonstrates how this improved method offers more precise and comprehensive flow data, facilitating the detailed analysis of fluid behavior in both academic research and industrial applications. This contribution to flow visualization enhances the capability of PTV as a tool for studying intricate flow phenomena and offers new avenues for future experimental techniques in fluid mechanics.

Riazanov et al. [36] present an in-depth study focused on the flow characteristics of coolant in a fuel rod bundle, specifically within the context of small modular reactors (SMRs). They employ advanced flow visualization techniques to capture the complex behavior of coolant as it interacts with the fuel rods, using CFD simulations alongside experimental methods. The study provides a detailed examination of flow patterns, temperature distributions, and pressure drops, which are crucial for the safe and efficient operation of nuclear reactors. By using sophisticated visualization tools, such as PIV and flow visualization in transparent models, the authors effectively highlight the intricacies of coolant behavior in a reactor environment. This work significantly contributes to the field of flow visualization by offering insights into the optimization of coolant flow management in nuclear reactors, an area where experimental techniques are key to ensuring safety and performance.

Looking forward, the future of flow visualization research will likely be driven by continued innovation in imaging technologies, computational methods, and interdisciplinary collaboration. As we move toward more sophisticated and dynamic experimental setups, it will be essential to focus on real-time data acquisition, multi-modal visualization, and developing hybrid techniques that combine traditional and modern approaches. Investigating the effects of flow visualization on different scales, from microfluidics to large-scale industrial applications, is also needed to provide a comprehensive understanding of fluid behavior [37–40]. Moreover, a deeper understanding of the fundamental fluid mechanics behind observed flow patterns will be critical for applying flow visualization to emerging environmental protection, sustainable energy, and biomechanics challenges.

It is worth noting that this Special Issue, entitled *Flow Visualization: Experiments and Techniques, 2nd Edition*, has launched in *Fluids*. One of the key themes addressed in this new Special Issue is the integration of machine learning and artificial intelligence into flow visualization workflows. These technologies offer new opportunities for automating data analysis, enhancing the resolution of flow features, and predicting flow behavior in previously intractable systems. This fusion of experimental techniques with computational power is set to revolutionize the field by enabling a more efficient and insightful analysis of complex fluid systems.

In conclusion, while substantial progress has been made, the field of flow visualization continues to evolve rapidly. Researchers are urged to explore new experimental techniques, integrate advanced data processing methods, and collaborate across disciplines to address the remaining gaps in the knowledge. This Special Issue provides a glimpse into the exciting future of flow visualization, where innovations in both experiments and techniques will offer new insights into the ever-complex world of fluid dynamics.

Conflicts of Interest: The authors declare no conflict of interest.

References

1. Weiskopf, D.; Gordon, E. Overview of flow visualization. In *Visualization Handbook*; Elsevier: Amsterdam, The Netherlands, 2005; pp. 261–278.
2. Smits, A.J. *Flow Visualization: Techniques and Examples*; World Scientific: Singapore, 2012.
3. Delmarcelle, T.; Lambertus, H. A unified framework for flow visualization. In *Computer Visualization*; CRC Press: Boca Raton, FL, USA, 2023; pp. 129–170.
4. Di Bella, B.; Khatamifar, M.; Lin, W. Experimental study of flow visualisation using fluorescent dye. *Flow Meas. Instrum.* **2022**, *87*, 102231. [CrossRef]
5. Raffel, M.; Christian, E.W.; Fulvio, S.; Christian, J.K.; Steve, T.W.; Jürgen, K. *Particle Image Velocimetry: A Practical Guide*; Springer: Berlin/Heidelberg, Germany, 2018.
6 Westerweel, J.; Gerrit, E.E.; Ronald, J.A. Particle image velocimetry for complex and turbulent flows. *Annu. Rev. Fluid Mech.* **2013**, *45*, 409–436. [CrossRef]
7. Mueller, T.J.; Brown, F.N.M. Flow visualization by direct injection. In *Fluid Mechanics Measurements*; Routledge: London, UK, 2017; pp. 367–450.
8. Etminan, A.; Muzychka, Y.S.; Pope, K.; Nyantekyi-Kwakye, B. Flow visualization: State-of-the-art development of micro-particle image velocimetry. *Meas. Sci. Technol.* **2022**, *33*, 092002. [CrossRef]
9. Lindken, R.; Rossi, M.; Große, S.; Westerweel, J. Micro-particle image velocimetry (μPIV): Recent developments, applications, and guidelines. *Lab Chip* **2009**, *9*, 2551–2567. [CrossRef] [PubMed]
10. Ge, M.; Petkovšek, M.; Zhang, G.; Jacobs, D.; Coutier-Delgosha, O. Cavitation dynamics and thermodynamic effects at elevated temperatures in a small Venturi channel. *Int. J. Heat Mass Transf.* **2021**, *170*, 120970. [CrossRef]
11. Ge, M.; Zhang, G.; Petkovšek, M.; Long, K.; Coutier-Delgosha, O. Intensity and regimes changing of hydrodynamic cavitation considering temperature effects. *J. Clean. Prod.* **2022**, *338*, 130470. [CrossRef]
12. Merola, F.; Memmolo, P.; Miccio, L.; Savoia, R.; Mugnano, M.; Fontana, A.; D'ippolito, G.; Sardo, A.; Iolascon, A.; Gambale, A.; et al. Tomographic flow cytometry by digital holography. *Light Sci. Appl.* **2017**, *6*, e16241. [CrossRef]
13. Meyers, J.; Charles, M. Flow visualization using momentum and energy transport tubes and applications to turbulent flow in wind farms. *J. Fluid Mech.* **2013**, *715*, 335–358. [CrossRef]
14. Aliseda, A.; Theodore, J.H. X-ray flow visualization in multiphase flows. *Annu. Rev. Fluid Mech.* **2021**, *53*, 543–567. [CrossRef]
15. Okhravi, S.; Radoslav, S.; Yvetta, V. Flow resistance in lowland rivers impacted by distributed aquatic vegetation. *Water Resour. Manag.* **2022**, *36*, 2257–2273. [CrossRef]
16. Laramee, R.S.; Bruno, J.; Helwig, H. Image space based visualization of unsteady flow on surfaces. In Proceedings of the IEEE Visualization 2003 (VIS 2003), Seattle, WA, USA, 19–24 October 2003; pp. 131–138.
17. Jiang, X.Y.; Lee, C.B.; Smith, C.R.; Chen, J.W.; Linden, P.F. Experimental study on low-speed streaks in a turbulent boundary layer at low Reynolds number. *J. Fluid Mech.* **2020**, *903*, A6. [CrossRef]
18. Krake, T.; Stefan, R.; Marcel, H.; Bernd, E.; Daniel, W. Visualization and selection of dynamic mode decomposition components for unsteady flow. *Vis. Inform.* **2021**, *5*, 15–27. [CrossRef]
19. Zhang, B.; Huang, Y.; Zhu, C. Flow–structure interaction mechanism under Coriolis Conditions. *J. Eng. Mech.* **2021**, *147*, 04021015. [CrossRef]
20. Perissinotto, R.M.; Verde, W.M.; Biazussi, J.L.; Bulgarelli, N.A.V.; Fonseca, W.D.P.; de Castro, M.S.; de Moraes Franklin, E.; Bannwart, A.C. Flow visualization in centrifugal pumps: A review of methods and experimental studies. *J. Pet. Sci. Eng.* **2021**, *203*, 108582. [CrossRef]
21. Ge, M.; Apte, D.; Wang, C.; Zhang, G.; Zhang, X.; Coutier-Delgosha, O. Slip velocity and field information of two-phase cavitating flows. *Phys. Fluids* **2024**, *36*, 093301. [CrossRef]

22. Ge, M.; Sun, C.; Zhang, G.; Coutier-Delgosha, O.; Fan, D. Combined suppression effects on hydrodynamic cavitation performance in Venturi-type reactor for process intensification. *Ultrason. Sonochem.* **2022**, *86*, 106035. [CrossRef]
23. Carvalho, V.; Maia, I.; Souza, A.; Ribeiro, J.; Costa, P.; Puga, H.; Teixeira, S.; Lima, R.A. In vitro biomodels in stenotic arteries to perform blood analogues flow visualizations and measurements: A review. *Open Biomed. Eng. J.* **2020**, *14*, 87–102. [CrossRef]
24. Tan, C.; Murai, Y.; Liu, W.; Tasaka, Y.; Dong, F.; Takeda, Y. Ultrasonic Doppler Technique for Application to Multiphase Flows: A Review. *Int. J. Multiph. Flow* **2021**, *144*, 103811. [CrossRef]
25. Bussière, M.; Bessa, G.M.; Koch, C.R.; Nobes, D.S. Application of a Combinatorial Vortex Detection Algorithm on 2 Component 2 Dimensional Particle Image Velocimetry Data to Characterize the Wake of an Oscillating Wing. *Fluids* **2024**, *9*, 53. [CrossRef]
26. Elaswad, R.; El-Sheimy, N.; Mohamad, A. Analyzing the Influence of Dean Number on an Accelerated Toroidal: Insights from Particle Imaging Velocimetry Gyroscope (PIVG). *Fluids* **2024**, *9*, 103. [CrossRef]
27. Shirinzad, A.; Jaber, K.; Xu, K.; Sullivan, P.E. An Enhanced Python-Based Open-Source Particle Image Velocimetry Software for Use with Central Processing Units. *Fluids* **2023**, *8*, 285. [CrossRef]
28. Colli, A.; Zanotti, A.; Gibertini, G. Wind Tunnel Experiments on Parallel Blade–Vortex Interaction with Static and Oscillating Airfoil. *Fluids* **2024**, *9*, 111. [CrossRef]
29. Jabado, A.H.; El Hassan, M.; Hammoud, A.; Sakout, A.; Assoum, H.H. Aeroacoustic Coupling in Rectangular Deep Cavities: Passive Control and Flow Dynamics. *Fluids* **2024**, *9*, 187. [CrossRef]
30. Mehta, Y.; Bhargav, V.N.; Kumar, R. Flowfield and Noise Dynamics of Supersonic Rectangular Impinging Jets: Major versus Minor Axis Orientations. *Fluids* **2024**, *9*, 169. [CrossRef]
31. Si, X.; Xi, J.S.; Talaat, M.; Park, J.H.; Nagarajan, R.; Rein, M.; Xi, J. Visualization and quantification of facemask leakage flows and interpersonal transmission with varying face coverings. *Fluids* **2024**, *9*, 166. [CrossRef]
32. Prisăcariu, E.; Prisecaru, T. Quantitative Color Schlieren for an H_2O_2 Exhaust Jet Developing in Air. *Fluids* **2024**, *9*, 19. [CrossRef]
33. Zhou, D.; Karatayev, K.; Fan, Y.; Straiton, B.; Marashdeh, Q. Experimental Study of Oil–Water Flow Downstream of a Restriction in a Horizontal Pipe. *Fluids* **2024**, *9*, 146. [CrossRef]
34. Jojo-Cunningham, Y.; Guo, X.; Zhou, C.; Liu, Y. Volumetric Flow Field inside a Gas Stirred Cylindrical Water Tank. *Fluids* **2023**, *9*, 11. [CrossRef]
35. Takeyama, M.; Fujiwara, K.; Hattori, Y. Improvement in the Number of Velocity Vector Acquisitions Using an In-Picture Tracking Method for 3D3C Rainbow Particle Tracking Velocimetry. *Fluids* **2024**, *9*, 226. [CrossRef]
36. Dmitriev, S.; Demkina, T.; Dobrov, A.; Doronkov, D.; Kuritsin, D.; Nikolaev, D.; Pronin, A.; Riazanov, A.; Solntsev, D. Modeling and Visualization of Coolant Flow in a Fuel Rod Bundle of a Small Modular Reactor. *Fluids* **2024**, *9*, 235. [CrossRef]
37. Liu, C.; Jiang, R.; Wei, D.; Yang, C.; Li, Y.; Wang, F.; Yuan, X. Deep learning approaches in flow visualization. *Adv. Aerodyn.* **2022**, *4*, 17. [CrossRef]
38. Ge, M.; Manikkam, P.; Ghossein, J.; Subramanian, R.K.; Coutier-Delgosha, O.; Zhang, G. Dynamic mode decomposition to classify cavitating flow regimes induced by thermodynamic effects. *Energy* **2022**, *254*, 124426. [CrossRef]
39. Dhruv, A.; Ge, M.; Coutier-Delgosha, O. Numerical investigation of a cavitating nozzle for jetting and rock erosion based on different turbulence models. *Geoenergy Sci. Eng.* **2023**, *231*, 212300.
40. Chatzimparmpas, A.; Martins, R.M.; Jusufi, I.; Kucher, K.; Rossi, F.; Kerren, A. The state of the art in enhancing trust in machine learning models with the use of visualizations. *Comput. Graph. Forum* **2020**, *39*, 713–756. [CrossRef]

fluids

MDPI

Article

Modeling and Visualization of Coolant Flow in a Fuel Rod Bundle of a Small Modular Reactor

Sergei Dmitriev, Tatiyana Demkina, Aleksandr Dobrov, Denis Doronkov, Daniil Kuritsin, Danil Nikolaev, Alexey Pronin, Anton Riazanov * and Dmitriy Solntsev

The Department of the Nuclear and Thermal Power Stations, Institute of Nuclear Power Engineeniring and Applied Physics n.a. Academican F.M. Mitenkov, Nizhny Novgorod State Technical University n.a. R.E. Alekseev, 603950 Nizhny Novgorod, Russia; dmitriev@nntu.ru (S.D.); tatyana07041998@gmail.com (T.D.); dobrov@nntu.ru (A.D.); nevid000@mail.ru (D.D.); d.d.kuritsin@yandex.ru (D.K.); nikolaevdanil_ft@mail.ru (D.N.); proninaleksei@mail.ru (A.P.); solntsev@nntu.ru (D.S.)

* Correspondence: riazanov.av@nntu.ru

Abstract: This article presents the results of an experimental study of the coolant flow in a fuel rod bundle of a nuclear reactor fuel assembly of a small modular reactor for a small ground-based nuclear power plant. The aim of the work is to experimentally determine the hydrodynamic characteristics of the coolant flow in a fuel rod bundle of a fuel assembly. For this purpose, experimental studies were conducted in an aerodynamic model that included simulators of fuel elements, burnable absorber rods, spacer grids, a central displacer, and stiffening corners. During the experiments, the water coolant flow was modeled using airflow based on the theory of hydrodynamic similarity. The studies were conducted using the pneumometric method and the contrast agent injection method. The flow structure was visualized by contour plots of axial and tangential velocity, as well as the distribution of the contrast agent. During the experiments, the features of the axial flow were identified, and the structure of the cross-flows of the coolant was determined. The database obtained during the experiments can be used to validate CFD programs, refine the methods of thermal-hydraulic calculation of nuclear reactor cores, and also to justify the design of fuel assemblies.

Keywords: coolant; hydrodynamic; fuel assembly; reactor core; small modular reactor

Citation: Dmitriev, S.; Demkina, T.; Dobrov, A.; Doronkov, D.; Kuritsin, D.; Nikolaev, D.; Pronin, A.; Riazanov, A.; Solntsev, D. Modeling and Visualization of Coolant Flow in a Fuel Rod Bundle of a Small Modular Reactor. *Fluids* **2024**, *9*, 235. https://doi.org/10.3390/fluids9100235

Academic Editors: Mingming Ge and Tomoaki Kunugi

Received: 4 September 2024
Revised: 4 October 2024
Accepted: 5 October 2024
Published: 8 October 2024

1. Introduction

Rosatom, the State Atomic Energy Corporation Rosatom, is currently in the final stages of developing a modernized ground-based nuclear power plant that will generate electricity for remote and decentralized areas. For the modernized power unit, specialists from Rosatom have developed the RITM-200S small modular reactor (Nizhniy Novgorod, Russia). One of the main components of the reactor is the core. It is subject to increased demands for durability, energy efficiency, reliability, and safety. The new core uses fuel assemblies with an increased active part height of 1650 mm to extend the service life. It also uses fuel rods with a thicker cladding made from a corrosion-resistant alloy [1–4]. New technical solutions require experimental and computational justification. One of the stages in the design process for the core includes an experimental study of the hydrodynamics of the coolant inside the fuel rod bundle within a fuel assembly.

The main results of the thermophysical experiments were obtained by scientific teams that are part of the State Atomic Energy Corporation Rosatom and are published in [5]. These studies present the results of both experimental and computational research on critical heat flows in RITM-200 reactor fuel assemblies used in ground-based small modular nuclear power plants. All experiments were conducted under normal operating conditions of the reactor.

Also, complex analytical work is being conducted in scientific organizations to analyze the design and operational features of modernized small modular reactors. The works in [6–9] present the results of analyzing the design of these reactors, identifying their weaknesses, analyzing their operating modes, and determining paths for their further modernization and development. It also includes recommendations for improving the design of the reactor core.

Scientists are mainly involved in the computational and experimental modeling of coolant hydrodynamics in PWR reactor cores. The main goal of these studies is to determine the influence of mixing grids and fuel bundle geometry on coolant flow processes. However, the influence of the geometry of mixing vanes on flow mixing and vortex formation in fuel bundles has not been thoroughly investigated, making it difficult to accurately assess the impact of these design features on flow patterns due to the complexity of processes in rod bundles, including the presence of secondary flows and turbulence anisotropy [10–14].

To identify the main patterns of coolant movement behind mixing grids, various approaches are used. These include the study of transverse mass transfer using tracer methods and laser Doppler technology, as well as the measurement of temperature fields in the coolant flow to gain a better understanding of dynamics [15,16].

In these studies, the focus is on the coolant flow within the fuel assembly, specifically the area behind the mixing grids. This area is represented by normal cells, formed by the arrangement of fuel rods, without any more complex cells such as peripheral regions, central displacer regions, or corner regions.

Accurate knowledge of the hydrodynamics of the coolant can help to improve the accuracy of results from thermohydraulic calculations for cores when justifying their reliability. Therefore, the aim of this work is to experimentally determine hydrodynamic characteristics of flow in fuel element bundles [17–26].

The results of experimental studies of the coolant hydrodynamics at the inlet and outlet, as well as in the fuel rod bundle of the fuel assembly for the core of the ground-based small modular reactor, are not available in the open press, which also confirms the relevance and scientific novelty of the results presented in this article.

2. Research Facility and Experimental Model

The flow of water coolant was simulated using airflow, based on the theory of hydrodynamic similarity, with the same Reynolds number in the range of 90,000 to 100,000 relative to the natural reactor unit. The open-circuit aerodynamic research facility consists of an airflow injection system and an experimental model equipped with a measurement system [27–31].

The experimental model is designed as a transparent hexagonal channel whose walls imitate the cover of a standard fuel assembly. The model's fuel bundle replicates the geometry of a fragment of a standard fuel assembly's fuel rod bundle and consists of fuel rod simulators and burnable absorber rods. In the fuel bundle, there are three spacer grid simulators installed. These simulators have a complex design, consisting of plates of different thicknesses and shapes. The central displacer simulator is located in the inner shells of the grids. The position of the grids in the model is based on their standard location, which has been enlarged by a scale factor. Six stiffening corners, which connect the grid simulators, are fixed to their outer shell and follow the geometry of the structural element of a standard fuel assembly (Figure 1). The scale factor for all elements in the experimental model is 5.8. By increasing the size of the design compared to the standard fuel assembly, we can reduce the impact of the probe sensors on the flow pattern. This solution also allows us to achieve a Reynolds number of 90,000 during low airflow velocities.

(**a**)

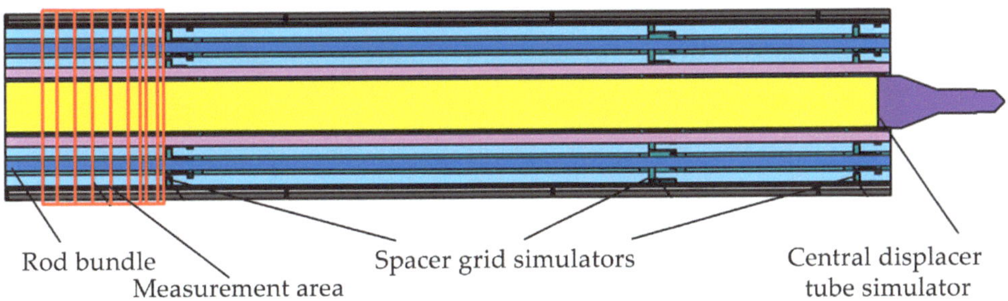

(**b**)

Figure 1. Model of a fuel rod bundle fragment: (**a**) External appearance of the experimental model, (**b**) Arrangement of structural elements in the experimental model channel.

3. The Representativeness of the Study

When performing hydrodynamic similarity, it is possible to choose the optimal design of the experimental model and recalculate the flow characteristics obtained during experiments for the natural conditions of coolant flow. This is achieved by ensuring geometric, kinematic, and dynamic similarity between the model and actual construction.

Geometric similarity is maintained by matching the geometric dimensions of the experimental model and the real object, taking into account scale factors.

The dynamic similarity of the flow of fluid in geometrically similar objects can be observed due to the close values of the Reynolds number. The maximum Reynolds number in the experimental model is 90,000, and in the reference object, it is 100,000.

The kinematic similarity of flows in geometrically similar objects is observed due to the proportionality of dimensionless velocities at the corresponding points of the experimental model and the real object.

The value of the Reynolds number, which was used in the research, corresponds to the range of self-similarity of flow, which allows us to use experimental data to study the flow of water coolant. The start of the self-similar zone is defined by a Reynolds number of 35,000. All measurements were taken at an average air velocity of 36.2 m/s at the entrance to the model with a Reynolds of 90,000

Additionally, the representativeness of the experiments is proved by the agreement between the hydraulic resistance coefficients of the test grids and those of standard design elements, which have identical Reynolds numbers.

4. Experimental Methodology

The hydrodynamic properties of the flow within a fuel bundle were studied using the pneumometry approach and the technique of impurity injection.

A multi-hole pressure probe has been used to measure the velocity vector in an experimental model. The sensitive part of the probe consists of five steel capillaries with a diameter of 0.8 × 0.1 mm, located in two perpendicular diametric planes. The rest of the capillaries are placed inside a steel tube that serves as the probe holder. The central capillary is cut at a 90-degree angle to its axis, while the four lateral capillaries are cut at a 45-degree angle (Figure 2).

Figure 2. Multi-hole pressure probe.

To measure the magnitude and direction of the velocity vector, a pneumatic probe is sequentially positioned at specific points on the experimental fuel assembly model using a coordinate system. In this process, pressure values in each capillary of the probe are recorded using piezoelectric pressure sensors.

Flow velocity components were measured with an accuracy of not more than 7.5% error.

The structure of transverse flows was investigated using the injection method with the addition of propane. This gas was injected at a specific flow rate into areas located in front of the third grate along the flow path. After injection, the distribution of the added substance in the volume of the model was monitored using an infrared gas analyzer. The gas flow rate was maintained with an accuracy of ±0.25%. The impurity concentration levels were determined using a gas analyzer with an error of ±1.5% or lower.

The size of the measuring area is determined by the large number of cells of various shapes and by the geometry of the spacers, which intersect in various ways. This area occupies one-third of the cross-sectional area of the fuel bundle and has a significant number of impurity entry points (Figure 3). The flow velocity components and impurity concentrations were measured across the entire area of interest using a uniform grid of measurement points. Along the length of the rod bundle, measurements were taken at 10 sections behind the simulator of the third spacer grid. The distance between these measurement points increased as the distance from the grid increased (Figure 1). This choice of measurement area size and placement of impurity injection points allows us to characterize the overall flow structure.

The analysis of the coolant flow was conducted based on contour plots of the dimensionless impurity concentration and the flow velocity. The values of the dimensionless axial (W_z/W_{av}) and tangential (W_{xy}/W_{av}) velocities were obtained by dividing their local values at a point by the value of the average flow velocity at the inlet of the experimental model (W_{av}). The values of the dimensionless impurity concentration (C/C_0) are calculated by dividing its local value (C) at a given point by the maximum value of the impurity concentration in the injection region immediately in front of the spacer grid (C_0). To estimate the flow disturbances caused by the structural elements in the experimental model, we used the value L/d_h, where L is the distance from the measurement point to the structure, and d_h is the hydraulic diameter of the model.

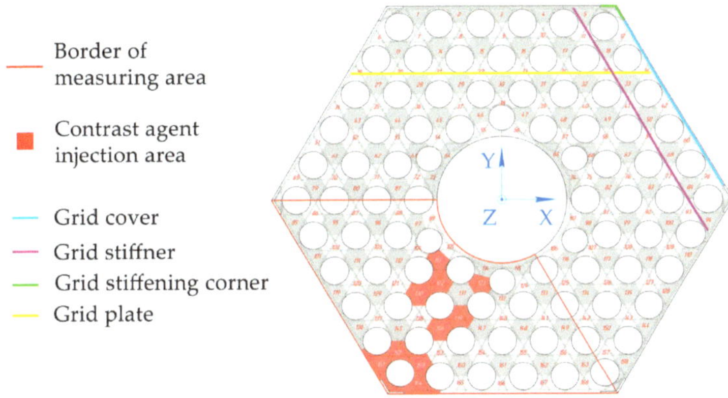

Figure 3. The location of the measuring area and contrast agent injection zones in the cross-section of the experimental model.

5. The Results of the Study on the Hydrodynamics of the Coolant Using the Pneumometry Technique

In the field of regular cells, transverse flows were formed behind the plates and grid stiffeners at the points of their contact with the fuel elements at a distance of $L/d_h \approx 1$ (Figure 4) from the spacer grid. Behind the stiffeners, the dimensionless tangential velocity was 0.17–0.22, and behind the plates, it was two times less, which is due to their smaller thickness. The transverse flow maintained its structure up to a distance of approximately three times the hydraulic diameter ($L/d_h \approx 3.1$) from the plates.

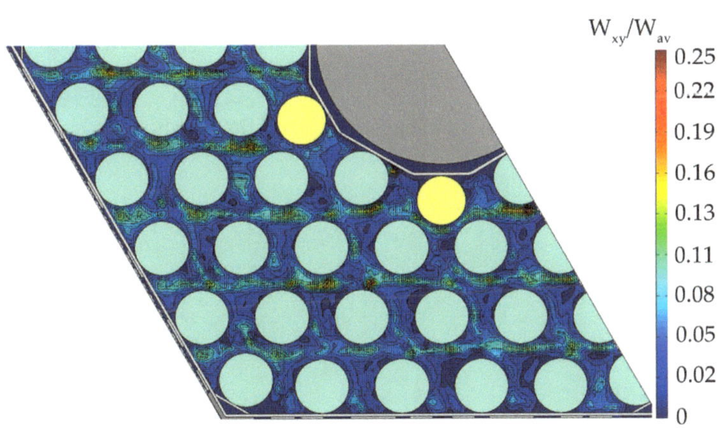

Figure 4. Dimensionless tangential flow velocity at a distance of $L/d_h \approx 1$ from the grid simulator plates.

At the faces of the fuel assembly cover and the stiffening corners, where there were no grid plates, the transverse flow was oriented from regular cells towards the outer edges. However, at the intersection of plates, the direction of transverse flow is reversed. Dimensionless tangential velocities ranged from 0.08 to 0.13 in the first case and 0.11–0.14 in the second case. In the case without plates in the cells, the transverse flow stopped at $L/d_h \approx 7.5$. In the case of the plates intersecting cells, it stopped at a distance of $L/d_h \approx 5$.

Near the central displacer, transverse flow moved across the surface of the displacer in a circular pattern, as well as within a field of regular cells located at a distance of $L/d_h \approx 2$ from the plates. These transverse flows were measured throughout the entire research area. In transverse flows moved around the central displacer, the dimensionless tangential velocity ranged from 0.13 to 0.16. In transverse flows directed into the field of regular cells, the dimensionless tangential velocity ranged from 0.11 to 0.13 (Figure 5).

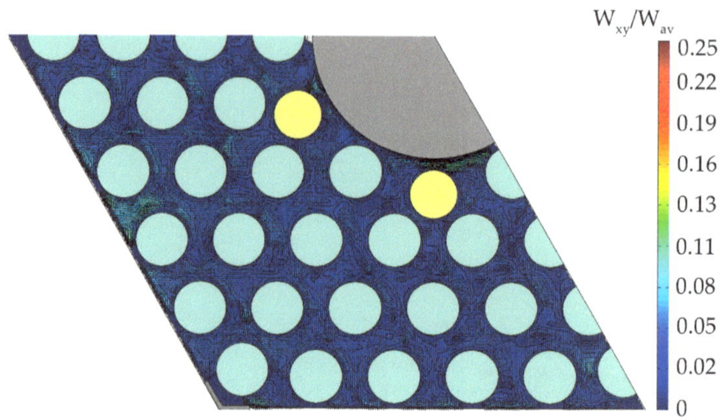

Figure 5. Dimensionless tangential flow velocity at a distance of $L/d_h \approx 2$ from the grid simulator plates.

In regular cells, with no grid plates in the center, the structure of the axial velocity profile had a uniform structure. The dimensionless axial velocity at distances $L/d_h \approx 1$ and $L/d_h = 10$ from the plate ranged from 1.1 to 1.2 and from 0.9 to 1, respectively. In regular cells with intersecting plates in the center, the axial flow was non-uniform. The dimensionless velocity ranged from 0.5 to 0.9. Behind the grid plate profiles, in areas where they contacted fuel elements and between fuel rods, axial flow velocity was minimal. The highest velocities were found in free areas of cells. At the points of the local minimum and maximum, the dimensionless axial velocity ranged from 0.5 to 0.7 and 0.8 to 0.9, respectively. Homogenization of the axial flow velocity in the field of regular cells occurred at a distance of $L/d_h \approx 10$ from the grid (Figures 6 and 7).

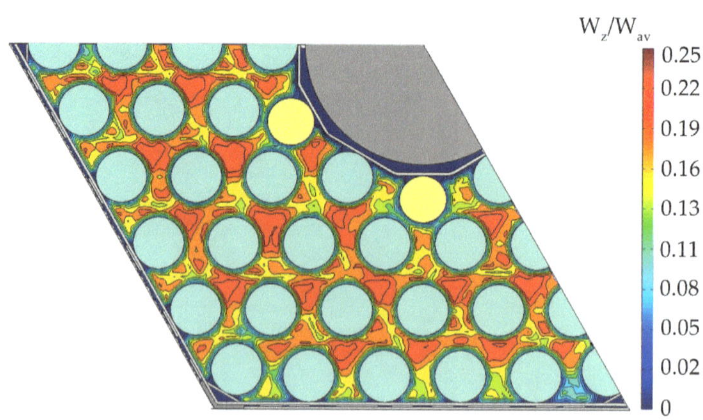

Figure 6. Dimensionless axial flow velocity at a distance of $L/d_h \approx 1$ from the grid simulator plates.

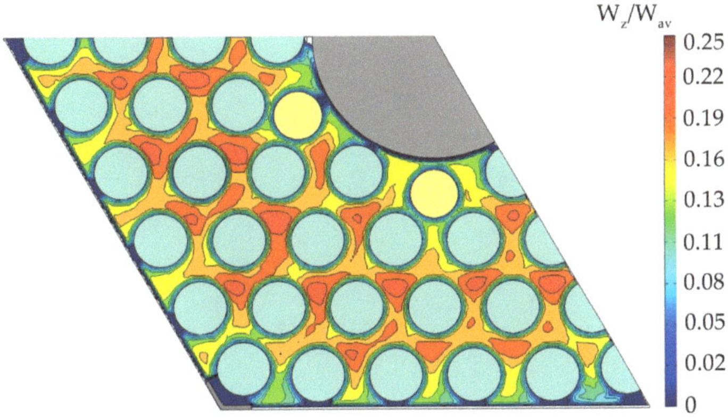

Figure 7. Dimensionless axial flow velocity at a distance of $L/d_h \approx 10$ from the grid simulator plates.

Near the fuel assembly cover, the flow velocity was lower than in the field of regular cells. The dimensionless velocity at a distance of $L/d_h \approx 1$ from the plates ranged from 0.3 to 0.9.

The axial flow velocity in the peripheral cells, remote from the stiffening corners, was influenced by the arrangement of the grid plates. At a distance of $L/d_h \approx 1$ from the grid, the dimensionless velocity in the cells without plates ranged from 0.8 to 0.9 and with plates from 0.4 to 0.75. At a larger distance of $L/d_h \approx 10$, velocities ranged from 0.9 to 1 without plates and from 0.5 to 0.8 with plates, indicating a lower intensity of flow velocity field equalization (Figures 6 and 7). The lowest flow velocity in the corner cells with grid plates in the center ranged from 0.3 to 0.6 at a distance of $L/dh \approx 1$ and from 0.4 to 0.7 at a distance of $L/d_h \approx 10$. In the corner cells without plates, the dimensionless axial flow velocity was higher. At distances of $L/d_h \approx 1$, it ranged from 0.6 to 0.8, and at distances of $L/d_h \approx 10$, it ranged from 0.8 to 0.9.

The speed of the axial flow was higher near the center of the displacer than at the periphery. Its dimensionless value ranged from 0.8 to 1.1 at distances $L/d_h \approx 1$ from the grid plates and from 0.75 to 0.9 at distances $L/d_h \approx 10$ (Figure 5).

6. The Results of the Study on the Hydrodynamics of the Coolant Using the Contrast Agent Injection Method

In the area of regular cells, which are located away from the central displacer tube and the fuel assembly cover, the transverse flow between the cells and the mixing of the coolant were low-intensity. The design of the spacer grid plates had a minimal impact on the process of cooling in regular cells. This is confirmed by the similarity in the distribution of impurities shown in Figure 8.

The flow in the area of regular cells near the central displacer tube is influenced by transverse flows from the displacer region. This is confirmed by the asymmetry in the distribution of the contrast agent, as shown in the contour plots in Figure 9.

In the peripheral cells, where the grid plates intersect, the transverse flow was oriented within the regular cells. There was no movement of coolant across the surface of the fuel assembly cover in a transverse direction. No reverse transverse flows were recorded from regular cells to peripheral cells (Figure 10). In peripheral cells without a plate intersection, transverse flows were oriented in the opposite direction from regular cells towards the periphery. Transverse flow of coolant along the surface of the fuel assembly cover was also observed. The presence of reverse flow from regular cells to the periphery has also been recorded (Figure 11).

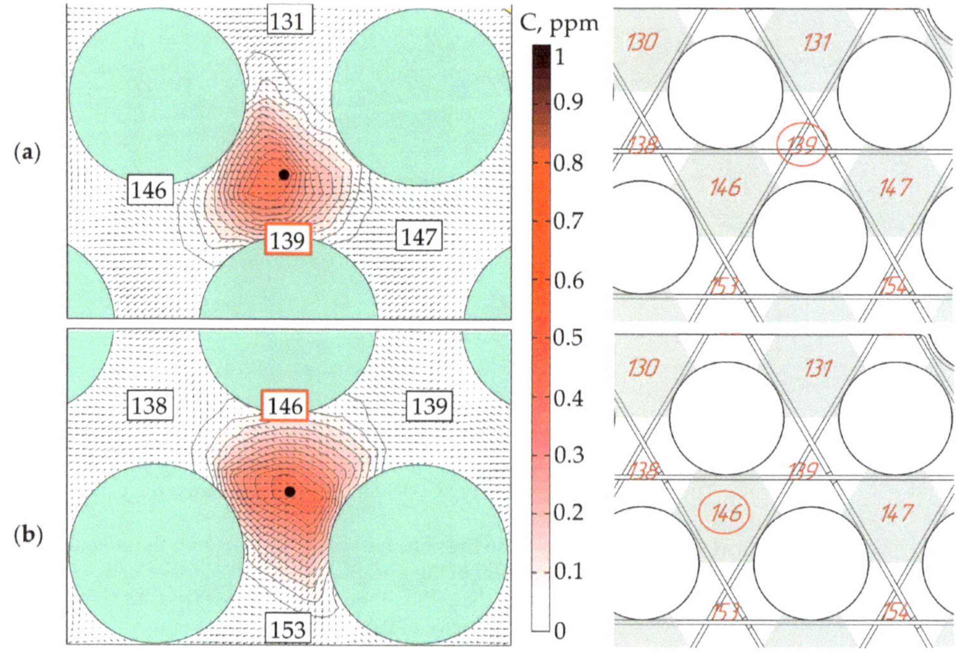

Figure 8. Contour plots of contrast agent distribution in regular cells: (**a**) Regular cell with plate intersection; (**b**) Regular cell without plate intersection.

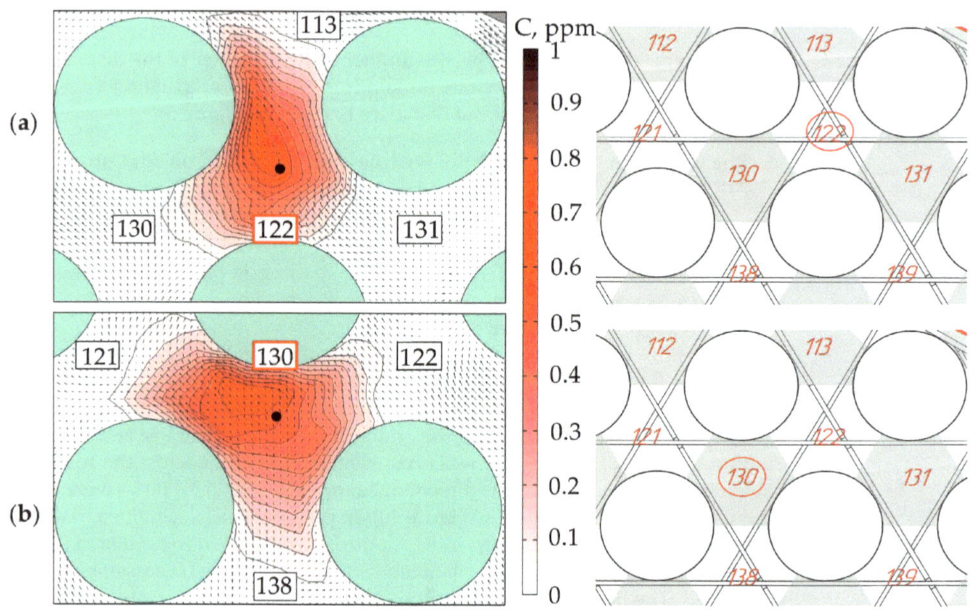

Figure 9. Contour plots of contrast agent distribution in regular cells: (**a**) Regular cell with plate intersection; (**b**) Regular cell without plate intersection.

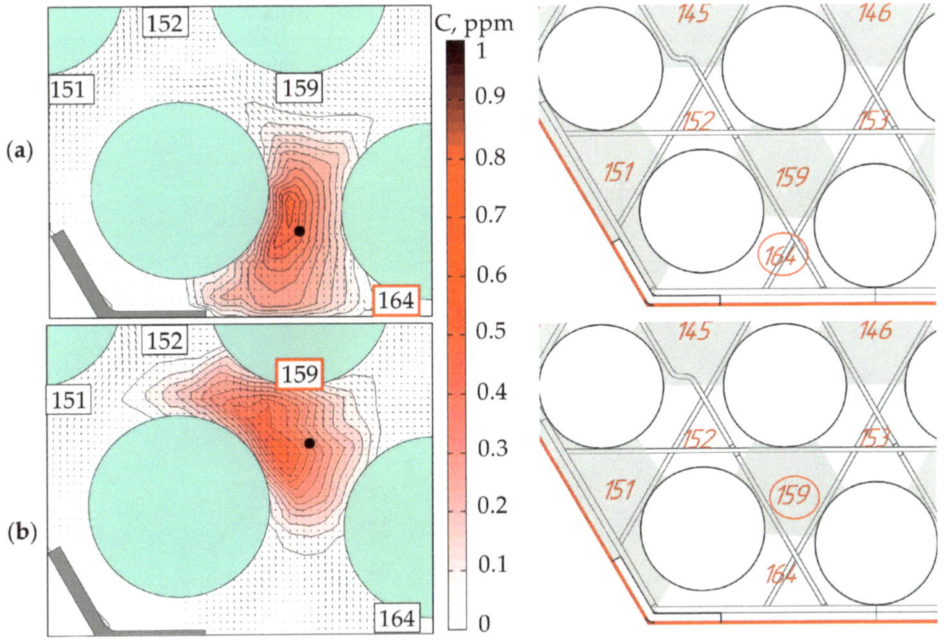

Figure 10. Contour plots of contrast agent distribution near the fuel assembly cover: (**a**) Cell with a plate intersection; (**b**) Cell without a plate intersection.

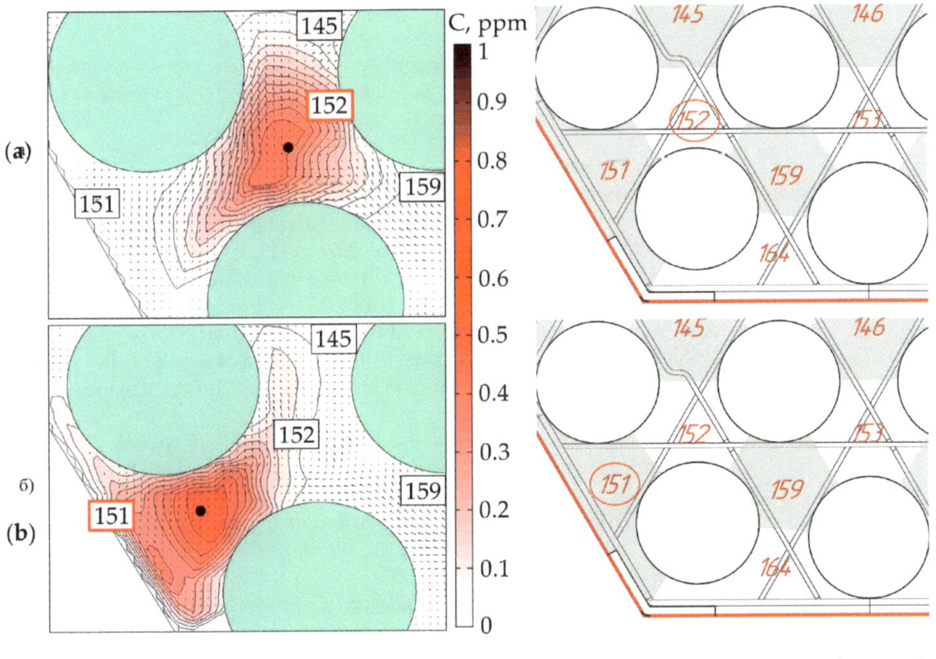

Figure 11. Contour plots of contrast agent distribution near the fuel assembly cover: (**a**) Cell with a plate intersection; (**b**) Cell without a plate intersection.

The flow of the coolant near the central displacer was determined by the transverse circular flow that was directed along its surface. This circular flow carried away some of the axial flow of the coolant from the cells near the displacer tube. The structure of the coolant flow is shown in the contour plot in Figure 12.

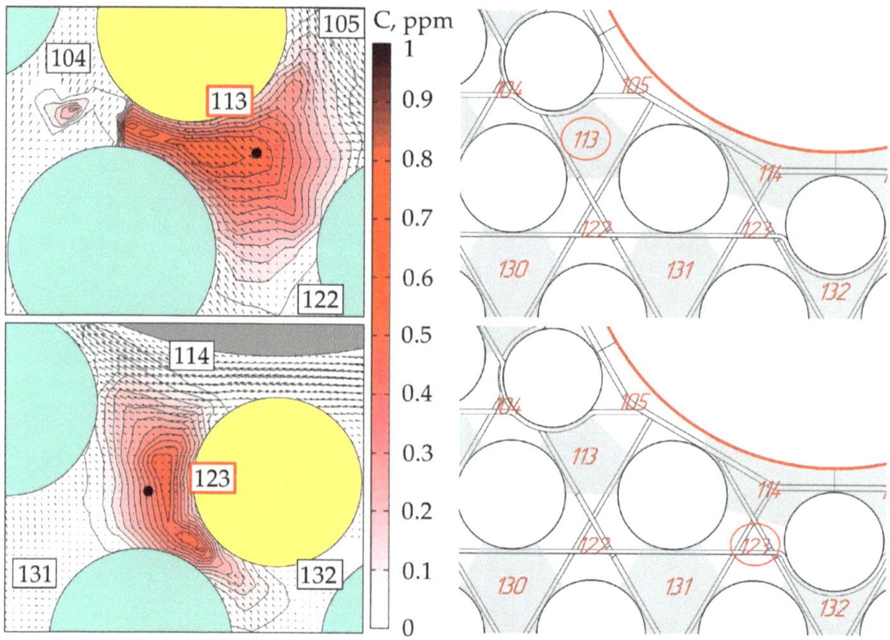

Figure 12. Contour plots of contrast agent distribution near the central displacer tube: (**a**) Cell with a plate intersection; (**b**) Cell without a plate intersection.

7. Discussion

In the fuel assembly, the coolant flow is significantly non-uniform due to the complex geometry of the fuel assembly bundle and spacer grids, as well as the presence of the fuel assembly cover, central displacer tube, and stiffening corners.

At the fuel assembly cover, the axial flow has the lowest velocities. The local minimum of the axial velocity is recorded in the areas around the stiffening corners. The average dimensionless velocity ranges from 0.3 to 0.7. The process of equalizing the axial flow velocity between the corner areas and the areas of regular cells is slow.

Behind the grid plates, the axial flow in the area of regular cells is non-uniform. The alignment of the axial flow structure in regular cells occurs at a distance of $L/d_h \approx 10$ from the grid plates.

At the central displacer tube, the axial flow velocity decreases, and its dimensionless value ranges from 0.75 to 0.9. This is due to the transverse flow directed towards the field of regular cells.

In the axial flow structure, three different regions can be identified with velocity values that differ by 25–30%. These are regular cells, the displacement region, and the peripheral region.

The arrangement of the grid plates has a significant effect on the structure of the axial flow in all areas of the fuel rod bundle and on the formation of transverse coolant flows only in the peripheral area near the fuel assembly cover.

The identified features of the coolant flow should be taken into account when justifying heat engineering reliability using one-dimensional calculation codes.

It is necessary to modernize one-dimensional calculation codes by increasing the number of cell types and considering the non-uniformity of flow through fuel bundle regions as well as different cell types. Additionally, it is important to update the coolant mixing matrix based on results from experimental studies using a contrast agent injection method.

Author Contributions: Conceptualization, S.D.; methodology, D.D. and A.P.; investigation, D.K. and D.N.; formal analysis, A.D. and A.R.; writing—original draft preparation, A.R.; writing—review and editing, D.D. and A.R.; visualization, T.D. and A.P.; software, A.D. and A.R.; supervision, S.D.; project administration, D.D.; funding acquisition, D.S. All authors have read and agreed to the published version of the manuscript.

Funding: This research was funded by the Ministry of Science and Higher Education of the Russian Federation (project No. FSWE-2024-0003).

Data Availability Statement: The original contributions presented in the study are included in the article, further inquiries can be directed to the corresponding author.

Conflicts of Interest: The authors declare no conflicts of interest.

References

1. Petrunin, V.V.; Sheshina, N.V.; Fateev, S.A.; Kurachenkov, A.V.; Shchekin, D.V.; BrykaloV, S.M.; Bezrukov, A.A. Scientific and technical aspects of developing a RITM-200N innovative reactor for SNPPs. *At. Energy* **2023**, *134*, 1–10. [CrossRef]
2. Filippov, S.P.; Veselov, F.V.; Pankrushina, T.G. Prospects of SNPPs in the electric power industry. *At. Energy* **2023**, *134*, 11–20. [CrossRef]
3. Zverev, D.L.; Fadeev, Y.P.; Pakhomov, A.N.; Galitskikh, V.Y.; Polunichev, V.I.; Veshnyakov, K.B.; Kabin, S.V.; Turusov, A.Y. Reactor Installations for Nuclear Icebreakers: Origination Experience and Current Status. *At. Energy* **2020**, *129*, 18–26. [CrossRef]
4. Belyaev, V.M.; Bol'shukhin, M.A.; Pakhomov, A.N.; Khizbullin, A.M.; Lepekhin, A.N.; Polunichev, V.I.; Veshnyakov, K.B.; Sokolov, A.N.; Turusov, A.Y. The World's First Floating NPP: Origination and Direction of Future Development. *At. Energy* **2020**, *129*, 27–34. [CrossRef]
5. Zakharychev, A.A.; Iksanova, G.S.; Kupriyanov, A.V.; Osin, A.B.; Petrunin, V.V.; Samoilov, O.B.; Shipov, D.L. Methodological Issues and Results of Experimental and Computational Studies of Critical Heat Flows in RITM-200 Reactor Fuel Assemblies for Small NPP. *At. Energy* **2021**, *130*, 63–68. [CrossRef]
6. Korolev, V.I. Analysis of New Technical Solutions for the RITM-200 Reactor Plant in Project 22220 Universal Nuclear Icebreakers. *Vestn. Gos. Univ. Morskogo Rechn. Flot. Im. Admirala S. O. Makarova* **2022**, *14*, 945–960. [CrossRef]
7. Tan, S.; Cheng, S.; Wang, K.; Liu, X.; Cheng, H.; Wang, J. The development of micro and small modular reactor in the future energy market. *Front. Energy Res.* **2023**, *11*, 1149127. [CrossRef]
8. Crețulescu, A. Small Modular Reactors in Romania's Energy Future. Two Essential Perspectives: Capital Costs and Public Perception. *Proc. Int. Conf. Bus. Excell.* **2024**, *18*, 1765–1775. [CrossRef]
9. Deng, C.; Zhu, S.; He, Y.; Wu, Y.; He, K.; Zhang, J.; Su, G. Analysis of Gas-Cooled Micro Modular Reactor (MMR) Fuel. *J. Nucl. Mater.* **2024**, *598*, 155191. [CrossRef]
10. Trupp, A.C.; Azad, R.S. The Structure of Turbulent Flow in Triangular Array Rod Bundles. *Nucl. Eng. Des.* **1975**, *32*, 47–84. [CrossRef]
11. Caraghiaur, D. *Experimental Study and Modeling of Spacer Grid Influence on Flow in Nuclear Fuel Assemblies*; Technical Report; KTH Stockholm: Stockholm, Sweden, 2009.
12. Bakosi, J.; Christon, M.A.; Lowrie, R.B.; Pritchett-Sheats, L.A.; Nourgaliev, R.R. Large-eddy simulations of turbulent flow for grid-to-rod fretting in nuclear reactors. *Nucl. Eng. Des.* **2013**, *262*, 544–561. [CrossRef]
13. Peña-Monferrer, C.; Muñoz-Cobo, J.L.; Chiva, S. CFD turbulence study of PWR spacer-grids in a rod bundle. *Sci. Technol. Nucl. Install.* **2014**, *2014*, 635651. [CrossRef]
14. Sergeev, D.A.; Kandaurov, A.A.; Troitskaya, Y.I. The particular use of PIV methods for the modelling of heat and hydrophysical processes in the nuclear power plants. *J. Phys. Conf. Ser.* **2017**, *891*, 012088. [CrossRef]
15. Yu, H.; Cai, J.; He, S.; Li, X. Analysis of neutron physics and thermal hydraulics for fuel assembly of small modular reactor loaded with ATFs. *Ann. Nucl. Energy* **2020**, *152*, 107957. [CrossRef]
16. He, S.; Cai, J.; Chen, Z.; Liu, R. Thermal hydraulic analysis of accident tolerant fuels for Reactivity-Initiated-Accident in PWR with different coolant channel geometries. *Nucl. Eng. Des.* **2019**, *351*, 131–142. [CrossRef]
17. Cong, T.; Zhang, X. Numerical Study of Bubble Coalescence and Breakup in the Reactor Fuel Channel with a Vaned Grid. *Energies* **2018**, *11*, 256. [CrossRef]
18. Tian, Z.; Yang, L.; Han, S.; Yuan, X.; Lu, H.; Li, S.; Liu, L. Numerical Investigation on the Flow Characteristics in a 17 × 17 Full-Scale Fuel Assembly. *Energies* **2020**, *13*, 397. [CrossRef]
19. Lee, I.S.; Yoon, D.H.; Bang, Y.S.; Kim, T.H.; Kim, Y.C. Assessment of Realistic Departure from Nucleate Boiling Ratio (DNBR) Considering Uncertainty Quantification of Core Flow Asymmetry. *Energies* **2021**, *14*, 1504. [CrossRef]

20. Vlasov, M.N.; Merinov, I.G. Application of an Integral Turbulence Model to Close the Model of an Anisotropic Porous Body as Applied to Rod Structures. *Fluids* **2022**, *7*, 77. [CrossRef]
21. Kim, K.; Kim, W.-S.; Choi, H.-S.; Seol, H.; Lim, B.-J.; Euh, D.-J. An Experimental Evaluation of the APR1000 Core Flow Distribution Using a 1/5 Scale Model. *Energies* **2024**, *17*, 2714. [CrossRef]
22. Beliavskii, S.; Alhassan, S.; Danilenko, V.; Karvan, R.; Nesterov, V. Effect of changing the outer fuel element diameter on ther-mophysical parameters of KLT-40S reactor unit. *Ann. Nucl. Energy* **2023**, *190*, 109877. [CrossRef]
23. Hidayatullah, H.; Susyadi, S.; Hadid Subki, M. Design and technology development for small modular reactors—Safety expectations, prospects and impediments of their deployment. *Prog. Nucl. Energy* **2015**, *79*, 127–135. [CrossRef]
24. Bhowmik, P.K.; Sabharwall, P.; Johnson, J.T.; Retamales, M.E.T.; Wang, C.; O'Brien, J.E.; Lietwiler, C.; Wu, Q. Scaling methodologies and similarity analysis for thermal hydraulics test facility development for water-cooled small modular reactor. *Nucl. Eng. Des.* **2024**, *424*, 113235. [CrossRef]
25. Ren, L.; Minjun, P.; Genglei, X.; Lin, S. The natural circulation flow characteristic of the core in floating nuclear power plant in rolling motion. *Ann. Nucl. Energy* **2020**, *142*, 107385. [CrossRef]
26. Alam, S.B.; de Oliveira, R.G.G.; Goodwin, C.S.; Parks, G.T. Coupled neutronic/thermal-hydraulic hot channel analysis of high power density civil marine SMR cores. *Ann. Nucl. Energy* **2019**, *127*, 400–411. [CrossRef]
27. Samoilov, O.B.; Shipov, D.L.; Kupriyanov, A.V.; Sholin, E.V.; Vishneva, T.Y.; Molodtsov, A.A.; Osin, A.B.; Dmitriev, S.M.; Khrobostov, A.E.; Doronkov, D.V.; et al. Efficiency studies of heat-transfer intensifier grids in TVSA-T model fuel assemblies. *At. Energy* **2020**, *128*, 17–23. [CrossRef]
28. Dmitriev, S.M.; Gerasimov, A.V.; Dobrov, A.A.; Doronkov, D.V.; Pronin, A.N.; Solntsev, D.N.; Khrobostov, A.E.; Shvetsov, Y.K.; Shipov, D.L. Hydrodynamics and mixing of a coolant in the core of the VVER with fuel assemblies of different designs. *Thermophys. Aeromech.* **2019**, *26*, 845–860. [CrossRef]
29. Gerasimov, A.V.; Dmitriev, S.M.; Dobrov, A.A.; Doronkov, D.V.; Pronin, A.N.; Ryazanov, A.V.; Solntsev, D.N.; Khrobostov, A.E. Computational-experimental investigations of the processes of motion of a coolant flow in the region of the guide channel behind mixing grids of fuel assemblies. *J. Eng. Phys. Thermophys.* **2020**, *93*, 145–154. [CrossRef]
30. Samoilov, O.B.; Noskov, A.S.; Shipov, D.L.; Dmitriev, S.M.; Dobrov, A.A.; Doronkov, D.V.; Legchanov, M.A.; Pronin, A.N.; Solntsev, D.N.; Sorokin, V.D.; et al. Hydrodynamic features of the flow downstream from the mixing spacer grid in a Kvadrat fuel assembly in PWRs. *Therm. Eng.* **2019**, *66*, 243–248. [CrossRef]
31. Dmitriev, S.M.; Dobrov, A.A.; Doronkov, D.V.; Doronkova, D.S.; Legchanov, M.A.; Pronin, A.N.; Ryazanov, A.V.; Khrobostov, A.E. Studies of the hydrodynamics of the coolant behind the mixing intensifier grids of the fuel assemblies of a PWR reactor. *High Temp.* **2022**, *60*, 358–365. [CrossRef]

fluids

MDPI

Article

Improvement in the Number of Velocity Vector Acquisitions Using an In-Picture Tracking Method for 3D3C Rainbow Particle Tracking Velocimetry

Mao Takeyama *, Kota Fujiwara and Yasuo Hattori

Meteorology and Fluid Science Division, Central Research Institute of Electric Power Industry, 1646 Abiko, Abiko-shi 270-1194, Chiba, Japan; yhattori@criepi.denken.or.jp (Y.H.)
* Correspondence: takeyama3826@criepi.denken.or.jp; Tel.: +81-70-6576-0603

Abstract: Particle image velocimetry and particle tracking velocimetry (PTV) have developed from two-dimensional two-component (2D2C) velocity vector measurements to 3D3C measurements. Rainbow particle tracking velocimetry is a low-cost 3D3C measurement technique adopting a single color camera. However, the vector acquisition rate is not so high. To increase the number of acquired vectors, this paper proposes a high probability and long-term tracking method. First, particles are tracked in a raw picture instead of in three-dimensional space. The tracking is aided by the color information. Second, a particle that temporarily cannot be tracked due to particle overlap is compensated for using the positional information at times before and after. The proposed method is demonstrated for flow under a rotating disk with different particle densities and velocities. The use of the proposed method improves the tracking rate, number of continuous tracking steps, and number of acquired velocity vectors. The method can be applied under the difficult conditions of high particle density (0.004 particles per pixel) and large particle movement (maximum of 60 pix).

Keywords: flow visualization; rainbow PTV; 3D3C measurement

Citation: Takeyama, M.; Fujiwara, K.; Hattori, Y. Improvement in the Number of Velocity Vector Acquisitions Using an In-Picture Tracking Method for 3D3C Rainbow Particle Tracking Velocimetry. *Fluids* **2024**, *9*, 226. https://doi.org/10.3390/fluids9100226

Academic Editors: Ashwin Vaidya and Tomoaki Kunugi

Received: 29 July 2024
Revised: 20 September 2024
Accepted: 28 September 2024
Published: 30 September 2024

1. Introduction

The present study aimed to increase the number of velocity vectors acquired from three-dimensional three-component (3D3C) velocity measurements made in rainbow particle tracking velocimetry (PTV) to enhance spatial resolution. To this end, we developed an in-picture tracking method adopting raw particle images in place of the conventional method of tracking 3D positions.

Particle image velocimetry (PIV) and PTV are common methods of measuring fluid velocities from images and have been used for a wide variety of objects, including multiphase flows [1,2] and microfluids [3]. These methods have evolved significantly, transitioning from 2D2C velocity vector measurements confined to the focal plane to full 3D3C measurements [4]. To enhance the dimensionality and components of these measurements, various techniques have been developed, including stereo PIV [5], scanning PIV [6,7], and tomographic PIV [8,9]. The 3D3C velocity measurements usually need to be made from multiple viewpoints using multiple cameras, such as when adopting scanning tomographic particle image velocimetry [9] and the shake-the-box method [10]. Rainbow PTV [11–16] is a low-cost method of obtaining 3D3C velocities with a single color camera and simple data processing. The flow volume is illuminated by multi-colored light that changes along the depth direction (z) using a liquid-crystal display projector. Particle positions along vertical and horizontal directions (x and y) in pictures are obtained in the same manner as in traditional PTV, whereas particle positions in the depth component are obtained from the particle colors. The degree of hue (H, 0–360) is calculated from RGB values of the particle and then converted to the depth position.

PTV provides Lagrangian velocity data at random locations. However, to calculate differential quantities such as vorticity, it is necessary to replace the data with grid data, such as the results of computational fluid dynamics analysis or particle image velocimetry. This requires the acquisition of spatially dense velocity vectors [17]. However, the number of vectors acquired in rainbow PTV is not high, as described below.

The accuracy of the depth position estimated from the particle color is not high. To improve accuracy, a multi-cycle color pattern was designed and used in [16]. In addition, recent methods based on machine learning still have a standard deviation error of $\pm 3\%$ of the total depth [14]. This error is too high relative to the positional accuracy of the x and y positions, which can be obtained at a sub-pixel level. The inaccuracy in the estimation of the depth position reduces not only the accuracy of velocity vectors but also the vector acquisition rate.

Low positional accuracy leads to a low vector acquisition rate in the following ways. As the velocity is calculated by dividing the particle movement by the time width of an image, it is necessary to link (i.e., track) the same particles in successive images. The tracking of particles adopts historical information on the particle position. Figure 1 shows moving particles and positional data, including errors for the z direction. Black circles indicate the correct positions, and red circles the measured positions with error. The particle position at time t + 1 is estimated from the particle positions at times t − 1 and t. The particle inside the estimated area (i.e., the dotted circle in Figure 1) is recognized as the same particle. However, a particle outside the search area cannot be tracked. A smaller movement in one-time step corresponds to a greater ratio of the position error to the amount of movement and a higher chance of tracking failure. Even if a slowly moving particle could be tracked, the accuracy of its velocity would be low. To reduce mistracking, in the example of the four-time-step method [18], a particle is judged to be the same if it is tracked in four consecutive images. However, although the four-time-step method, velocity gradient tensor method [19], and other methods are used to track particles, they cannot track particles in the described error-prone situations.

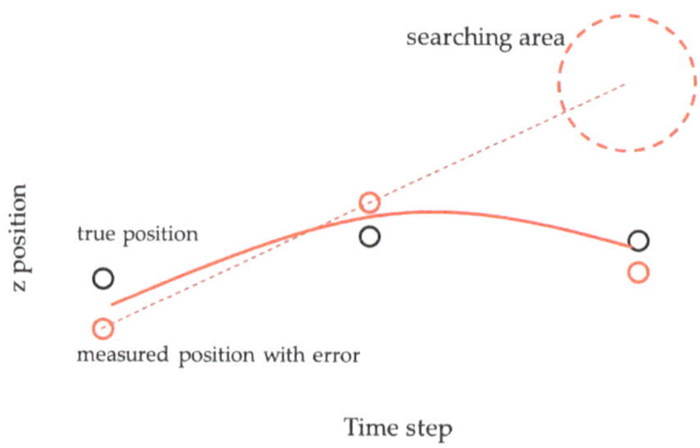

Figure 1. Schematic showing particle tracking. Black circles indicate true positions, red circles indicate measured positions with error, the red dashed circle is the search area, and the red curve is the fitted curve.

In addressing the above problem, the nearest-neighbor method was adopted to track particles in the picture instead of the 3D space [14]. However, using the nearest-neighbor method, the amount of movement of a particle in one step has to be sufficiently small for the particle to be tracked, which requires a sufficiently high sampling rate. Achieving a higher dynamic range of velocity requires that particles be tracked over a large range of movement.

Such a 3D tracking method requires a long-time tracking fitting process after tracking. Figure 1 shows that, even if tracking is possible, the accuracy of the velocity obtained in the z direction is low. It is expected that positional data containing zigzag errors in the depth direction, as shown in Figure 1, can be supplemented with smooth motion (the red curve in Figure 1) through filter processing; see, for example, [14,20]). This operation requires continuous tracking over a long period of time without losing the particle.

This paper thus proposes a high-probability and long-term tracking method that can be applied under conditions of large particle movement and high particle density by means of two main features.

First, particles are tracked in a raw picture, which is in contrast with the conventional method of tracking particles in 3D space. Using color information as an aid for tracking, it is possible to track a particle with high probability even under conditions of high particle density and large movement in a single time step. In other words, whereas conventional 3D tracking applies equal weights to x, y, and z positions (estimated from H), the proposed tracking process strongly weights the x and y positions, which have higher accuracy, and uses H from particle color to improve the tracking rate.

Second, even if the tracking of a particle fails temporarily, the tracking is reestablished in a later step. When two or more particle images overlap, which often happens under high particle density conditions, the particles are not recognized correctly, and the particle being tracked is temporarily lost. By compensating the positional information of untracked particles at the time step where particle tracking has failed, it is possible not only to increase the tracking rate but also to extend the period of continuous tracking. This increases the number of vectors acquired because the subsequent fitting process requires continuous tracking.

The present work evaluates the proposed method for flow under a rotating disk. Specifically, the work demonstrates the processes of the proposed method, the possibility of improving the number of vector acquisitions, and the application of the method under conditions of high particle density and large particle movement.

2. Data Processing

2.1. In-Picture Tracking

Figure 2 presents flowcharts of data processing. The left flowchart is for the conventional 3D tracking method, whereas the right flowchart is for the in-picture tracking method proposed in this paper. First, the background is subtracted from the measured image, and the particles are identified on the grey scale. Then, particles are identified from the raw image, and their xy position in the pixel and particle color space are extracted. The particle color is converted from RGB to the hue degree and then to the normalized z position using a calibration curve. The hue degree (H) is calculated from the values of the three color channels [11] according to

$$H = \begin{cases} 60 \times \frac{G-B}{max(R,G,B)-min(R,G,B)} & (\text{if } max(R,G,B) = R) \\ 60 \times \frac{B-R}{max(R,G,B)-min(R,G,B)} + 120 & (\text{if } max(R,G,B) = G) \\ 60 \times \frac{R-G}{max(R,G,B)-min(R,G,B)} + 240 & (\text{if } max(R,G,B) = B). \end{cases} \quad (1)$$

The calibration curve is obtained by adopting the in situ slit calibration method [14]. The operation up to this point is the same for the two methods.

In the conventional method, the xy position in millimeters is obtained from the z position because the viewing angle varies with depth. In addition, the normalized z is converted to the z position in millimeters. Tracking based on the 3D positions is then performed to obtain the velocity vector.

In contrast, the proposed in-picture tracking method tracks particles in the raw picture. Particles are tracked by the xy position in pix units, and normalized z is used as an aid. The xy position is known more accurately than the z position, and particles can thus be tracked at a higher rate than for tracking in the 3D domain. The detailed process of tracking and particle linking across time steps is explained in Section 2.2. Inaccurate z positions are compensated for by fitting the tracked passage in the z direction with a polynomial equation. Finally, the x, y, and z positions are converted from pixel units to millimeter units to obtain velocity vectors.

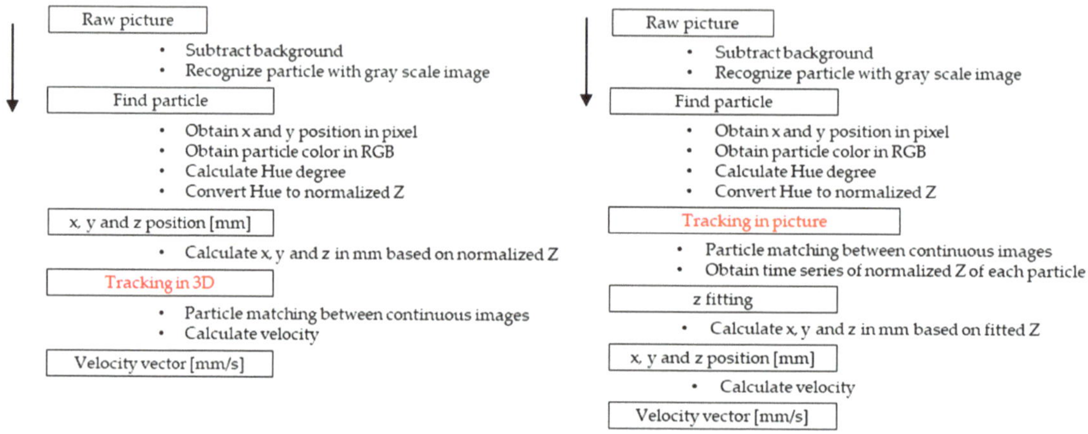

Figure 2. Flowcharts of data processing. The left flowchart is for the conventional 3D tracking method, whereas the right flowchart is for the proposed in-picture tracking method.

2.2. Detailed Process of Continuous Long-Time Tracking

One of the challenges in PTV is tracking particles when they overlap in an image. Since the particles are distributed in three dimensions, it is common for them to overlap when viewed in 2D, which can cause the tracking system to lose track of individual particles. This is a limiting factor of the particle density in PTV. Overlapping particles can be recognized as single particles of color different from the colors of the overlapping particles, such that the particles are lost. However, two overlapping particles have different motions in the xy plane because of their different z positions, and their overlap is thus a temporary event. Therefore, even if a particle is temporarily lost, it can be tracked for an extended period of time through compensation of the particle position at the time of overlap. To address this issue, our method includes a process to predict and recover the positions of lost particles, allowing for continuous tracking over a longer period.

Initially, the system predicts the position of a particle at a future time step by considering its velocity and acceleration from previous time steps. If the particle is not found in its predicted location at the next time step, the system anticipates where it might appear in subsequent frames. When the particle reappears, the system compensates for the missed positions by creating "virtual" particles at the points where the particle was temporarily lost. This compensation process allows the tracking to remain continuous, even when particles are lost for up to two consecutive time steps. This allows for continuous long-term tracking. Basically, there is no sudden disappearance of a particle except when a particle travels beyond the field of view.

The detailed tracking process at time t is described below and shown in Figure 3. Basically, the velocity is calculated from the positions at two consecutive times, and the acceleration is calculated from the positional information at three consecutive times, i.e., the velocities at two times. Tracked particles are assigned an ID that shows that they are the same particles in different images to continuously track temporarily lost particles.

The process involves four operations, as shown in Table 1. First, process 1, as the basic tracking process, predicts the position at $t + 1$, assuming that the velocity and acceleration at $t - 1$ are maintained at t, from the positional information at $t - 2$ and $t - 1$ for the particle that has been assigned an ID at t. The particles within the search area are filtered using the dimensionless z positional information obtained from the color information, e.g., by taking $\Delta Z < 0.025$. If a particle exists within the predicted range, the particle is linked as the same particle and given an ID, and the velocity and acceleration at time t are calculated.

Filtering by z is performed in the same way in subsequent processes. This threshold value of ΔZ must be greater than the distance moved in the depth direction for one step plus the error of color-depth transformation. This is to be the same as setting the search range in the xy direction. In previous studies, the error width (95th percentile width) has been around ±2%, and research is underway to reduce this further [16]. So, the threshold value should thus be set according to the velocity of the measurement target and sampling rate.

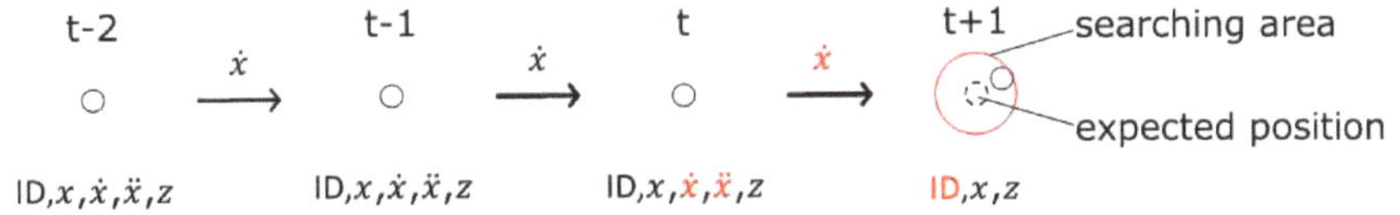

Figure 3. Basic tracking process. Black letters are known values; red letters are values obtained after tracking.

Table 1. Tracking processes for continuously tracked and temporarily lost particles. ○ means tracked, and × means untracked.

	t − 2	t − 1	t	Operation
Process 1	○	○	○	Basic tracking
Process 2	○	○	×	Compensate time t
Process 3	○	×	×	Compensate time t − 1 and t
Process 4	×	×	×	Four-time step method

Next, process 2 targets particles that have an ID at t − 1 but are not found at t and are thus temporarily lost. Assuming that the velocity and acceleration at t − 2 are maintained, the position at t + 1 is predicted, and if there is a particle within the estimated search area that has not yet been assigned an ID, it is judged to be the same particle and assigned an ID. The acceleration at t − 1 can be calculated from the positions at times t − 2, t − 1, and t, and the acceleration at time t can also be calculated from the positions at times t − 1, t, and t + 1. The virtual particle at time t is placed at the position where the acceleration at both times is constant.

Process 3 targets particles that have an ID at t − 2 but are not found at t − 1 and t. The same operation as in process two above is performed at the next time t + 1, and if a particle is found at t + 1, it is given an ID, and a virtual particle is placed at two times, t − 1 and t.

Finally, process 4 applies the four-step method to particles that have not yet been assigned an ID. This process is also applied to the initial state of the tracking process. For particles that have been successfully tracked four consecutive times, new IDs are assigned, and velocities and accelerations are calculated. When assigning new IDs, it is necessary to impose strict requirements that ensure that no mistakes are made.

Once a particle is given an ID, it is allowed to be lost through particle overlap up to two consecutive times, and it continues to be tracked until it moves out of the field of view.

The ability to track over long periods of time allows for polynomial fitting. The fitting function is not specified, but as in previous studies [14,20], a cubic equation is used in the present study. However, using a low number of points (e.g., four points) for fitting is inappropriate, and therefore, only particles that have been tracked more than 10 consecutive times are used in the velocity calculation. In addition, the long-time motion of particles having a short rotational period, such as particles in fine vortices, cannot be fitted. Therefore, initially, a fitting interval or curve order should be set for each particle. However, in this case, we uniformly set the maximum interval at 100 steps.

2.3. Calibration Curve

The calibration curve used in the conversion from the hue degree to the dimensionless z position is obtained by adopting in situ slit calibration [14]. The schematic of the slit calibration method is shown in the lower graph of Figure 4. Here, the color pattern from the projector falls incident on a slit. Only a portion of the color pattern passes through the slit, and the rest of the color pattern is blocked. The slit is moved in the z direction, and a curve is created by acquiring the hue of the particles at each depth position. The slit width is set at 1/30 of the full scale (i.e., 64 pix relative to 1920 pix and 6 mm relative to 180 mm). The slit is positioned such that half of its width overlaps with the previous measurement position. Values are obtained for 59 steps, and interior interpolation is performed between the measurements.

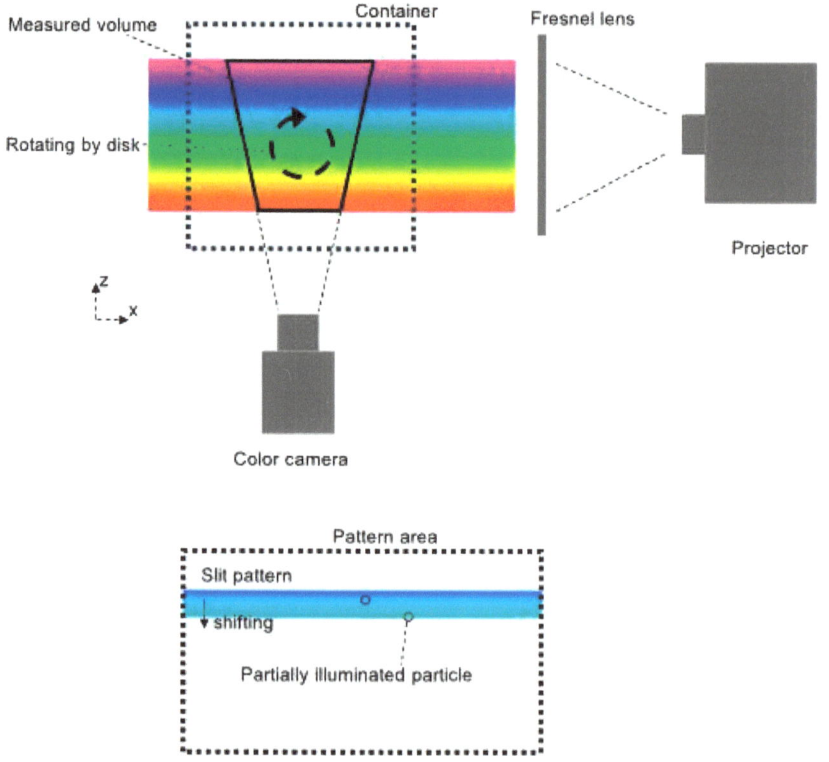

Figure 4. Top view of the experimental setup. The upper section shows a hole setup, a projector, a camera, a lens, and a test section. Lower shows illumination of slit calibration and sample images of illuminated particles.

There are two problems with slit calibration: the color projected from the projector at both ends of the slit is not the intended color, and the projected color pattern partially hits the particles at both ends of the slit, resulting in a blurred image. Therefore, only the central 50% range of hue degrees is adopted, and the color of the particles obtained through the slit is analyzed, ignoring the extreme.

3. Experimental Setup

Figure 4 is a schematic of the top view of the experimental setup, showing the camera's field of view and the optical arrangement. A VPL-HW60 liquid-crystal display projector (Sony) is used as the light source to output the color pattern. The projector has a brightness of 1800 lm and a resolution of 1920 × 1080 pix and provides illumination in the x direction. The image data (.tiff) of the color pattern is projected by the projector. The gamma value has a default value of 1.8. A Fresnel lens having a focal length of 350 mm is set next to the projector to collimate the expanding projection light. The illumination width in the z direction is 180 mm.

To increase the color detection accuracy, an AP3200T-PMCL-1 3CMOS camera (JAI) and VS-1218/3CMOS lens (VS Tech) are used in place of the typically used single CMOS color camera. For a single CMOS camera, a Bayer filter with R, G, and B colors and a 1:2:1 ratio covers a CCD/CMOS array where cells of the same color do not adjoin each other [21]. In contrast, the 3CMOS camera has three CMOS sheets and a dichroic mirror to separate light. Thus, R, G, and B values are obtained for the same cell of an image sensor array.

The distance between the center of the measured volume and the lens of the color camera is 610 mm. The field of view in the x direction ranges from 149 to 187 mm, and the spatial resolution ranges from 72 to 91 μm/pix, changing in the depth direction. The depth of view is sufficient to capture the particle image in the full depth range. The sampling rate is 40 fps, and the shutter speed is 1 ms. Generally, the sensitivity of channel G of a color camera is higher than the sensitivities of the other two channels. In this study, the sensitivity ratio of the camera is 1.6. To align the ranges of the three channels, the gains of the R and B channels are each set at 1.6 times the gain of channel G.

An acryl container having inner dimensions of 300 × 300 × 300 mm^3 is filled with particle extended water. The Diaion HP20 tracer particles (Mitsubishi Chemi) have a density of 1.01 g/cm^3, and their diameter is controlled to be within the range of 350–600 μm through sieving. The particle number density (particles per pixel, ppp) is evaluated from the captured image. The density cannot be correctly evaluated under dense conditions because overlapping particles cannot be accurately recognized. Therefore, we adopt a dilute condition with few overlapping particles as a standard, and we control the particle number density under a dense condition by mixing several times the dilute number of particles into the particle mixture. The particle concentrations are 0.001, 0.002, 0.003, and 0.004 ppp. Sample images for the different conditions are shown in Figure 5. A rotating disk with a diameter of 200 mm generates a circular flow in a container. The rotational speeds are 60, 120, 180, and 240 rpm, and the corresponding maximum movements in the image (i.e., the initial search areas) are 25, 40, 45, and 60 pix. The search area for the predicted position is uniformly set at 5 pixels. Incidentally, most particles are less than 10 pixels in diameter. Moreover, tracking in 3D space is almost impossible under such strict conditions. The search range for the predicted position is, therefore, set at one-fifth of the distance moved (in millimeters) in the previous step.

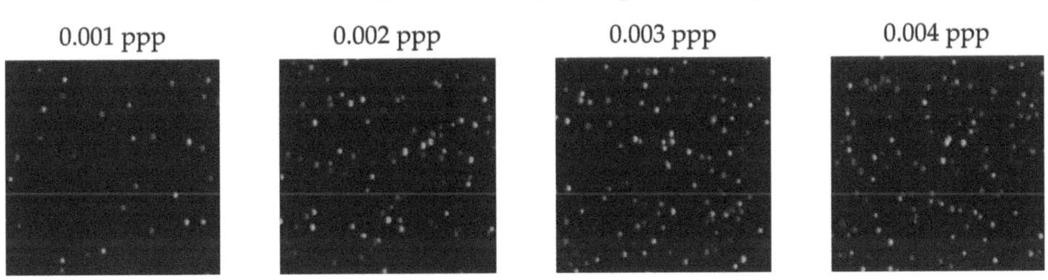

Figure 5. Example of images for different particle number densities. Brightness values are tripled for visibility.

4. Results

4.1. Demonstration of In-Picture Tracking

Figure 6 shows an example of the compensation process for a lost particle. The figure shows a cropped image of 100 × 100 pix, with two overlapping particles at the center indicated by a white circle. The overlapping particles are seen as a single particle having a color different from that of the particles before and after, and the tracking is thus disconnected. The blue circles indicate the positions of the two particles, five steps before and five steps after. The positions of the two particles at the overlap time are estimated from the particle positions at the times before and after. The IDs of the particles are taken from before and after the overlap and considered to label the same particles.

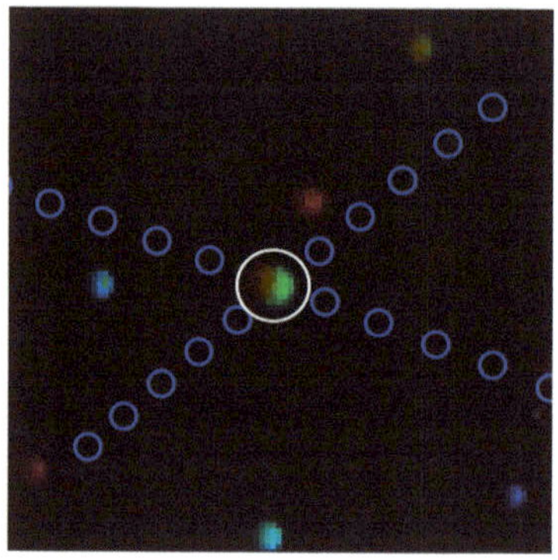

Figure 6. Example of an image with particle overlapping. Brightness values are tripled for visibility.

This processing compensates for lost particles up to two consecutive times, such that the tracking fails when the overlap continues for more than three consecutive times. Although it is possible to extend the complementation interval beyond three-time steps as a solution, the best solution is to properly recognize each overlapped particle. Analysis of the best solution is left as a future topic of study.

Next, an example of complementation by polynomial approximation is shown, where the low accuracy of the depth position estimated from particle colors is complemented by approximating the particle trajectory with a third-order polynomial, as in [14,20].

Here, only particles that can be tracked for more than 10 steps are considered because the approximation does not work well when the number of successive tracking steps is too small. In addition, if the tracking interval is too long or if the particle has many rounds of vortex motion during the interval of interest, the approximation does not work well with the third-order polynomial. In this study, the approximation interval is limited to 100 time steps for a large circular flow. Although there is no method at this time to determine the fitting function and tracking interval for the various flows, it needs to be adjusted according to the experimental results.

In Figure 7, the left panel shows an example time series of the z position of a particle for 100-time steps and the trajectory obtained by approximation, and the right panel shows the time series of the z-directional velocity and the velocity obtained by approximation. As shown in Figure 3, the xy position in 3D space changes the viewing angle depending on the depth, such that the z position also affects the xy position. Looking at the particle positions alone, the positional deviation before and after fitting does not appear to be large, but the velocity deviation is very large, indicating that fitting is essential. This velocity deviation is mitigated as the particle movement in a one-time step increases, but conversely, continuous particle tracking becomes more difficult.

Figure 8 shows sample results of the particle trajectory of rotational motion in 3D, XZ, and ZY 2D coordinates as the demonstration. The particle number density is 0.003 ppp, and the disk rotating speed is 180 rpm. Results are shown for 100-time steps but with a moderately reduced number of particles for visibility.

The average velocity field in the y-section of the same results shown in Figure 8 is illustrated in Figure 9. The velocities that can be measured for each $10 \times 10 \times 10$ mm^3 voxel are temporally averaged at a height of 30 mm < y < 40 mm. The velocity in the depth direction is measured with sufficient accuracy to determine the overall flow.

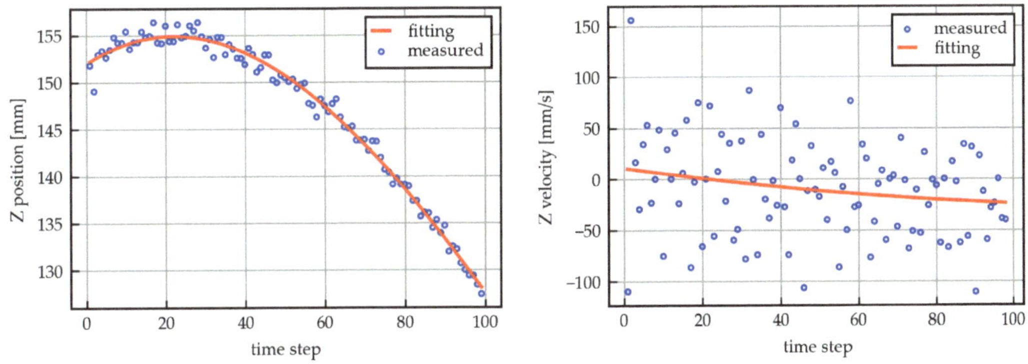

Figure 7. Results of the fitting process. Left: depth position, right: depth velocity.

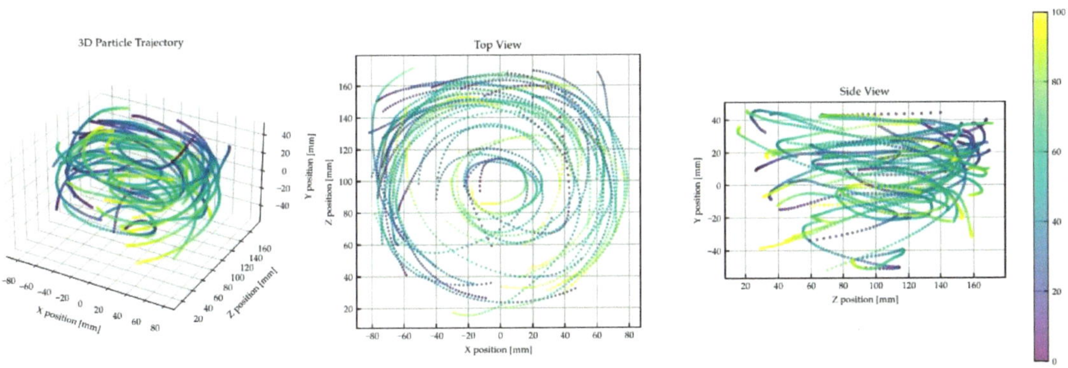

Figure 8. Particle trajectory in 3D and 2D coordinates.

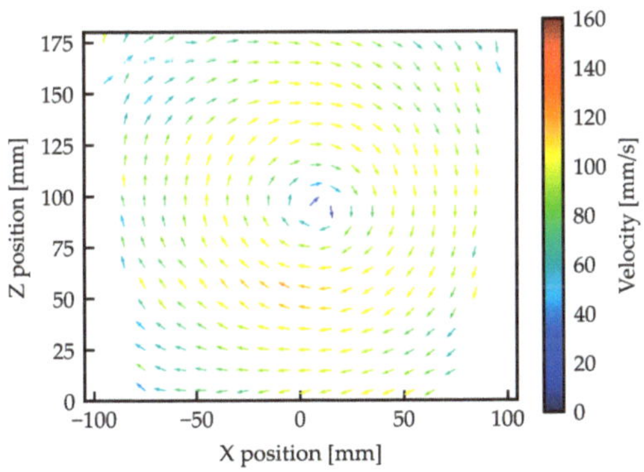

Figure 9. Time-averaged velocity field in XZ coordinates.

4.2. Improvement in Velocity Vector Acquisition

Whereas the tracking rate in three-dimensional space is around 30% at the particle number density of 0.001 ppp and the disc rotation speed of 60 rpm condition, in-picture tracking achieved around 90%, and furthermore, almost 100% of particles are successfully tracked by the lost particle compensation process for temporarily lost particles. In both methods, the search area for particles is set to 1/5 of the maximum particle moving distance (mm or pix) in one step for each rotational speed condition. In tracking in three-dimensional space, an increase in the search range can improve the tracking rate, but the probability of making a mistake in linking a particle to another particle increases.

Figure 10 shows the particle tracking rates for different particle densities and disk rotational speeds. Tracking rates achieved using the in-picture method are indicated by the colored sections of bars, and further increases in the tracking rate due to compensation are shown by the black sections of the bars. If there is particle overlap, the number of particles cannot be accurately determined from an image. Therefore, as there is almost no overlap at a particle density of 0.001 ppp, the number of particles is estimated by suspending an equal number of particles based on the number of particles recognized under this condition.

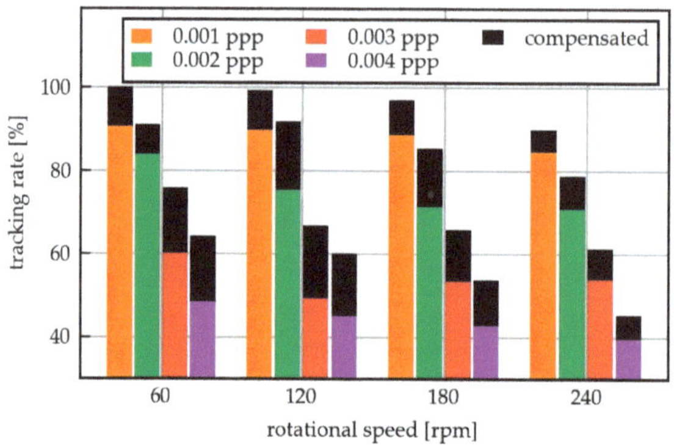

Figure 10. Particle tracking rates for different disk rotating speeds and particle densities.

At a disk rotation speed of 60 rpm, almost all particles are tracked, with the tracking rate slightly exceeding 100%. The tracking rate is calculated by dividing the number of tracked particles by the number of recognized particles. The tracking rate exceeds 100% because the number of tracked particles exceeds the number of recognized particles due to the compensation process.

Increasing the particle density and particle movement in one step increases the number of candidate particles to be tracked, making tracking more difficult. An increase in particle density increases the probability of particle overlap and, therefore, the number of lost particles. In this method, if a particle is not recognized for three consecutive steps, compensation is not possible, and the velocity vector cannot be obtained.

An increase in particle movement in one step decreases the tracking rate, as particles are more likely to be outside the measurement range. It also increases the number of candidate particles within the searching area, which also makes tracking more difficult.

In the present study, more than 40% of the particles in the image can be tracked, and velocity vectors are obtained at 0.004 ppp and 240 rpm conditions. If each overlapping particle could be recognized individually, the number of vectors obtained would be expected to increase dramatically.

A polynomial fitting process is required to obtain the velocity vectors, and therefore, particles that are tracked continuously for a certain number of steps are needed. Figure 11 shows the frequency of successive tracking steps before and after the compensation process. The sample data are the same as those in Figures 8 and 9 under the condition of 0.003 ppp and 180 rpm. In this study, the continuous track required for fitting is 10 steps. The compensation process reduces the frequency of occurrences below 10 steps and generally increases the number of continuous tracking steps.

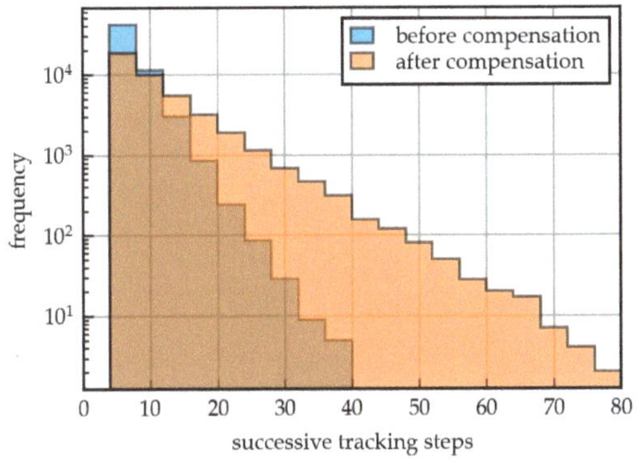

Figure 11. Improvements in the number of successive tracking steps due to the compensation process.

Figure 12 shows the velocity vector acquisition number for different particle densities and disk rotational speeds. As in Figure 11, the results obtained using the in-picture method are indicated by the colored sections of bars, and the improvements due to compensation are shown by black sections.

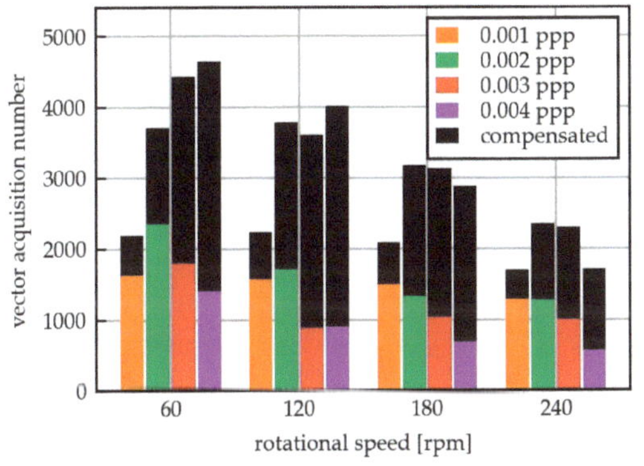

Figure 12. Vector acquisition numbers for different disk rotating speeds and particle densities.

Here, only particles that can be tracked continuously for more than 10 steps are used. The improvement in the tracking rate and the number of consecutive tracking steps due to compensation for lost particles significantly increases the number of acquisition vectors. For example, two sets of particles that are tracked for only five consecutive steps cannot be processed for fitting, so no velocity vectors are obtained. However, if they are concatenated as identical particles by the compensation process, the number of acquisition velocity vectors increases by 10. This greatly increases the number of vector acquisitions. Under some conditions, the increase is several times.

At a disk rotational speed of 60 rpm, where the movement per step is small, the number of vectors is highest at a particle density of 0.004 ppp, which corresponds to the highest number of particles. In contrast, at a disk rotational speed of 240 rpm, where the movement is large, an intermediate number of particles gives the highest number of vector acquisitions.

5. Conclusions

This paper proposed a high-probability and long-term tracking method with which to increase the number of vector acquisitions for 3D3C color PTV. Particles are tracked in a raw picture instead of in 3D space and by using color information as an aid. A particle that temporarily cannot be tracked is then compensated for using location information at the times before and after.

The proposed method is evaluated for flow under a rotating disk. The adoption of the proposed method improves the tracking rate, continuous tracking period, and a number of acquired velocity vectors. This methodology can be applied to the difficult conditions of high particle density (0.004 ppp) and large particle movement (maximum 60 pix).

The results of the present study have increased the number of vector acquisitions and improved the continuous particle tracking time, which have important implications for the post-processing of the measurements. Increased vector acquisition density leads to higher quality results in methods that reconstruct grid velocity data and pressure and vorticity fields from Lagrangian data [17]. Long-term tracking is also an important factor in improving the result of the method that reconstructs the velocity data as it is Lagrangina data [22].

Author Contributions: Conceptualization, methodology, and writing—original draft preparation, M.T.; software, M.T. and K.F.; writing—review and editing, project administration, Y.H. All authors have read and agreed to the published version of the manuscript.

Funding: This research received no external funding.

Data Availability Statement: The experimental data is available upon request.

Conflicts of Interest: The authors declare no conflict of interest.

References

1. Neogi, A.; Mohanta, H.K.; Sande, P.C. Particle Image Velocimetry Investigations on Multiphase Flow in Fluidized Beds: A Review. *Flow Meas. Instrum.* **2023**, *89*, 102309. [CrossRef]
2. Seyfi, S.; Karimpour, S.; Balachandar, R. Simultaneous Flow and Particle Measurements for Multiphase Flows in Hydraulic Engineering: A Review and Synthesis of Current State. *Flow Meas. Instrum.* **2024**, *99*, 102666. [CrossRef]
3. Etminan, A.; Muzychka, Y.S.; Pope, K.; Nyantekyi-Kwakye, B. Flow Visualization: State-of-the-Art Development of Micro-Particle Image Velocimetry. *Meas. Sci. Technol.* **2022**, *33*, 092002. [CrossRef]
4. Discetti, S.; Coletti, F. Volumetric Velocimetry for Fluid Flows. *Meas. Sci. Technol.* **2018**, *29*, 042001. [CrossRef]
5. Prasad, A.K. Stereoscopic Particle Image Velocimetry. *Exp. Fluids* **2000**, *29*, 103–116. [CrossRef]
6. Hori, T.; Sakakibara, J. High-Speed Scanning Stereoscopic PIV for 3D Vorticity Measurement in Liquids. *Meas. Sci. Technol.* **2004**, *15*, 1067–1078. [CrossRef]
7. Burgmann, S.; Brücker, C.; Schröder, W. Scanning PIV Measurements of a Laminar Separation Bubble. *Exp. Fluids* **2006**, *41*, 319–326. [CrossRef]
8. Elsinga, G.E.; Scarano, F.; Wieneke, B.; Van Oudheusden, B.W. Tomographic Particle Image Velocimetry. *Exp. Fluids* **2006**, *41*, 933–947. [CrossRef]
9. Casey, T.A.; Sakakibara, J.; Thoroddsen, S.T. Scanning Tomographic Particle Image Velocimetry Applied to a Turbulent Jet. *Phys. Fluids* **2013**, *25*, 2. [CrossRef]
10. Schanz, D.; Gesemann, S.; Schröder, A. Shake-The-Box: Lagrangian Particle Tracking at High Particle Image Densities. *Exp. Fluids* **2016**, *57*, 70. [CrossRef]
11. Watamura, T.; Tasaka, Y.; Murai, Y. LCD-Projector-Based 3D Color PTV. *Exp. Therm. Fluid Sci.* **2013**, *47*, 68–80. [CrossRef]
12. Park, H.J.; Saito, D.; Tasaka, Y.; Murai, Y. Color-Coded Visualization of Microbubble Clouds Interacting with Eddies in a Spatially Developing Turbulent Boundary Layer. *Exp. Therm. Fluid Sci.* **2019**, *109*, 109919. [CrossRef]
13. Park, H.J.; Yamagishi, S.; Osuka, S.; Tasaka, Y.; Murai, Y. Development of Multi-Cycle Rainbow Particle Tracking Velocimetry Improved by Particle Defocusing Technique and an Example of Its Application on Twisted Savonius Turbine. *Exp. Fluids* **2021**, *62*, 71. [CrossRef]
14. Noto, D.; Tasaka, Y.; Murai, Y. In Situ Color-to-Depth Calibration: Toward Practical Three-Dimensional Color Particle Tracking Velocimetry. *Exp. Fluids* **2021**, *62*, 131. [CrossRef]
15. Noto, D.; Tasaka, Y.; Murai, Y. Low-Cost 3D Color Particle Tracking Velocimetry: Application to Thermal Turbulence in Water. *Exp. Fluids* **2023**, *64*, 92. [CrossRef]
16. Takeyama, M.; Suto, H.; Hattori, Y. 3D3C Rainbow Particle Tracking Velocimetry: Improving Depth Resolution and Velocity Vector Acquisition Rate by Using Color Space for a Multi-Cycle Rainbow Pattern. *J. Vis.* **2024**. [CrossRef]
17. Schneiders, J.F.G.; Scarano, F. Dense Velocity Reconstruction from Tomographic PTV with Material Derivatives. *Exp. Fluids* **2016**, *57*, 139. [CrossRef]
18. Malik, N.A.; Dracos, T.; Papantoniou, D.A. Particle Tracking Velocimetry in Three-Dimensional Flows. *Exp. Fluids* **1993**, *15*, 279–294. [CrossRef]

19. Ishikawa, M.; Murai, Y.; Wada, A.; Iguchi, M.; Okamoto, K.; Yamamoto, F. A Novel Algorithm for Particle Tracking Velocimetry Using the Velocity Gradient Tensor. *Exp. Fluids* **2000**, *29*, 519–531. [CrossRef]
20. Aguirre-Pablo, A.A.; Aljedaani, A.B.; Xiong, J.; Idoughi, R.; Heidrich, W.; Thoroddsen, S.T. Single-Camera 3D PTV Using Particle Intensities and Structured Light. *Exp. Fluids* **2019**, *60*, 25. [CrossRef]
21. Pick, S.; Lehmann, F.O. Stereoscopic PIV on Multiple Color-Coded Light Sheets and Its Application to Axial Flow in Flapping Robotic Insect Wings. *Exp. Fluids* **2009**, *47*, 1009–1023. [CrossRef]
22. Li, L.; Pan, Z. Three-Dimensional Time-Resolved Lagrangian Flow Field Reconstruction Based on Constrained Least Squares and Stable Radial Basis Function. *Exp. Fluids* **2024**, *65*, 57. [CrossRef]

fluids

MDPI

Article

Aeroacoustic Coupling in Rectangular Deep Cavities: Passive Control and Flow Dynamics

Abdul Hamid Jabado [1], Mouhammad El Hassan [2,*], Ali Hammoud [1], Anas Sakout [3] and Hassan H. Assoum [1,3]

1 Mechanical Engineering Department, Beirut Arab University, Tripoli 1300, Lebanon; ajabado10@gmail.com (A.H.J.)
2 Mechanical Engineering Department, Prince Mohammad Bin Fahd University, Al Khobar 34218, Saudi Arabia
3 LASIE UMR CNRS 7356, La Rochelle University, 17000 La Rochelle, France
* Correspondence: melhassan@pmu.edu.sa

Abstract: Deep cavity configurations are common in various industrial applications, including automotive windows, sunroofs, and many other applications in aerospace engineering. Flows over such a geometry can result in aeroacoustic coupling between the cavity shear layer oscillations and the surrounding acoustic modes. This phenomenon can result in a resonance that can lead to significant noise and may cause damage to mechanical structures. Flow control methods are usually used to reduce or eliminate the aeroacoustic resonance. An experimental set up was developed to study the effectiveness of both a cylinder and a profiled cylinder positioned upstream from the cavity in reducing the flow resonance. The cavity flow and the acoustic signals were obtained using particle image velocimetry (PIV) and unsteady pressure sensors, respectively. A decrease of up to 36 dB was obtained in the sound pressure levels (SPL) using the passive control methods. The profiled cylinder showed a similar efficacy in reducing the resonance despite the absence of a high-frequency forcing. Time-space cross-correlation maps along the cavity shear layer showed the suppression of the feedback mechanism for both control methods. A snapshot proper orthogonal decomposition (POD) showed interesting differences between the cylinder and profiled cylinder control methods in terms of kinetic energy content and the vortex dynamics behavior. Furthermore, the interaction of the wake of the control device with the cavity shear layer and its impact on the aeroacoustic coupling was investigated using the POD analysis.

Keywords: cavity flow; passive control; PIV; POD; vortex dynamics

Citation: Jabado, A.H.; El Hassan, M.; Hammoud, A.; Sakout, A.; Assoum, H.H. Aeroacoustic Coupling in Rectangular Deep Cavities: Passive Control and Flow Dynamics. *Fluids* **2024**, *9*, 187. https://doi.org/10.3390/fluids9080187

Academic Editors: Guangjian Zhang and Pier Marzocca

Received: 27 June 2024
Revised: 8 August 2024
Accepted: 15 August 2024
Published: 17 August 2024

1. Introduction

Cavity flow is found in various industrial applications, and numerous researchers have utilized various experimental and numerical methods to investigate the mechanism involved in such flows since the 1950s [1,2]. The aeroacoustic mechanism in such flows results from a coupling between the surrounding acoustic modes and the aerodynamic modes. The flow instability present in the cavity shear layer generates pressure waves upon its impact on the trailing corner of the cavity. The generated pressure waves travel the flow upstream (forming a feedback loop) to control the shear layer near its formation at the cavity's leading corner. The high acoustic levels inside a cavity, or resonance, and the flow fluctuations are caused by the coupling between periodic oscillations of the flow shear layer and the acoustic modes of the surrounding geometry [3–10]. Therefore, several approaches have been attempted to explore the impact of these fluctuations on flow properties, including drag and heat transfer [11,12].

The highly unsteady flow field, the oscillating shear layer, and the production of acoustic waves are fundamental characteristics of cavity flows. However, cavities were classified in different ways, based on the incoming flow, their geometry, or even by the static pressure distribution. Cavity flow has been observed over a wide variety of geometries, as well as Reynolds and Mach numbers. This led to one of the initial classifications, proposed by Charwat et al. [13], who demonstrated that cavities can be categorized as either open or closed. Rossiter [14] added a third category known as transitional. This classification was based on the differences in aspect ratio and the static pressure distribution for each category. However, open cavities can be categorized as deep and shallow cavities. Based on earlier experiments [15–17], deep cavities have a depth greater than their length (L/H < 1), while shallow cavities have aspect ratio greater than 1 (L/H > 1). This paper focuses on the open deep cavity type, characterized (aspect ratio L/H < 10 [15]). All cavities within this range share a common characteristic; they have a boundary layer separating at the upstream corner and reattaching at the downstream edge.

Rockwell et al. [18] stated that both shallow and deep cavities exhibit fluid oscillations know as self-sustaining oscillations. For deep cavities, a forcing mechanism acts in the shear layer with resonant waves propagating in the transverse direction. This may induce further aeroacoustic couplings that lead to resonance [19–21]. On the other hand, shallow cavities are associated with longitudinal acoustic standing waves [22]. Experiments have also shown that propagating disturbances also affect the shear layer of shallow cavities at low subsonic speeds [23].

A reference model for predicting self-sustained cavity oscillations was constructed by Rossiter [24], who developed a semi-empirical analytical model to evaluate the proper frequency of the flow. The production of a high level of noise is often the result of coupling between an acoustic resonance and a certain periodic instability in the flow. Indeed, a portion of the energy is extracted from the flow to sustain the acoustic oscillation. The model proposed by Rossiter evaluates the fundamental frequency of the flow over a cavity, based on a comprehensive description of the interaction between the mixing layer and the acoustic waves. The expression of the Rossiter model is given by

$$S_t = \frac{f_r \times L}{U_0} = \frac{n - \alpha}{M + \frac{1}{k}}, \tag{1}$$

where S_t is the Strouhal number, f_r is the acoustic frequency, n is the cavity mode, L is the length of the cavity, U_0 is the free stream velocity, M represents the Mach number, $k = \frac{u_c}{U_0}$ represents the ratio between the velocities of the external flow and that of the structures convection in the shear layer, and α represents the delay in time between the impingement of the vortices and the creation of an acoustic disturbance. The parameter α is considered an empirical value and is corrected based on experiments.

In addition to analyzing cavity flow properly, it is highly important to propose control methods that can eliminate resonance without significant energy cost or a substantial increase in drag. Therefore, numerous control methods have been proposed by various authors [25]. These methods include passive control techniques, which involve altering the cavity geometry [26,27] or adding external devices, particularly on the leading edge (such as spoilers [24] or cylinder [28,29], etc.). These methods have shown their effectiveness in eliminating resonance and modifying the flow behavior in the shear layer. Furthermore, some limitations of passive control in certain applications have drawn attention to active control techniques, which involve the use of devices requiring external energy. Active control techniques can be classified as either open-loop or closed-loop, and they demonstrate significant potential in achieving attenuation [30].

In this study, we utilized the same configurations used by El Hassan et al. [28], where two passive control techniques were employed: the use of a cylindrical cylinder and the use of a profiled cylinder. The reduction of noise level using such a device has been studied by Stanek et al. [31–33] for subsonic and supersonic flows. These authors propose the following explanation for the effectiveness of the rod: the high-frequency forcing stabilizes the hydrodynamic stability of the flow and thus reduces the pressure levels inside the cavity. However, this is just one hypothesis among others, as the addition of the rod involves complex, highly nonlinear physics, as well as several mechanisms involved in the suppression of pressure fluctuations. Both passive and active control methods were used to attenuate cavity resonance [34]. The control of a cavity flow has been studied to a limited extent for a deep cavity (El Hassan et al. [29]). In most studies, the flow velocity was high (applied to military aviation). At relatively low velocities, the control of deep cavity flow finds application in the automotive and railway fields.

Proper orthogonal decomposition (POD), presented by Lumley in 1967 [35], involves decomposing the random vector field into a set of functions that effectively represent the turbulent motion and organization of the flow through POD modes. It enables the capture of the flow's total fluctuating kinetic energy [36]. Over the years, various POD methods have been developed for numerous fluid dynamic applications [37–40], including cavity flow [41–43]. Overall, these studies highlight the diverse applications of POD in analyzing cavity flows, showcasing its efficacy in capturing dominant flow features, identifying coherent structures, and investigating flow instability. By extracting crucial information on flow features such as frequency, amplitude, and spatial distribution, POD enables researchers to understand the impact of design parameters on these flow characteristics. However, one of the main objectives of utilizing POD in the case of deep cavity flows is to gain insights into the flow physics of the complex flow phenomena. This involves analyzing and identifying coherent structures and recirculating zones that play an important part in describing the flow behavior. Understanding the flow mechanism leading to flow separation, turbulence, and drag is of utmost importance in deep cavity flow analysis.

In this study, a circular cylinder and a profiled cylinder were employed to alter the aeroacoustic resonance within a deep, large cavity subjected to low subsonic flow. Hot wire and pressure measurements were conducted to investigate the acoustic resonance of the cavity. The particle image velocimetry (PIV) technique is employed to investigate the flow dynamics in both cases of the circular and profiled cylinders. Examination of the spatiotemporal development of vortical structures is derived from consecutive snapshots. Statistical support for interpreting the primary mechanisms is provided through both spatiotemporal cross-correlation maps and proper orthogonal decomposition (POD).

2. Materials and Methods

2.1. Cavity and Control Mechanism

A deep cavity inside a closed-circuit wind tunnel of a cross-sectional area of 2×2 m^2 and a length of 10 m (test section) and allowing a maximum velocity of 60 m/s was used to conduct our experiments. In this study, a freestream velocity of U0 = 43 m/s was used [28].

The geometry of the considered cavity is defined as follows: length (L) = 10.4 cm, depth (H) = 52 cm, and width (W) = 200 cm. The cavity is located on the lateral wall of the working area of the wind tunnel. Its leading corner is 8 m far from the inlet of the test section. To assess the boundary layer characteristics, velocity profiles were measured using hotwire measurements immediately upstream from the leading corner of the cavity. In the experimental setup, a cylinder of 0.6 cm in diameter was located 3 cm upstream from the cavity leading corner in the transverse direction. The cylinder was placed at a vertical position of yc = 10 mm from the wall. This specific positioning was selected to achieve an effective control of the resonance of the cavity. The same position was used for the profiled cylinder which has the dimensions shown in Figure 1.

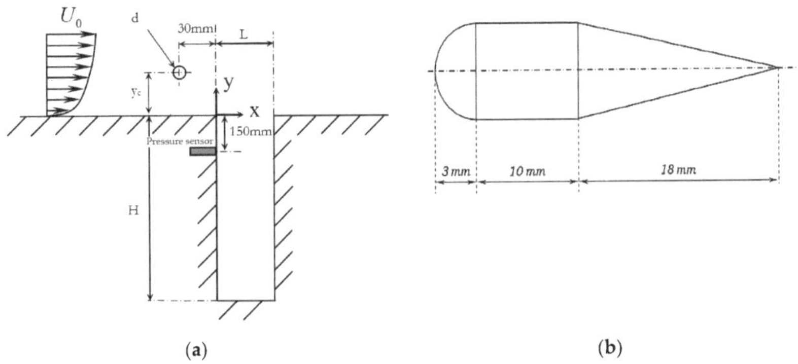

Figure 1. Experimental configuration: (**a**) rectangular cavity; (**b**) control device (profiled cylinder).

2.2. Hot-Wire Measurements

A hot wire probe (Dantec55P15, manufactured by Dantec Dynamics, Skovlunde, Denmark) is located at x/L = 0.1 from the leading corner of the cavity. Acquisition and storage of C.T.A. signals (Constant Temperature Anemometry, DANTEC 90C10, manufactured by Dantec Dynamics, Skovlunde, Denmark) were conducted thanks to the software "Streamline 3.0" from DANTEC Dynamics, Skovlunde, Denmark. Through this procedure, we obtain a voltage signal related to the flow velocity signal at the sensor location.

2.3. Acoustic Measurements

The nominal sensitivity of Kulite sensors used for this study is 275 mV/bar. For each sensor, the output was connected to a multi-channel conditioner that allows for the adjusting of the gain while keeping an average around zero. At the output of the conditioner, the pressure signal is transferred to an analog-to-digital acquisition card with a resolution of 12 bits. The chosen sampling frequency was 6 KHz, and the number of samples was 180,000 per channel, corresponding to an acquisition time of 30 s. A low-pass filter (cut-off at 3 KHz) was implemented to eliminate the aliasing effect.

2.4. PIV Velocity Measurements

Particle image velocimetry (PIV) offers non-invasive and highly accurate results for several flow configurations, making it an attractive choice for researchers investigating complex fluid phenomena in numerous applications. The PIV system is based upon a time resolved PIV Dantec "DynamicStudio" system, including a 2 × 10 mJ dual YAG laser and a 3 kHz Photron Ultima APX-RS Camera (1024 × 1024 pixels), used at 2 kHz frequency (1 kHz vector map). The time delay between two laser pulses was 30 µs. The PIV camera was mounted on a traversing system, perpendicular to the light sheet plane of the laser. PIV measurements were taken in a streamwise plane (x, y), normal to the wall.

In this study, the PIV technique was employed at a sampling rate of 15 Hz, while the freestream velocity was maintained at 43 m/s. PIV differs from other flow measurement and visualization techniques in several ways. It enables the instantaneous determination of the kinematic field, offering good spatial coverage in a 2D or 3D plane but with limited temporal resolution ranging from 1/15 s to 1/1000 s, depending on the experimental setup. PIV works by seeding the flow with particles and capturing images then analyzing these images to determine the velocity field of the fluid.

A typical PIV system includes a laser sheet that illuminates a plane in the flow, and the particles in this plane are captured by a camera pre-processed to improve the data quality, including filtering and other processing measures. Multiple particle images are then created, and these particles are tracked to determine the displacement of all particles, providing information about the flow velocity.

PIV is subjected to uncertainties in velocity measurements due to multiple sources, such as camera calibration or an incorrect choice of the time interval between two images, among other factors. This makes uncertainty analysis important for minimizing errors through the quantification of uncertainty using statistical methods.

The complexity of the PIV system generates measurement errors related to several parameters. Some PIV measurement errors, such as those due to camera calibration, can be minimized by refining the settings, but other errors cannot be eliminated and must be estimated. The main PIV measurement uncertainties can be summarized as follows.

Errors Related to Particle Size and Displacement: The precision of PIV measurements is primarily related to particle size, density, and their average displacement relative to the size of the interrogation zones. When the ratio of the particle image diameter (dpar) to the pixel size (dpix) is less than 2, the energy spectrum is overestimated due to the "peak-locking" phenomenon [44]. Prasad et al. [45] recommend an optimal particle image size according to the relation 3 < dpar/dpix < 4. In this case, the measurement uncertainty is estimated at about 1/10 to 1/20 of the particle image diameter.

In our case, PIV image magnification shows that the average particle diameter is approximately 28 to 32 μm, implying a dpar/dpix ratio of 3.1 to 3.6. Consequently, pixel resolution is adequate, and the uncertainty in particle displacement measurement is approximately 1/15 of the particle image size. Normalizing this uncertainty with the average particle displacement results in a relative error of 1.6%.

The bias error, related to the accuracy of sub-pixel interpolation (Gaussian interpolation), is estimated in our case to be 1/10 of a pixel size with 8-bit images [46]. Thus, the velocity bias error is calculated using the equation:

$$U_{err} = \frac{0.1 \times T_{pixel} \times M_{ech}}{\Delta t}, \tag{2}$$

where T_{pixel} is the pixel size (9 μm), M_{ech} is the image scale factor, and Δt is the time between two laser pulses, with $M_{ech} = 3$ and $\Delta t = 66$ ms; the bias error is approximately 0.04×10^{-3} m/s.

Errors Due to the Number of Samples: A sufficient number of samples, N, must be taken when calculating the average velocity fields. The larger the number N, the closer the velocity distribution function, approximates a Gaussian function. Theoretically, the mean velocity value approaches its exact value as N tends to infinity. The central limit theorem shows that the uncertainty in the mean velocity, ϵ_n, is σ/N (where σ is the standard deviation of the measured velocities). In our case, the standard deviation is obtained with a 95% confidence interval, so $\epsilon_n = 1.96\sigma/\sqrt{N}$. For our study, the highest value for this type of error does not exceed 10^{-2} U_0.

3. Results

3.1. Acoustic Field

The primary objective if the present work is to reduce significantly or eliminate the resonance resulting from the aeroacoustic coupling. Therefore, a simple control method that consists of two mechanisms of control—a cylindrical rod and a profiled cylinder placed transversely upstream of the cavity—is proposed (Figure 1).

Stanek et al. [31] linked the performance of the cylinder to the high-frequency forcing phenomenon induced by the vortex shedding after the cylinder. He suggested that the frequency of shedding must be much higher (at least 10 times) than that of cavity modes. Illy et al. [47] varied the cylinder diameter to observe the impact of the vortex shedding frequency on the control effectiveness, and they found contradiction with Stanek's hypothesis.

Thus, the idea behind implementing a profiled cylinder is to have a body similar to the cylinder but with a profile that prevents the formation of high-frequency vortex shedding in order to explore the effectiveness of such a control and its relationship with the vortex shedding behind the cylinder.

Figure 2 shows the spectrum of the sound pressure level (SPL) of the pressure signal in the absence of control. In the same figure, we plotted the spectrum of the normalized velocity acquired through the hot wire technique. One can notice that both the normalized velocity and SPL present a peak at a frequency of 155 Hz, indicating a strong relation between the shedding vortices and the acoustic field produced in such a flow. This result confirms the existence of a self-sustained loop, where the vortices impinge on the trailing corner of the cavity, thus creating a pressure wave at the same frequency under optimal conditions of energy transfer from the aerodynamic field to the acoustic one.

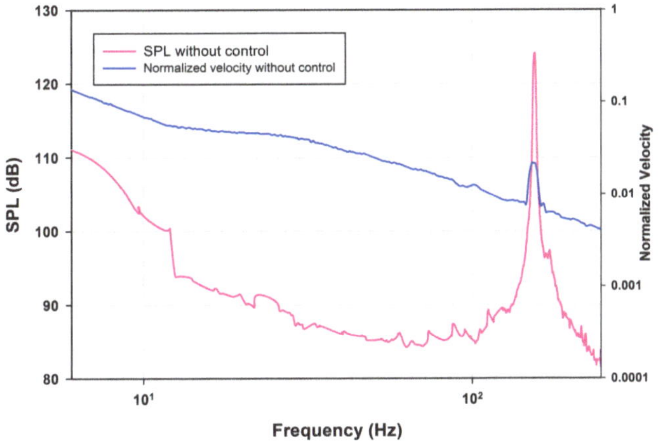

Figure 2. Spectrum of the sound pressure level (SPL) and that of normalized velocity without control.

In the pressure level spectrum Figure 3a, it is noticeable that in the presence of the profiled cylinder, the noise reduction is almost identical to that obtained with the cylinder (About 30 dB of reduction). In the spectrum of the normalized velocity obtained from hot-wire measurements (Figure 3b), it can be noted that the peak of the Rossiter mode has indeed disappeared with the use of the streamlined cylinder, as well as the vortex shedding behind the cylinder. It can also be noted that the energy distribution has shifted towards higher frequencies to a lesser range when the profiled cylinder is employed. This result confirms that the noise reduction is not due to high-frequency forcing.

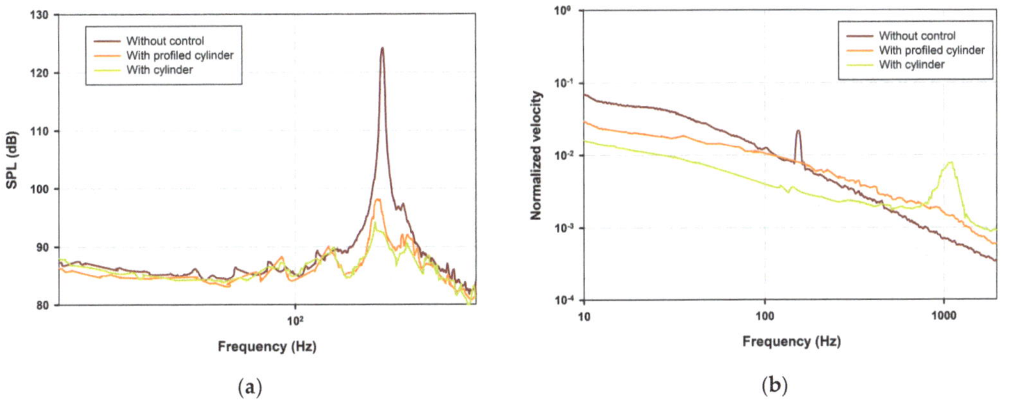

(**a**) (**b**)

Figure 3. Spectrum of (**a**) the sound pressure level (SPL) and (**b**) the normalized velocity for different configurations.

3.2. Kinematic Field

Velocity fields were derived from PIV snapshots to describe the flow. PIV enabled the measurement of mean velocity distribution for both longitudinal and transverse velocity components. Figure 4 illustrates the mean longitudinal velocity distribution, and Figure 5 illustrates the vorticity magnitude, offering valuable information on the flow dynamics, for all three studied cases. In Figure 4a, it is observed that the flow leaving the leading edge undergoes a transition from a boundary-layer flow to a shear layer flow. Consequently, the shear layer, which develops after the boundary layer separation at the cavity leading corner, expands and thickens as it progresses from the separation point to the trailing corner of the cavity. Figure 4b,c show a decrease in velocity in the wake zone of both control mechanisms (cylinder and profiled cylinder). This decrease leads to the thickening of the shear layer in the presence of the cylinders. Similar findings were reported by Illy et al. [48], who considered the impact of a cylinder on the flow of a deep cavity (L/H = 0.42).

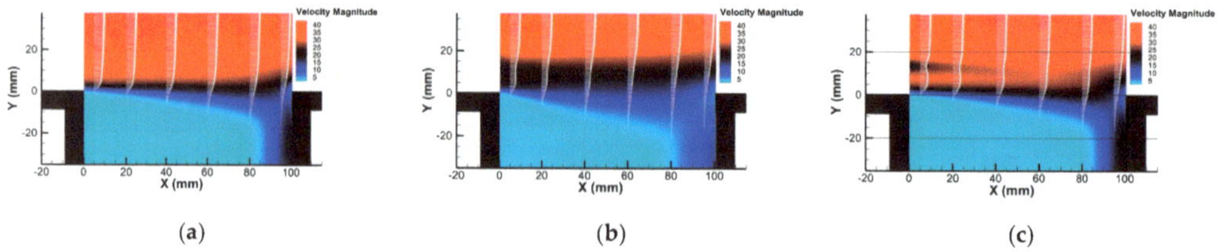

Figure 4. Mean longitudinal velocity distribution: (**a**) without control; (**b**) with cylinder; (**c**) with profiled cylinder.

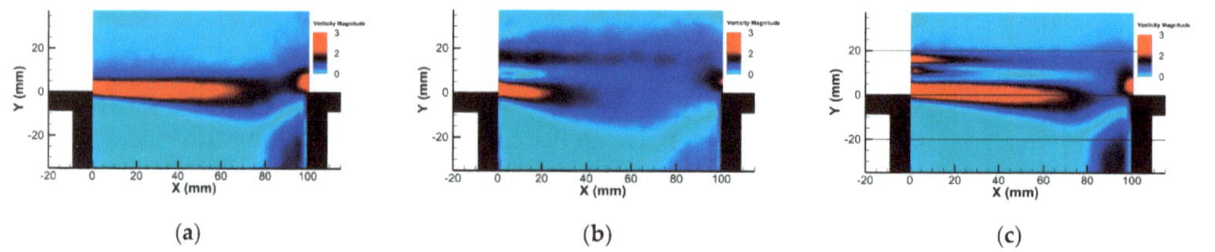

Figure 5. Vorticity magnitude: (**a**) without control; (**b**) with cylinder; (**c**) with profiled cylinder.

Figure 5a shows the presence of high vorticity magnitude at the point of flow separation and along the cavity shear layer. It also shows a region of deficiency (70 mm < X < 93 mm) just before the flow impinges on the cavity trailing corner, where vorticity rises once more. In Figure 5b, it is shown that the streamwise vorticity produced by the cylinder has considerably reduced within the shear layer of the cavity. This indicates the suppression of large-scale vortices in the wake of the cylinder. In Figure 5c, one can note that for the profiled cylinder, the vorticity distribution within the shear layer of the cavity closely resembles that observed without control. This distribution suggests that the profiled cylinder induces a minimal disturbance within the shear layer of the flow.

The normalized turbulent kinetic energy (TKE) is plotted in Figure 6. In absence of control (Figure 6a), elevated values of TKE are exhibited in the downstream region of the cavity shear layer, attributed to the development of large-scale vortices as they travel toward the cavity trailing corner. Introducing the cylinder (Figure 6b) results in a significant rise of TKE in the upstream region. Two peaks in the TKE distribution can be distinguished (at X = 40 mm) and are associated with the shedding from the cylinder. With the profiled cylinder (Figure 6c) compared to the case with cylinder, one can notice that the wake of the profiled cylinder presents dramatic decrease in the TKE. The TKE values in most of the cavity shear layer are even lower than that without control (Figure 6a). This could be related to a configuration where the aeroacoustic coupling is not optimized or suppressed when the profiled cylinder is used despite the elimination of high-frequency forcing (control with cylinder). When the cylinder is employed, it is interesting to observe lower TKE magnitudes in the downstream part of the shear layer as compared to the case without control. These effects will be further investigated using the analysis of the instantaneous velocity fields.

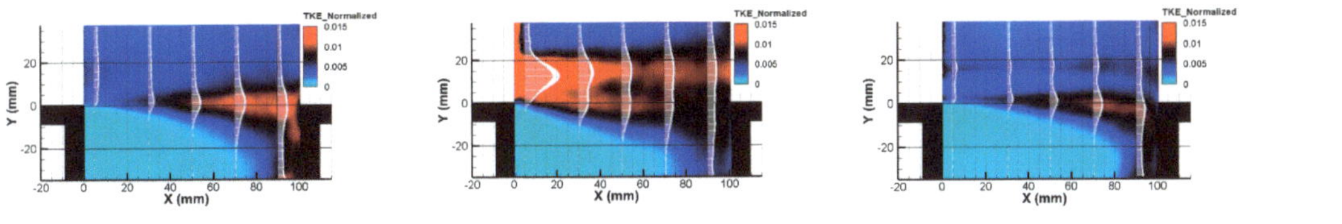

(a) (b) (c)

Figure 6. Normalized total kinetic energy: (**a**) without control; (**b**) with cylinder; (**c**) with profiled cylinder.

For vortex identification purpose, the Lambda-2 criterion, which detects the center of rotating structures by assuming the minimum local pressure at the vortex centre, is used [49]. Essentially, it is a mathematical tool based on the eigenvalues of the velocity gradient tensor that helps in identifying of coherent structures exhibiting strong rotational behavior. Lambda-2 criterion is favored since it gives clearer visualizations of purely rotational structures without capturing regions of high strain, and it does not require the complex calculations involved, respectively, in the Q-criterion and Delta criterion methods. In Figure 7, Lambda-2 fields are displayed for successive phases, following the presentation method used by Assoum (FDR 2013). This presentation reveals the vortex patterns along the shear layer of the cavity flow. Observing Figure 7a for the case without control, one can notice a single pattern of vortices which is dominated by that of the shear layer of the cavity flow. In the case of control with a cylinder (Figure 7b), two main vortex paths are observed: the first path is linked to the vortices in the cavity shear layer, and the second path in the upper part of this figure (10 < Y < 20) corresponds to the shedding vortices from the cylinder. One can also observe a lower concentration of large-scale vortices near the cavity downstream corner when the cylinder is used. In Figure 7c, when a profiled cylinder is used, one can see that the upper path of vortices observed when a cylinder was used almost disappears.

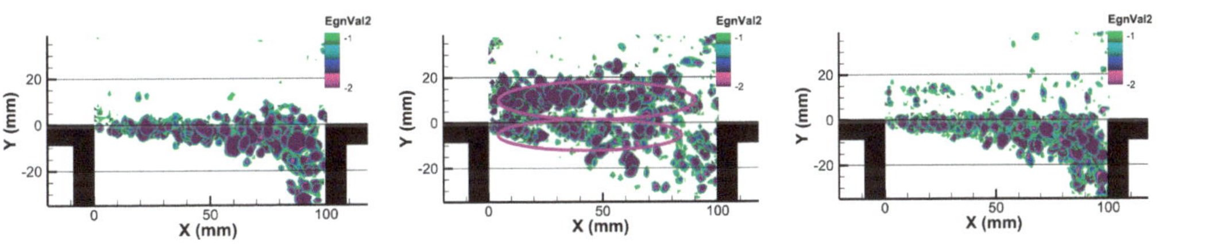

(a) (b) (c)

Figure 7. Turbulent structures using Lambda-2 criterion: (**a**) without control; (**b**) with cylinder; (**c**) with profiled cylinder.

Using both control methods, a spatial shift of the cavity shear layer towards inside the cavity is obtained, with a more pronounced shift in the middle region of the cavity shear layer, especially for the cylinder case. The interaction between the wake of the control devices and the cavity shear layer seem to weaken the impact of the large scale vortical structures on the downstream edge, thus dramatically reducing the aeroacoustic coupling and thus the generated noise.

To better apprehend these complex flows, time–space cross-correlation maps are established to provide more characterization of the cavity–wake interaction.

3.3. Cross-Correlation Maps

In order to acquire more understanding of the cavity–wake interaction, we propose time–space cross-correlation maps. Fluctuation transverse velocity at a specific point chosen as a reference undergoes cross-correlations across the whole field. The cross-correlation coefficient is given by

$$R_{v'v'} = \frac{\frac{1}{N}\sum_1^N v'(x,y,t)v'(x_0,y_0,t)}{RMS(v'(x,y)) \times RMS(v'(x_0,y_0))}. \tag{3}$$

In this investigation, many points were considered as reference points starting from the cavity leading corner (just downstream from the location of the control devices) towards the impingement zone. These points can be observed in Figure 8 at the positions where $R_{v'v'} = 1$ (which corresponds to an autocorrelation). These points were chosen in order to observe the relationship between the cylinder's wake and the cavity shear layer.

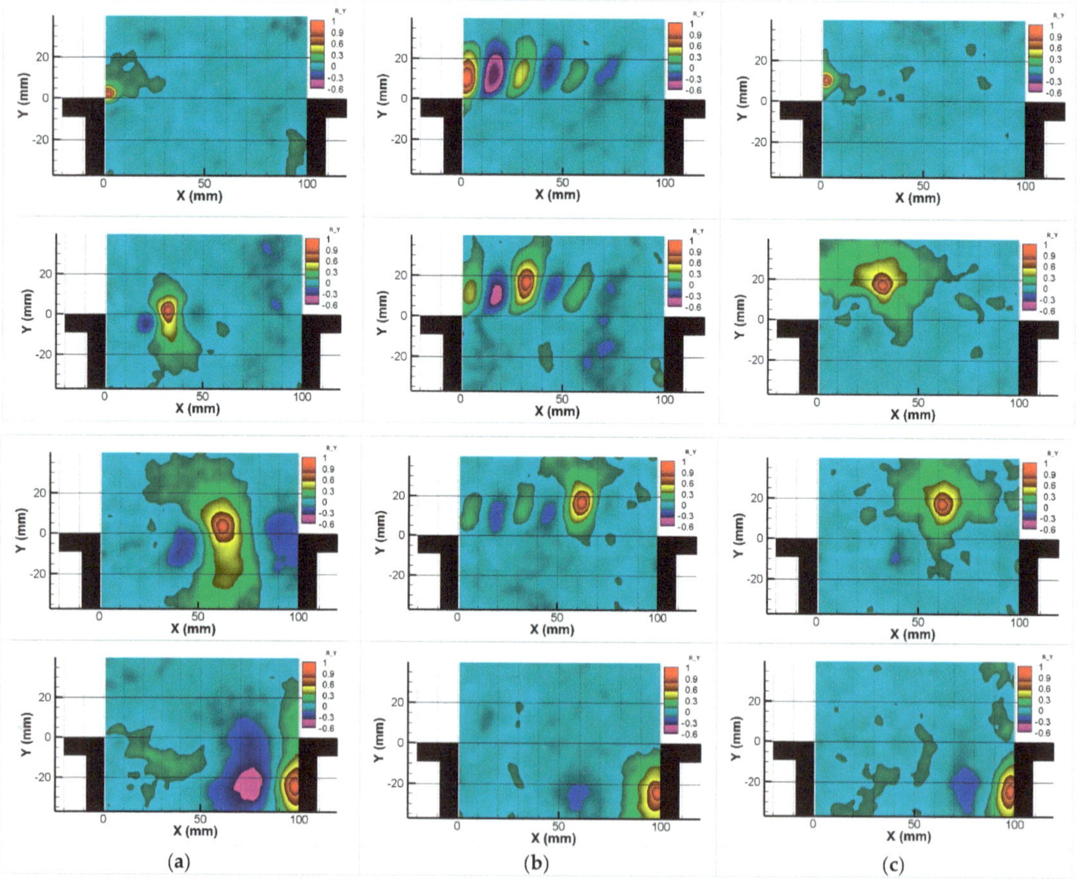

Figure 8. Cross correlation maps: (**a**) without control; (**b**) with cylinder; (**c**) with profiled cylinder.

As compared to the case of controlled flow with a cylinder, it is obvious that the profiled cylinder does not have high-frequency shedding. This confirms that an effective control of the aeroacoustic coupling does not require high-frequency forcing.

In the case of no control, it is interesting to see a correlation between the impingement zone (below the downstream corner) and the location where the vortices are generated after the boundary layer separation, just downstream from the cavity leading corner. Such a mechanism (explained by the feedback loop) is absent for the cylinder and profiled cylinder.

3.4. Proper Orthogonal Decomposition (POD)

In confined deep cavity flow, the interactions between the fluid and the cavity walls can lead to complex fluid motion. Velocity fields data can be analyzed using the POD to identify the most frequent and influential fluid patterns in the flow. POD is recognized as a valuable method in providing insights and improving the understanding of fluid flow dynamics. It consists of finding an optimal basis that represents the main flow features. In the current study, the snapshot POD was performed using 550 snapshots taken at regular intervals (acquisition frequency of 15 Hz). Figure 9 displays the cumulative energy sum over the POD modes and the energy fraction for each mode.

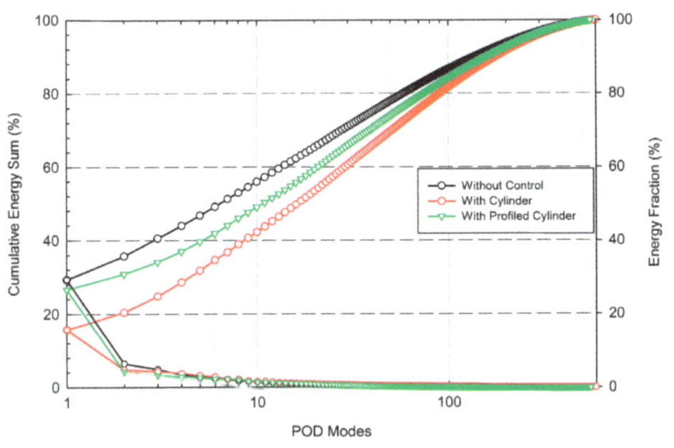

Figure 9. Energy portion and cumulative energy of the POD modes for the longitudinal velocity (U).

In case of no control, the first POD mode contains almost 30% of the flow kinetic energy (KE), and the first five POD modes contain nearly 50% of the total KE. When the profiled cylinder was used, it was found that the first mode contains 27% of the KE, whereas the cumulative KE of the first five POD modes is 40% of the total KE. This result is in agreement with Figure 6a,c, where a small effect of the profiled cylinder on the TKE of the cavity flow was observed. It should, however, be noted that only 15% of the KE is contained in the first POD mode when the cylinder is used, and the cumulative energy for the first five modes is reduced to 31% (Figure 9). This result may be affirmed by Figure 5b, where the vorticity magnitude was found to be significantly decreased in the shear layer of the flow; therefore, the vortices are weakened. Thus, the POD projection attributes less energy to the first POD modes since they are related are related to these coherent structures.

The five most energetic spatial POD modes obtained from longitudinal velocities are illustrated in Figure 10 for controlled and non-controlled flows. Without control, the most energetic POD modes are linked to the advection of the large vortices in the downstream flow of the cavity. The profiled cylinder case presents some similarities with the case without control. However, one can distinguish the higher KE content shifted toward the modes where the vortices are oriented upward near the cavity trailing corner. This might, in part, explain the weak and less organized pressure waves following the impingement of the vortices in the case of the profiled cylinder when compared to the non-controlled flow. With the cylinder, it is noted that the first two POD modes are linked to the cylinder shedding, whereas the three other modes reflect the interaction of the cylinder wake with the cavity shear layer.

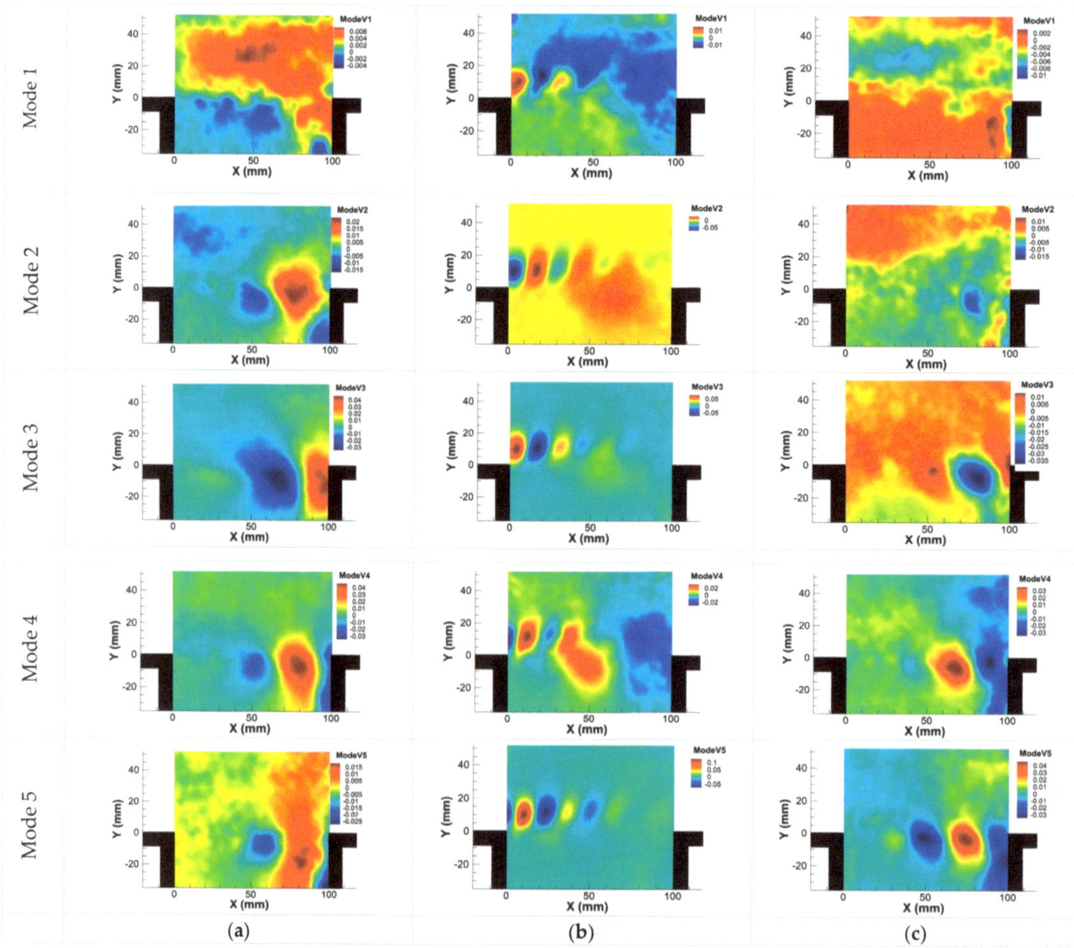

Figure 10. POD spatial modes: (**a**) without control; (**b**) with cylinder; (**c**) with profiled cylinder.

4. Conclusions

The state-of-the-art experimental techniques used in this study provide accurate qualitative and quantitative descriptions (within the uncertainties presented in the experimental section) of the aeroacoustics, which is the focus of the present work. The complex studied flow dynamics (the fully turbulent incoming boundary layer, its separation at the cavity leading edge, and the development of cavity shear layer) make it very difficult to predict using analytical methods. To the best of the authors' knowledge, no analytical predictions exist in the literature for the same cavity geometry and operating flow regimes. However, the aeroacoustic characteristics of similar flows were compared to both semi-empirical and analytical predictions, which were already discussed in the study of El Hassan et al. [21].

Experimental characterization of the interaction between the cavity vortex system and the wake of a control mechanism consisting of a cylinder and a profiled cylinder has been conducted. Different experimental methods were employed to establish this analysis. The main conclusions can be summarized as follows:

- Achieving a similar significant reduction in noise with both the cylindrical and profiled cylindrical configurations confirms that this reduction is not attributable to high-frequency forcing.
- Vorticity distribution suggests that the presence of the profiled cylinder introduces minimal disturbance within the cavity shear layer compared to the circular cylinder;
- The distribution of normalized turbulent kinetic energy (TKE) significantly decreases with the profiled cylinder, indicating lower TKE compared to the uncontrolled flow.
- Lambda-2 fields reveal two primary vortex paths in the case of control with a cylinder: one linked to vortices in the cavity shear layer and the other related to the cylinder shedding.
- Cross-correlation maps validate that effective control of aeroacoustic coupling does not necessitate high-frequency forcing.
- Snapshot POD analysis indicates that in the absence of control, and with control using a profiled cylinder, the first POD mode contains nearly 30% of the flow's KE, while only 15% of the KE is contained in the first POD mode when using the standard cylinder. This aligns with the decreased vorticity distribution in the shear layer of the flow controlled with the cylinder, resulting in weakened vortical structures. Consequently, the POD projection attributes less energy to the first POD modes since they are associated with these coherent structures.
- Spatial modes reveal that the higher KE content shifts towards modes where vortical structures are oriented upward near the cavity trailing edge. This may partially explain the weak and less-organized pressure waves following the impingement of vortices with the controlled flow compared to the uncontrolled case.
- Many perspectives could be proposed:
- In this study, the focus was on reducing noise near deep cavities. It is worth noting that this noise reduction, achieved by adding a cylinder upstream of the cavity, may lead to an increase in drag and could therefore be detrimental to aerodynamic efficiency. Measurement of friction (estimation of drag) should be considered with the aim of optimizing both aerodynamic and aeroacoustic aspects.
- Investigating energetic transfers from the aerodynamic field towards the acoustic field with and without the control mechanisms would be of interest in a future investigation.

Author Contributions: A.H.J.: writing—original draft preparation, Data analysis. M.E.H.: Conceptualization, writing—review and editing. A.H.: Supervision. A.S.: Resources and funding acquisition. H.H.A.: Supervision, writing—review and editing. All authors have read and agreed to the published version of the manuscript.

Funding: This research received no external funding.

Data Availability Statement: Data available on request due to restrictions (e.g., privacy, legal or ethical reasons). The data presented in this study are available on request from the corresponding author (accurately indicate status).

Conflicts of Interest: The authors declare no conflict of interest.

References

1. Powell, A. On Edge Tones and Associated Phenomena. 1953. Available online: https://www.ingentaconnect.com/search/article?option2=author&value2=Powell,+Alan&pageSize=10&index=7 (accessed on 28 February 2020).
2. Krishnamurty, K. *Acoustic Radiation from Two-Dimensional Rectangular Cutouts in Aerodynamic Surfaces*; NTRS: Washington, DC, USA, 1955.
3. Assoum, H.H.; Hamdi, J.; Hassan, M.; Abed-Meraim, K.; Kheir, M.E.; Mrach, T.; Asmar, S.E.; Sakout, A. Turbulent kinetic energy and self-sustaining tones: Experimental study of a rectangular impinging jet using high Speed 3D tomographic Particle Image Velocimetry. *J. Mech. Eng. Sci.* **2020**, *14*, 6322–6333. [CrossRef]
4. Hamdi, J.; Assoum, H.H.; Abed-Meraïm, K.; Sakout, A. Volume reconstruction of a plane jet impinging on a slotted plate using the phase averaging technique. *Energy Procedia* **2017**, *139*, 404–409. [CrossRef]
5. Assoum, H.H.; El Hassan, M.; Abed-Meraim, K.; Sakout, A. The vortex dynamics and the self sustained tones in a plane jet impinging on a slotted plate. *Eur. J. Mech.-B/Fluids* **2014**, *48*, 231–235. [CrossRef]

6. El Zohbi, B.; Assoum, H.H.; Alkheir, M.; Afyouni, N.; Meraim, K.A.; Sakout, A.; El Hassan, M. Experimental investigation of the Aero-Acoustics of a rectangular jet impinging a slotted plate for different flow regimes. *Alex. Eng. J.* **2024**, *87*, 404–416. [CrossRef]
7. Afyouni, N.E.; Alkheir, M.; Assoum, H.; El Zohbi, B.; Abed-Meraim, K.; Sakout, A.; El Hassan, M. Effect of a Control Mechanism on the Interaction between a Rectangular Jet and a Slotted Plate: Experimental Study of the Aeroacoustic Field. *Fluids* **2023**, *8*, 309. [CrossRef]
8. Assoum, H.H.; El Kheir, M.; Afyouni, N.E.; El Zohbi, B.; Meraim, K.A.; Sakout, A.; El Hassan, M. Control of a rectangular impinging jet: Experimental investigation of the flow dynamics and the acoustic field. *Alex. Eng. J.* **2023**, *79*, 354–365. [CrossRef]
9. Alkheir, M.; Assoum, H.H.; Hamdi, J.; Mrach, T.; Hassan, M.; Sakout, A. Experimental study of the vortex organization in a rectangular impinging jet in the presence of self-sustained tones. *Energy Rep.* **2021**, *8*, 1486–1492. [CrossRef]
10. Assoum, H.H.; Hamdi, J.; Alkheir, M.; Meraim, K.A.; Sakout, A.; Obeid, B.; El Hassan, M. Tomographic Particle Image Velocimetry and Dynamic Mode Decomposition (DMD) in a Rectangular Impinging Jet: Vortex Dynamics and Acoustic Gen-eration. *Fluids* **2021**, *6*, 429. [CrossRef]
11. Jabado, A.H.; Assoum, H.H.; Hammoud, A.; Meraim, K.A.; Sakout, A.; El Hassan, M. Review-Heat Transfer Inside Cavity Flows Trends. *IOP Conf. Series Earth Environ. Sci.* **2022**, *1008*, 012002. [CrossRef]
12. Jabado, A.H.; El Hassan, M.; Assoum, H.H.; Hammoud, A.; Meraim, K.A.; Sakout, A. A review of cavity heat transfer under separated/reattached flow conditions. *Energy Rep.* **2021**, *8*, 949–956. [CrossRef]
13. Charwat, A.; Roos, J.; Dewey, F., Jr.; Hitz, J. An investigation of separated flows-Part I: The pressure field. *J. Aerosp. Sci.* **1961**, *28*, 457–470. [CrossRef]
14. Rossiter, J. The Effects of Cavities on the Buffeting of Aircraft. RAE Technical Memorandum No. Aero 754. 1962. Available online: https://www.semanticscholar.org/paper/The-effects-of-cavities-on-the-buffeting-of-Rossiter/6643bd3131210206868827364f0cb8cd5ecf40ae (accessed on 14 August 2024).
15. East, L. Aerodynamically induced resonance in rectangular cavities. *J. Sound Vib.* **1966**, *3*, 277–287. [CrossRef]
16. Plumblee, H.; Gibson, J.; Lassiter, L. *A Theoretical and Experimental Investigation of the Acoustic Response of Cavities in an Aerodynamic Flow*; Flight Dynamics Laboratory, Aeronautical Systems Division, Air Force Systems: Wright-Patterson AFB, OH, USA, 1962.
17. Heller, H.; Holmes, D.; Covert, E. Flow-induced pressure oscillations in shallow cavities. *J. Sound Vib.* **1971**, *18*, 545–553. [CrossRef]
18. Rockwell, D.; Naudascher, E. Review—Self-Sustaining Oscillations of Flow Past Cavities. *J. Fluids Eng.* **1978**, *100*, 152–165. [CrossRef]
19. El Hassan, M.; Labraga, L.; Keirsbulck, L. Aero-Acoustic Oscillations inside Large Deep Cavities. 2007. Available online: https://espace.library.uq.edu.au/data/UQ_120829/Hassan_afmc_16_07.pdf?Expires=1723967529&Key-Pair-Id=APKAJKNBJ4MJBJNC6NLQ&Signature=GeOqc~Fgk-VhHIpIcKlbwYVBOvgOUEaGSVAleFBUb8Y-hXBY9EDHReUO8D0ifQH5lmgpI3XWPujvibc8P3MEtaHuogmkfxrP-jpntFHGom4bhc2bNeAxwFM9m2JUiJ9uDgQaA5-Hn6VOhvHcuYTgtvbe8nd4Ru2PbhMF-yJpGJalPybwbQRoVZVVcdVAUmS5tCSCW~wIcugfNJaMbczTxxPskxgJG0l~MOGl6dg6WclsKRGCDjG2xA0rkppW5jzK0NQbPpH0f0jVFvS1cd-k1vG4DvOT6Q5YQI80fPpGnkqAM1EDL9PJHCIn7QO5lvga3QvxRVZ9KaddBKiy2NeRQA__ (accessed on 15 August 2024).
20. El Hassan, M.; Keirsbulck, L.; Labraga, L. Non-oscillating/Oscillating Shear Layer over a Large Deep Cavity at Low-Subsonic Speeds. *Flow, Turbul. Combust.* **2008**, *82*, 359–374. [CrossRef]
21. El Hassan, M.; Keirsbulck, L.; Labraga, L. Aero-Acoustic Coupling Inside Large Deep Cavities at Low-Subsonic Speeds. *J. Fluids Eng.* **2008**, *131*, 011204. [CrossRef]
22. Sinha, J.; Arora, K. Review of the flow-field analysis over cavities. In Proceedings of the 2017 International Conference on Infocom Technologies and Unmanned Systems (Trends and Future Directions) (ICTUS), Dubai, United Arab Emirates, 18–20 December 2017; pp. 870–876.
23. Sarohia, V. Experimental investigation of oscillations in flows over shallow cavities. *AIAA J.* **1977**, *15*, 984–991. [CrossRef]
24. Rossiter, J. *Wind-Tunnel Experiments on the Flow over Rectangular Cavities at Subsonic and Transonic Speeds*; Cranfield University: Bedford, UK, 1964.
25. Lawson, S.; Barakos, G. Review of numerical simulations for high-speed, turbulent cavity flows. *Prog. Aerosp. Sci.* **2011**, *47*, 186–216. [CrossRef]
26. Vikramaditya, N.; Kurian, J. Pressure oscillations from cavities with ramp. *AIAA J.* **2009**, *47*, 2974–2984. [CrossRef]
27. Kok, J.C.; Soemarwoto, B.I.; van der Ven, H. X-LES simulations using a high-order finite-volume scheme. *Notes Numer. Fluid Mech. Multidiscip. Des.* **2008**, *97*, 87.
28. El Hassan, M.; Keirsbulck, L. Passive control of deep cavity shear layer flow at subsonic speed. *Can. J. Phys.* **2017**, *95*, 894–899. [CrossRef]
29. El Hassan, M.; Keirsbulck, L.; Labraga, L.; Lippert, M. Control of deep cavity tones using a spanwise cylinder at low-subsonic speeds. *HEFAT* **2008**, *2008*, 5.
30. Cattafesta, L.N., III; Song, Q.; Williams, D.R.; Rowley, C.W.; Alvi, F.S. Active control of flow-induced cavity oscillations. *Prog. Aerosp. Sci.* **2008**, *44*, 479–502. [CrossRef]
31. Stanek, M.; Ross, J.; Odedra, J.; Peto, J. High Frequency Acoustic Suppression-The Mystery of the Rod-in-Crossflow Revealed. In Proceedings of the 41st Aerospace Sciences Meeting and Exhibit, Reno, NV, USA, 6–9 January 2003.
32. Stanek, M.; Raman, G.; Kibens, V.; Ross, J.; Odedra, J.; Peto, J. Control of cavity resonance through very high frequency forcing. In Proceedings of the 6th Aeroacoustics Conference and Exhibit, Hartford, CT, USA, 4–6 June 1980.

33. Stanek, M.; Raman, G.; Ross, J.; Odedra, J.; Peto, J.; Alvi, F.; Kibens, V. High Frequency Acoustic Suppression—The Role of Mass Flow & The Notion of Superposition. In Proceedings of the 8th AIAA/CEAS Aeroacoustics Conference & Exhibit, Breckenridge, CO, USA, 17–19 June 2002.
34. Shaw, L.; Clark, R.; Talmadge, D. F-111 generic weapons bay acoustic environment. *J. Aircr.* **1988**, *25*, 147–153. [CrossRef]
35. Lumley, J.L. The structure of inhomogeneous turbulent flows. *Atmos. Turbul. Radio Wave Propag.* **1967**, 166–178.
36. Weiss, J. A Tutorial on the proper orthogonal decomposition. In Proceedings of the AIAA Aviation 2019 Forum, Dallas, TX, USA, 17–21 June 2019.
37. Berkooz, G.; Holmes, P.; Lumley, J.L. The proper orthogonal decomposition in the analysis of turbulent flows. *Annu. Rev. Fluid Mech.* **1993**, *25*, 539–575. [CrossRef]
38. Ribau, M.; Gonçalves, N.D.; Ferrás, L.L.; Afonso, A.M. Flow Structures Identification through Proper Orthogonal Decompo-sition: The Flow around Two Distinct Cylinders. *Fluids* **2021**, *6*, 384. [CrossRef]
39. Delville, J. Characterization of the organization in shear layers via the Proper Orthogonal Decomposition. *Flow, Turbul. Combust.* **1994**, *53*, 263–281. [CrossRef]
40. Hamdi, J.; Assoum, H.; Abed-Meraïm, K.; Sakout, A. Volume reconstruction of an impinging jet obtained from stereoscopic-PIV data using POD. *Eur. J. Mech.-B/Fluids* **2018**, *67*, 433–445. [CrossRef]
41. Podvin, B.; Fraigneau, Y.; Lusseyran, F.; Gougat, P. A Reconstruction Method for the Flow Past an Open Cavity. *J. Fluids Eng.* **2005**, *128*, 531–540. [CrossRef]
42. Rowley, C.; Colonius, T.; Murray, R. POD based models of self-sustained oscillations in the flow past an open cavity. In Proceedings of the 6th Aeroacoustics Conference and Exhibit, Hartford, CT, USA, 4–6 June 1980.
43. Lawson, S.; Barakos, G.; Simpson, A. Understanding cavity flows using proper orthogonal decomposition and signal pro-cessing. *J. Algorithms Comput. Technol.* **2010**, *4*, 47–69. [CrossRef]
44. Westerweel, J. *Digital Particle Image Velocimetry*; Delft University Press: Delft, The Netherlands, 1993.
45. Prasad, A.; Adrian, R.; Landreth, C.; Offutt, P. Effect of resolution on the speed and accuracy of particle image ve-locimetry interrogation. *Exp. Fluids* **1992**, *13*, 105–116. [CrossRef]
46. Raffel, M.; Willert, C.E.; Kompenhans, J. *Particle Image Velocimetry: A Practical Guide*; SPIE Optical Engineering Press: Bellingham, WA, USA, 1998.
47. Illy, H.; Geffroy, P.; Jacquin, L. Control of cavity flow by means of a spanwise cylinder. In Proceedings of the 21th International Congress of Theoretical and Applied Mechanics, IUTAM, Warsaw, Poland, 15–21 August 2004.
48. Illy, H.; Geffroy, P.; et Jacquin, L. Contrôle des oscillations de cavité au moyen d'un cylinder placé transversalement l'écoulement. In Proceedings of the 17me Congrès Français de mécanique, Troyes, France, 29 August–2 September 2005.
49. Jeong, J.; Hussain, F. On the identification of a vortex. *J. Fluid Mech.* **1995**, *285*, 69–94. [CrossRef]

fluids

MDPI

Article

Flowfield and Noise Dynamics of Supersonic Rectangular Impinging Jets: Major versus Minor Axis Orientations †

Yogesh Mehta [1,*], Vikas N. Bhargav [2] and Rajan Kumar [3]

1 Olympic College, Washington State University, Bremerton, WA 98337, USA
2 Aerospace Engineering, Auburn University, Auburn, AL 36849, USA
3 Florida Center for Advanced Aero-Propulsion, Department of Mechanical Engineering, FAMU-FSU College of Engineering, Tallahassee, FL 32310, USA
* Correspondence: ymehta257@gmail.com

† This paper is an extended version of paper published in the AIAA AVIATION 2022 Forum, Chicago, IL, USA, 20 June 2022; p. 3896.

Abstract: The current study explores the flowfield and noise characteristics of an ideally expanded supersonic (Mach 1.44) rectangular jet impinging on a flat surface. The existing literature is primarily concentrated on axisymmetric jets, known for their resonance dominance, pronounced unsteadiness, and acoustic signatures. In contrast, non-axisymmetric jets remain relatively less understood, particularly those impinging on a ground surface. By employing Schlieren imaging, high-frequency pressure measurements using high-bandwidth transducers, and particle image velocimetry (PIV), this research comprehensively examines the flow-acoustic phenomena. Schlieren imaging revealed distinct, coherent structures and strong acoustic waves, while pressure measurements at the impingement surface exhibited high-amplitude fluctuations, peaking at approximately 186 dB. Acoustic analysis identified multiple high-amplitude tones with unique directional characteristics, suggesting the potential for multiple simultaneous modes in rectangular jets. Furthermore, the PIV data elucidated differences in the jet shear layer and wall jet development attributed to the nozzle orientation. These findings contribute to a deeper understanding of non-axisymmetric jet behavior, offering insights relevant to fundamental flow physics and practical applications such as vertical takeoff and landing aircraft.

Keywords: supersonic jets; flowfield; jet noise; impinging jets

Citation: Mehta, Y.; Bhargav, V.N.; Kumar, R. Flowfield and Noise Dynamics of Supersonic Rectangular Impinging Jets: Major versus Minor Axis Orientations. *Fluids* **2024**, *9*, 169. https://doi.org/10.3390/fluids9080169

Academic Editor: Robert Martinuzzi

Received: 18 June 2024
Revised: 11 July 2024
Accepted: 20 July 2024
Published: 24 July 2024

1. Introduction

Axisymmetric supersonic impinging jets under specific operating conditions are known to exhibit high unsteadiness and pronounced acoustic fluctuations due to the feedback mechanism [1–7]. This dynamics is largely attributed to a mechanism where Kelvin–Helmholtz instabilities within the jet shear layer give rise to vortical structures which grow as they convect downstream. Upon impinging on a surface, these structures induce significant pressure fluctuations, which manifest as upstream-traveling acoustic waves that subsequently excite the shear layer at the nozzle lip, thus perpetuating the feedback loop. The extent of unsteadiness and noise generated by these jets is modulated by various parameters, which include the distance of the nozzle to the ground plane, presence of adjacent jets, momentum, Mach number, nozzle pressure ratio(NPR), and temperature ratio(TR) of the jet [8–11]. Figure 1 describes the two modes of feedback loop mechanisms associated with supersonic jets. Powell [12,13] developed a theoretical model to predict the frequencies associated with the feedback loop of round impinging jets.

Despite extensive investigations into axisymmetric jets, the study of non-axisymmetric jets, particularly those emanating from rectangular nozzles, remains relatively sparse. Rectangular jets are distinguished by their enhanced mixing rates with the ambient environment compared with their axisymmetric counterparts, attributed to their distinct geometric configuration [14–18]. Previous research has delved into the flowfield evolution and three-dimensional characteristics of rectangular jets, identifying three primary regions within these jets: a potential core, a self-similar decay region along the minor axis, and an axisymmetric decay region where the velocity decay is inversely proportional to the axial distance. Notably, non-axisymmetric jets exhibit a phenomenon known as axis switching [15,18,19], where the size of the jet along its minor axis surpasses the one along its major axis at a certain downstream distance. Previous studies utilizing particle image velocimetry (PIV) and acoustic measurements in rectangular jets have revealed the impact of corner and streamwise vortices on the generation of self-induced excitation, commonly referred to as the screech tone [18,20–22]. Similar findings have also been reported in various numerical studies (see, for example, [23,24]).

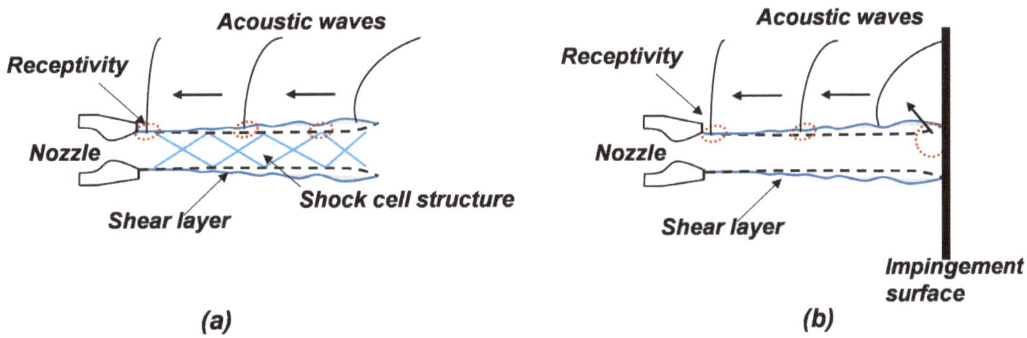

Figure 1. Feedback loop mechanism in a supersonic (**a**) screeching jet and (**b**) impinging jet.

The interaction of such jets with a surface introduces additional noise components, characterized by both discrete (impingement tones) and broadband elements. Early foundational work by researchers, such as Krothapalli [25], shed light on the fundamental features of rectangular impinging jets, identifying the dominant tones associated with screech and impingement. Subsequent studies [26–30], including detailed investigations into acoustic behavior such as multiple tones and flow oscillations of rectangular jets, have furthered our understanding of these complex phenomena.

However, much of the existing literature has focused on off-design, screech-dominated conditions, with limited exploration of the impact of the impingement distance on the flowfield evolution at the ground surface and its consequent acoustic signature. In addressing this gap, the present study aims to elucidate the effects of the impingement distance on ground-level unsteady pressures and nearfield acoustics as well as examine variations in the flow and acoustic fields along the major and minor axes of the rectangular nozzle. Through this investigation, we seek to answer two pivotal scientific questions. (1) What is the effect of the impingement distance on the unsteady pressures on the ground and the nearfield acoustics of jets exhausting from rectangular nozzles? (2) How do the flow and acoustic fields vary about the two axes (i.e., major and minor axes) of the rectangular nozzle?

2. Experimental Set-up and Conditions

2.1. Test Facility

All experiments were carried out in the short take-off and vertical landing (STOVL) facility at the Florida Center for Advanced Aero Propulsion located at the FAMU-FSU College of Engineering. The compressed air for this facility was supplied through a set of tanks with a capacity of 110 m^3 which could supply air at a maximum pressure of 3450 kPa. The facility has the capability of supplying a maximum absolute pressure of 827 kPa and stagnation temperature of 750 K, which are controlled by high-pressure valves and an inline 192 kW electric heater to achieve the desired nozzle pressure ratio and temperature ratio, respectively.

A series of honeycomb straighteners and meshes were installed upstream of the nozzle to streamline and condition the flow. A ground plane, which was positioned perpendicular to the nozzle axis, was attached to a traverse mechanism. This mechanism was connected to a stepper motor to simulate various distances between the nozzle and the ground. A circular impingement plate insert 266.7 mm in diameter was flush-mounted on the rectangular ground plane to study the impinging jet flows. The impingement plate insert was designed to accommodate unsteady pressure transducers in an array. The plate insert could also be rotated about its axis to change the orientation of the pressure transducer array.

The experiments were conducted with a converging–diverging rectangular nozzle with a design Mach number of 1.44. The nozzle had an aspect ratio of 4:1 with the minor axis of the nozzle (h = 10 mm), while the major axis was w = 40 mm. The converging part of the nozzle was designed using a fifth-order polynomial, and the diverging section was designed based on the method of characteristics (MOC), which existed only along the minor axis. The major axis was kept straight from the throat to the exit.

2.2. Measurement Techniques and Instrumentation

2.2.1. Schlieren

The flowfield was visualized using a Z-type Schlieren technique. White light from a Luminus Devices 7000K 3A LED source (pulse duration of 1 µs) was focused onto a pinhole using collimating optics. The beam emerging from the pinhole was directed to a concave mirror (focal length of 2.54 m, $f_{\#} = 8$) through a planar folding mirror. The concave mirror generated a collimated beam of light, while the plane folding mirror was used to overcome spatial constraints within the test facility. The collimated beam subsequently illuminated the region of interest and then traveled to a second identical concave mirror on the other side of the test section. This beam was captured using a Photron FASTCAM SAZ camera after being reflected by a second planar folding mirror. A vertical knife edge was placed at the focal point of the converging beam from the concave mirror, and 500 instantaneous images were acquired at a rate of 60 Hz for the current set of experiments.

2.2.2. Nearfield Acoustics

Nearfield acoustic measurements were performed using two Brüel & Kjaer 4939 free field microphones 6.35 mm in diameter. They were positioned in such a way that they were perpendicular to each other, corresponding to the major and minor axes, and 25.4 h upstream of the plane containing the nozzle exit. A schematic of the microphone locations about the nozzles is shown in Figure 2. Microphones Mic 1 and Mic 2 were placed at a radial distance of r = 38 h with respect to the nozzle axes. The microphone signals were first amplified through a B&K 2960 C signal conditioning amplifier at a sensitivity of 3.16 mV/Pa. The signals were then passed through low-pass analog filters (Stanford Systems SR 650) with a cutoff frequency of 50 kHz. Before each set of tests, the microphones were calibrated using a B&K 4228-type pistonphone at a frequency of 250 Hz and an amplitude of 124 dB. Nearby metal surfaces were covered with acoustic foam to minimize acoustic reflections, as the tests were conducted in a non-anechoic environment.

2.2.3. Surface Unsteady Pressure Measurements

The flow-induced pressure fluctuations on the impingement plate were measured with two high-frequency Kulite pressure transducers 1.6 mm in diameter (K1 and K2) (model no. XCE-062-100A) separated by a distance of 54.61 mm (5.46 h) such that K1 was located exactly beneath the impingement point, and K2 was situated in the wall jet, as shown in Figure 2. The pressure transducers were carefully calibrated using a Druck DPI 610 pressure calibrator before the tests.

All unsteady pressure measurements and acoustics signals were synchronized and sampled at a rate of 102.4 kHz for 1 sec at each nozzle-to-ground plate distance. The data were acquired through a National Instruments PCI-4472 acquisition card and were monitored using LabView software. A frequency resolution of 50 Hz was achieved, and the results were post-processed using the 'Pwelch' function on the MATLAB computing platform.

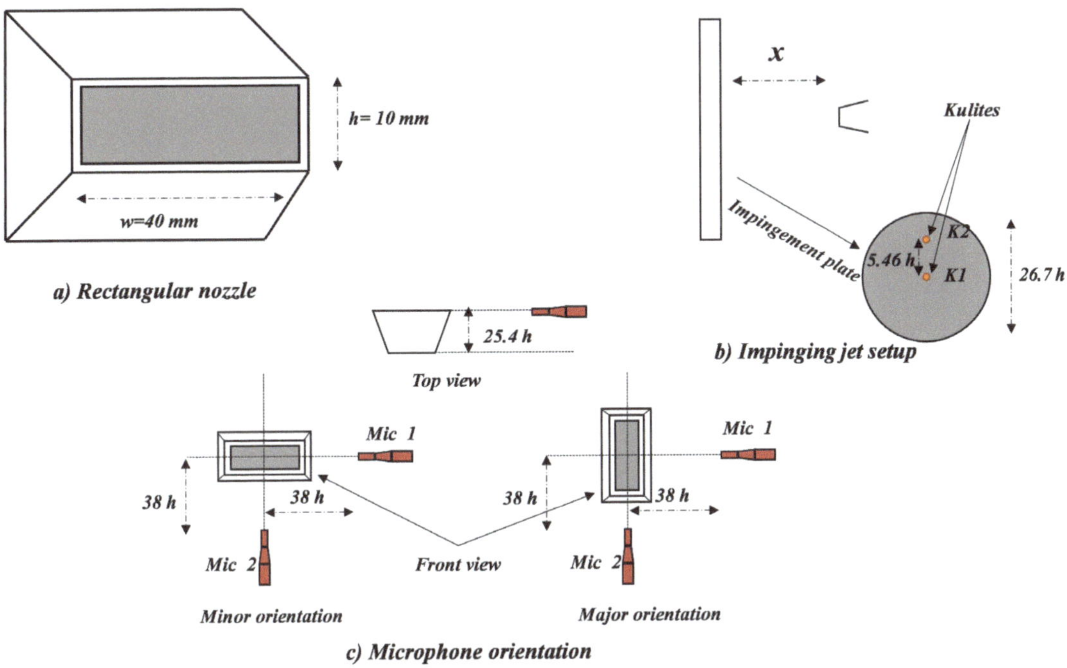

Figure 2. Schematics of the rectangular nozzle, impingement surface, and instrumentation setup.

2.2.4. Planar Particle Image Velocimetry (PIV)

PIV was carried out to quantify the velocity field. The flow was seeded by sub-micron size avocado oil using a Laskin nozzle seeder. A double-pulsed Nd:YAG (200 mJ/pulse) laser was used for illumination of the seed particles. A light sheet with a thickness of about 1 mm was created using a suitable combination of spherical and cylindrical lenses. The time between the two laser pulses was set to 1.8 µs. Images were acquired at a rate of 15 Hz using a 5.5 megapixel, 2560 × 2160 pixel sCMOS camera. The processing employed a multi-pass approach. During the initial two passes, an interrogation window of 64 × 64 pixels with 50% overlap was utilized. In the final pass, an interrogation window of 32 × × 32 pixels with 75% overlap was applied. For each case, 500 instantaneous image pairs were acquired, and the image pairs were processed using LaVision DaVis software.

2.3. Test Conditions and Measurement Uncertainties

The experiments were performed at a nozzle pressure ratio of 3.37, corresponding to the ideally expanded condition. All experiments were performed at a temperature ratio of 1. The nozzle-to-impingement-plate distance was varied from 3 to 20 h in steps of 0.25 h for the acoustics and pressure measurements. The Schlieren, acoustics, pressure transducers, and PIV set-ups were held fixed, and only the nozzle was rotated by 90 degrees to obtain measurements in both (major and minor) orientations. Table 1 summarizes the test conditions for the current set of tests. The NPR and TR values were accurate within ±0.0375 and ±0.04, respectively. The ground-plane-to-nozzle distance uncertainty was ±3 mm. The data acquisition system used for acoustic and unsteady pressure measurements had a 24 bit analog-to-digital-converter resolution. The measurement uncertainty in the unsteady pressure signals provided by the manufacturer was ±0.5% of the full-scale output. The velocities measured through planar PIV were found to have uncertainties within 2% of the jet's fully expanded exit velocity in the jet cores and within 5% of the jet velocity in the shear layer. These values were calculated using DaViS software.

Table 1. Test matrix.

Measurement	TR	NPR	x/h	Measurement Location
Unsteady pressure	1.0	3.37	3–20	0–5.46 h
Nearfield acoustics	1.0	3.37	3–20	38 h
Schlieren	1.0	3.37	5, 17, free jet	....
PIV	1.0	3.37	5, 17	....

3. Results and Discussion

The objective of this study was to characterize the aeroacoustic properties of rectangular impinging jets. As mentioned in the previous section, the results were examined at various impingement distances. However, only pertinent results are included herein. The global flowfield was visualized utilizing the Schlieren technique. Unsteady pressure measurements and nearfield acoustics were employed to characterize the flow's unsteadiness. Finally, the flowfield was quantified and elucidated by employing PIV.

3.1. *Qualitative Flow Visualization*

The conventional Schlieren technique was employed to visualize the flowfield associated with the impinging jet, with the findings compared against those of free jets under analogous conditions. Instantaneous Schlieren images for three cases are shown in Figure 3 for both the major and minor orientations. The results of the minor axis are illustrated in Figure 3a–c for x/h of 5, 17, and the free jet condition, respectively. At $x/h = 5$ (Figure 3a), the supersonic jet was observed to exit the nozzle, advect downstream, and approach the impingement plate, where it decelerated abruptly. The jet subsequently altered its trajectory, owing to the surface-imposed zero penetration boundary conditions. Post impingement, a wall jet formed and diffused at a certain distance from the impingement point. One can visualize the coherent structures within the jet shear layer, forming a comparatively subdued shock cell structure, and various acoustic wave patterns were discernible. Cylindrical wave patterns were also observed to originate from the jet shear layer. Despite maintaining nozzle pressure conditions conducive to an ideally expanded jet, minor pressure variances, coupled with nozzle imperfections and a finite lip thickness, resulted in a faint shock cell structure.

One can also visualize several waves originating near the impingement point, which correspond to the high-amplitude impingement tones. These tones directly resulted from the aeroacoustic coupling, a self-sustained phenomenon initiated at the nozzle lip by the acoustic waves. As briefly described, these resulted in instabilities close to the nozzle lip, which were amplified and manifested into large-scale coherent structures as they convected downstream. These structures subsequently impinged upon the surface. The resulting impingement led to the generation of pressure fluctuations, leading to high-intensity acoustic waves. The phase relationship of the feedback loop determine a specific frequency to be satisfied. In the noise spectrum, these discrete high-amplitude tones manifest as impingement tones. Detailed analyses of the noise measurements are in subsequent sections. At an increased impingement distance of $x/h = 17$, as depicted in Figure 3b, features such as the shear layer, wall jet, coherent structures, and acoustic waves persisted, albeit with diminished intensity. Previous studies on axisymmetric nozzles suggest that the flowfield at shorter impingement distances is predominantly dominated by resonance [9,10]. In contrast, when the impingement plate is removed, resulting in a free jet scenario, as shown in Figure 3c, the flowfield underwent significant alteration, notably the absence of wall jets and acoustic waves. The jet displayed a flapping mode, yet wave patterns were absent, indicating a weakened feedback loop (screech) between the shock structures and the shear layer.

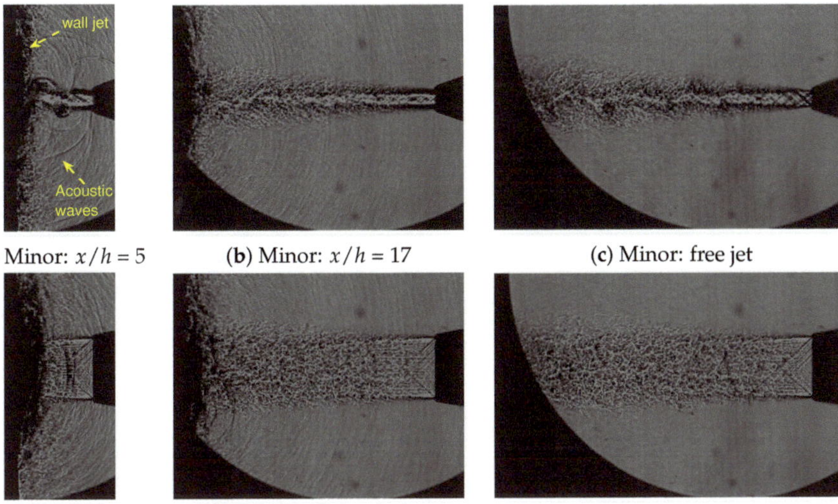

(**a**) Minor: $x/h = 5$ (**b**) Minor: $x/h = 17$ (**c**) Minor: free jet

(**d**) Major: $x/h = 5$ (**e**) Major: $x/h = 17$ (**f**) Major: free jet

Figure 3. Instantaneous Schlieren images at different nozzle heights for major and minor axes (flow from right to left).

When the nozzle was oriented such that the major axis aligned with the direct line of sight in Schlieren imaging, both similar and distinct features were discerned in comparison with those observed with the minor axis orientation. As depicted in Figure 3d, the presence of a weak shock structure, impingement tones, and unsteady wall jets, akin to those observed in the minor axis configuration, can be identified. An increase in the impingement distance, as illustrated in Figure 3e, led to diminished intensity for the impingement tones, expansion of the shear layer, and persistence of the unsteady wall jets. Comparable to the minor axis scenario, the major axis configuration also revealed analogous characteristics for the free jet condition, as shown in Figure 3f. While the overarching features of the major axis orientation bore resemblance to those of the minor axis orientation, several nuanced distinctions were also observed. Notably, the pronounced vortical structures evident in the minor axis (Figure 3a) at $x/h = 5$ were absent in the major axis configuration. Concurrently, the intensity of the acoustic waves was somewhat reduced, indicating the influence of the nozzle orientation on the behavior of non-axisymmetric jets. Figure 4 delineates the root mean square (RMS) characteristics of the Schlieren images, derived from post-processing 100 instantaneous Schlieren images, to accentuate the flow's fluctuating nature. Figure 4a vividly illustrates the jet's interaction with the impingement surface, showcasing a bright, dense plate shock indicative of a high-velocity jet impinging upon the plate, leading to the typical spread observed in wall jet formation. A comparative analysis across all orientations and impingement distances revealed more pronounced shear layer growth in the minor axis configuration. These observations will be quantitatively elaborated upon in subsequent sections. Overall, the Schlieren findings offer invaluable insights into the dynamics of a supersonic jet, elucidating the influence of the nozzle orientation and standoff distance on the jet spread, shock intensity, and flow structure. These distinctions play a pivotal role in comprehending the jet's acoustic and dynamic behaviors, with significant ramifications for applications in jet propulsion and noise mitigation. The observed non-uniformity in coherent structures and wave emission in rectangular nozzles necessitate further investigation through additional pressure and acoustic measurements, as detailed in the following subsections.

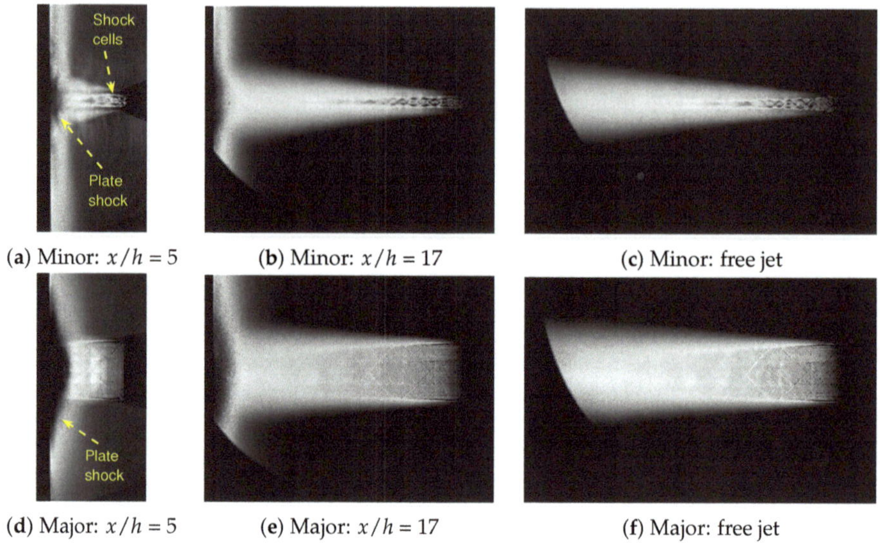

Figure 4. RMS Schlieren images at different nozzle heights for major and minor axes (flow from right to left).

3.2. *Unsteady Surface Pressure and Nearfield Acoustics*

Unsteady surface pressure and nearfield acoustic measurements were performed using two Kulite pressure transducers flush-mounted on the impingement plate and free field microphones (Mic 1 and Mic 2), respectively, as described in the experimental section (Figure 2). Pressure and acoustics measurements were synchronized to better understand and quantify the flow unsteadiness. The results were obtained for both axes (major and minor) and at different impingement heights as shown in Figure 5. We begin by presenting the overall sound pressure level (OASPL) for two microphones when the nozzle was in the major axis orientation (Figure 5a). Mathematically, the OASPL on a dB scale is defined by Equation (1):

$$OASPL(dB) = 20 \times log_{10}\left(\frac{P_{rms}}{P_{ref}}\right) \quad (1)$$

where P_{ref} is 20 µPa and P_{rms} is the fluctuating pressure component of the pressure. For Microphone 1, it is apparent that the OASPL distribution demonstrated distinct local maxima and minima as a function of the impingement distance, indicating a pronounced dependence on the nozzle standoff distance. An OASPL of 147 dB was recorded at the minimal impingement distance of 3 h. Notably, the OASPL values reached their peak within a 5–6 h distance from the nozzle, suggesting this region was indicative of strong resonance-based impingement tones which substantially contributed to the sound pressure levels. It is widely recognized that such intermediate distances typically exhibit resonance-dominated behavior in axisymmetric impinging jets [1,10]. Interestingly, analogous characteristics were observed in the present study of non-axisymmetric impinging jets. A more detailed analysis of resonance-dominated impingement tones will be conducted subsequently. Beyond $x/h = 6$, the OASPL diminished with an increasing impingement distance, displaying smaller peaks and troughs, which implies that a weaker resonance was maintained even at greater impingement distances. The OASPL value decreased to 137 dB at the maximum distance from the nozzle. Subsequently, the nearfield noise levels measured by Microphone 2 are presented and contrasted with those of Microphone 1.

The trend of the OASPL for Microphone 2 closely mirrored that of Microphone 1, with OASPL values peaking at a specific nozzle height before diminishing markedly as the distance between the nozzle and the impingement plate increased. Although the qualitative trends showed minimal variation, quantitative differences between the two microphones' results are evident. Primarily, the OASPL magnitude for Microphone 2 was lower than that for Microphone 1 at most locations, indicating a pronounced directivity in non-axisymmetric jets. Previous studies, such as those conducted on axisymmetric single impinging jets, have typically reported symmetric sound emission. However, the current findings indicate a unique directivity pattern in nearfield noise from rectangular impinging jets, which may also correlate with the Schlieren images previously discussed, where the nozzle exhibited a preferential flapping mode along one axis. Notably, the disparity in OASPL amplitudes was more pronounced at shorter impingement distances, diminishing with increased distance from the nozzle and eventually nullifying at $x/h = 20$. This trend could be associated with the attenuation of coherent structures at larger distances, leading to a weakened feedback loop mechanism.

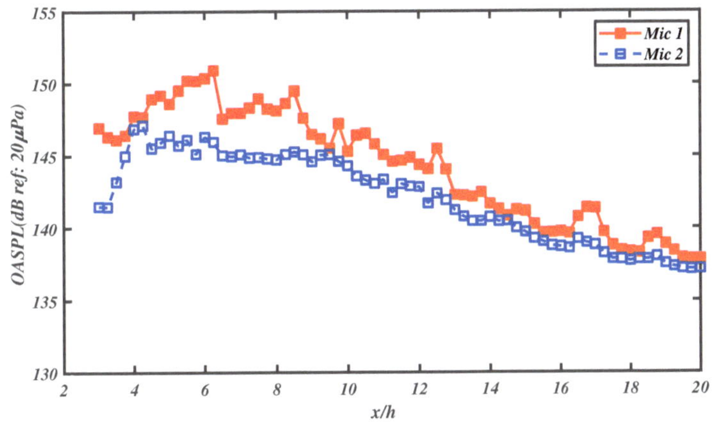

(**a**) Overall sound pressure level (major axis orientation)

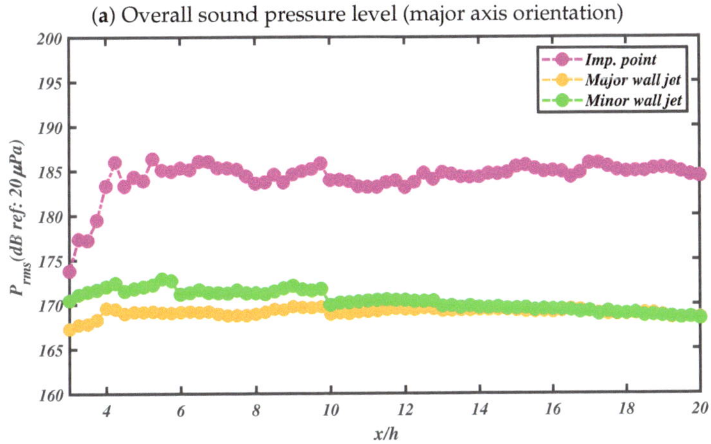

(**b**) Pressure fluctuations on the impingement plate

Figure 5. Unsteady characteristics of a rectangular impinging jet.

Further examination was directed toward the unsteady pressures measured at the impingement(Imp.) point and within the wall jets along both the major and minor axes on the ground plane, as depicted in Figure 5b. To assess the wall jet in a different orientation, the nozzle was rotated by 90 degrees, transitioning from the major to the minor view plane. The unsteady pressures, reaching as high as 185 dB at the impingement point, peaked near

$x/h = 5$ and 6, aligning with the resonance-dominated distances akin to the acoustic findings. Beyond this point, the pressure fluctuations surprisingly remained relatively constant, a phenomenon consistent with prior studies on impinging jets from circular nozzles, where impingement point pressure fluctuations reach a peak and then stabilize for most distances. Such elevated pressure fluctuations are commonly observed in supersonic jets. Interestingly, until $x/h = 13$, the unsteadiness in the wall jet differed between the major and minor axis orientations, with higher pressures observed in the minor axis orientation. This could be attributed to the more pronounced coherent structures and rapid shear layer growth in the minor axis orientation, as evidenced in the Schlieren images. Beyond $x/h = 13$, the unsteadiness equalized between the two axes, potentially due to the impingement surface extending beyond the potential core of the jet. Another contributing factor might be the axis-switching phenomenon frequently observed in rectangular jets, where at sufficient distances, the minor axis of the jet begins to expand more rapidly than the major axis [18]. It is hypothesized that at approximately $x/h = 13$, the jet may undergo an axis switch, leading to a more axisymmetric jet shape prior to their eventual convergence further downstream. Comparing nearfield acoustics and unsteady pressure fluctuations revealed that the unsteadiness was global and resonance-dominated, also being influenced by the jet's axis orientation.

We further investigated the acoustic spectra derived from Microphones 1 and 2 at two distinct nozzle standoff distances ($x/h = 5$ and 17), as depicted in Figure 6, with the nozzle oriented along its major axis. The frequency is represented by the x axis, while the y axis denotes the sound pressure level (SPL), calculated with a reference pressure of 20 μPa. As illustrated in Figure 6a, the spectra from Microphone 1 distinctly exhibited multiple discrete high-amplitude impingement tones and their harmonics, indicative of flow-acoustic coupling. Contrarily, previous studies on axisymmetric nozzles typically reported a singular dominant peak or a series of peaks corresponding to either the axisymmetric or non-axisymmetric modes and their harmonics [1,10]. The observed multiplicity of modes and harmonics at a resonant distance of $x/h = 5$ in this study suggests the coexistence of a lower-frequency tone, associated with the symmetric feedback instability wave mode, and a higher-frequency tone, linked to the purely antisymmetric feedback instability wave mode of the jet. Similar findings have been reported in experimental studies by [26,28]. The frequencies of additional peaks, beyond the symmetric and asymmetric modes, equated to the summation and differences, thereby representing combinations of the two fundamental tones. In this study, these fundamental tones were identified at frequencies of roughly 3850 kHz and 8300 kHz. The presence of these additional tones in the spectra is typically attributed to nonlinearities within the jet flow, including nonlinear oscillatory motions of the jet and its interaction with the impingement surface.

Conversely, for Microphone 2, as shown in Figure 6b, there was a notable reduction in the intensity of the impingement tones and a decrease in the number of distinct tones, aligning with the overall reduction in the SPL observed in Figure 5. Furthermore, a diminution in low-frequency broadband noise levels was observed for Microphone 2, indicative of hydrodynamic noise. It is postulated that this variation in noise levels may be related to differences in the shear layer thickness between the two axes, a hypothesis which may be supported by Schlieren imaging analysis. Figure 6b also presents the acoustic spectra for Microphone 1 at an axial distance of 17 h from the nozzle, revealing a significant decrease in the number of impingement tones, akin to observations from axisymmetric studies in which impingement tones were markedly attenuated at greater heights. Additionally, a closer examination suggests alterations in the fundamental frequencies of base tones, likely due to changes in the feedback loop mechanism caused by the varying absolute distances between the initial perturbation location (i.e., nozzle lip) and the impingement plate. There was an observable reduction in the amplitude of the impingement tones. These acoustic spectra outcomes highlight the unique directivity of noise levels in rectangular impinging jets, particularly regarding the tonal content across the two axes.

The narrowband pressure spectra for the unsteady pressures measured at the ground plane at $x/h = 5$ and 17 are illustrated in Figure 7. As shown in Figure 7a, strong multiple impingement tones and harmonics, mirroring those seen in the acoustic spectra, are evident at a nozzle standoff distance of $x/h = 5$. While the amplitude of these tones varied from the acoustic spectra, the fundamental frequencies remained relatively constant across most tones. Additionally, the low-frequency unsteadiness was heightened due to the direct impingement of the jet on the pressure transducer, contributing to significant unsteady loads on the impingement plate. The wall jet, observed in both major and minor axes, exhibited notable differences in the broadband and discrete tonal contents. Increasing the impingement distance to 17 h, as shown in Figure 7b, resulted in significant attenuation of the impingement tones, consistent with the acoustic findings. Intriguingly, the low-frequency broadband noise increased significantly compared with the shorter impingement distance, potentially due to the increased jet thickness and hydrodynamic noise, which contributed to the low-frequency content. For wall jets, a similar spectral content was observed in both axes across most frequencies, aligning with the unsteady pressure levels depicted as *Prms* values in Figure 5b.

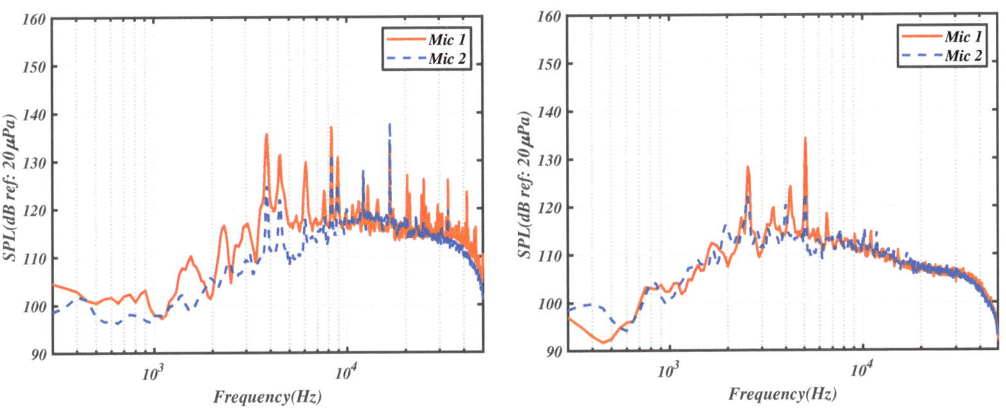

(**a**) $x/h = 5$ (**b**) $x/h = 17$

Figure 6. Nearfield acoustic spectra (major axis orientation).

Numerous researchers have extensively documented the feedback loop-related resonance characteristics of high-speed impinging jets for circular nozzles, highlighting a phase-locked process between the nozzle lip and the impingement surface which results in discrete, high-amplitude impingement tones [1,10,31]. These processes are known to be sensitive to variations in the boundary conditions, such as the temperature, pressure, and impingement distance. Moreover, a well-documented staging phenomenon has been observed in axisymmetric impinging jets, where the frequency of the tones gradually decreases with an increase in the impingement distance up to a certain threshold, beyond which any further increase in the impingement distance leads to an abrupt increase in frequency, signifying the transition to a new stage.

The resonance characteristics and staging phenomena for rectangular impinging jets were subsequently examined, as depicted in Figure 8a,b. The spectrogram plots the frequency on the y axis against the impingement distance on the x axis, with the z-axis representing the nearfield acoustics quantified by sound pressure level. In the spectrogram, a lighter grayscale denotes broadband noise or tones of lower magnitudes, while darker lines indicate high-amplitude tones and their harmonics. The frequency of the impingement tones was also estimated using Powell's equation [12], as expressed in Equation (2):

$$\frac{N+p}{f_n} = \frac{h'}{C_{ac}} + \int_0^{h'} \frac{dh'}{C_{st}} \tag{2}$$

where f_n represents the predicted frequency of an impingement tone, h' is the distance between the nozzle and the plate, C_{ac} is the ambient speed of upstream traveling acoustic waves (340 m/s), and C_{st} is the velocity of coherent structures (0.63*U_j), while N is an arbitrary integer denoting different frequency modes and p represents the phase lag, with $p = -0.30$ providing the best approximation for the current tests.

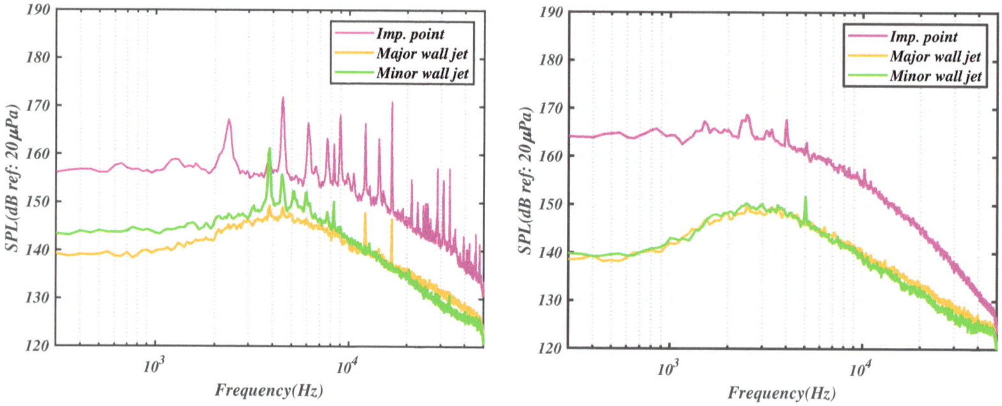

(**a**) $x/h = 5$ (**b**) $x/h = 17$

Figure 7. Unsteady pressure measurements at impingement plate (nozzle rotated to obtain both major and minor wall jets).

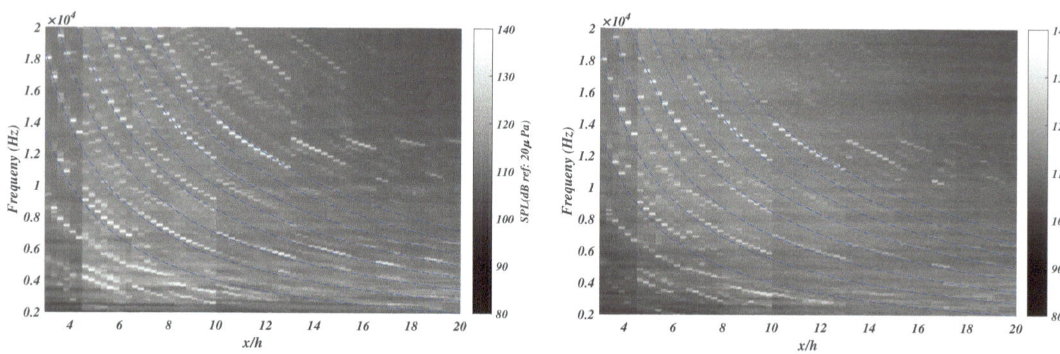

(**a**) Mic 1 (**b**) Mic 2

Figure 8. Comparison of impingement tones with Powell's prediction (blue lines) (major axis orientation).

As observed in Figure 8a, the spectrogram displays a plethora of high-amplitude tones across various frequencies and amplitudes. Similar to the findings from axisymmetric studies, the staging phenomenon is evident, with tone frequencies generally decreasing as the impingement height increased to a certain limit, after which the tones transitioned to higher frequencies. The spectrum exhibited a higher number of tones up to $x/h = 10$, beyond which the quantity of tones typically diminishes, although the spectrogram remained rich in tonal content even at larger impingement distances. These experimental findings are in alignment with the theoretical predictions of impingement tone frequencies estimated by Powell's equation, which are superimposed on the spectrogram as a blue dotted line. Similarly, Figure 8b showcases a rich array of impingement tones for Microphone 2, exhibiting similar staging phenomena and in line with the theoretical estimations. While the overall features of the spectrogram remained consistent, Microphone 2 captured significantly fewer and weaker tones compared with Microphone 1. This discrepancy is briefly discussed in the context of the varying shear layer growth across the two axes, potentially resulting in different mode shapes and thus causing azimuthal variations in the impingement tones.

3.3. *Quantitative Flow Visualization*

3.3.1. Main Jet Column Properties

Figure 9 shows the mean velocity results as derived from PIV flowfield measurements. The velocity results were non-dimensionalized with the jet exit velocity. Furthermore, due to the surface reflections, some of the data (bad vectors) near the nozzle exit and the impingement plate were trimmed. Starting from the left, the first column represents the contours of the axial velocity, wherein the first two rows (starting from the top) are the results of the major and minor axes orientations at $x/h = 5$. The results in Figure 9a distinctly capture the supersonic jet profile from the nozzle exit extending toward the vicinity of the impingement plate. Notably, in close proximity to the impingement plate, a unique parabolic velocity profile indicative of reduced velocities was observed, which was attributed to the plate shock phenomenon. This phenomenon typically occurs when supersonic jets encounter a solid surface situated within one of its shock cell regions. At this impingement distance, the shear layer growth was not discernible on either the major or minor axes (Figure 9a,d). However, at an increased impingement distance of $x/h = 17$ (Figure 9g,j), the shear layer profiles became evident in both orientations, with the minor axis demonstrating greater growth compared with the major axis. Despite the increased distances, the flow remained supersonic, and the plate shock typically observed at lower heights was absent.

The second column investigates the transverse velocity component for both orientations at varying heights. At an impingement distance of $x/h = 5$, as depicted in Figure 9b,e, the positive and negative velocities indicate the flow direction. The contours clearly differentiated the wall jet from the main jet, with noticeable differences in the thickness of the wall jets between the two orientations, potentially correlating with the variations in wall jet strength. These results suggest that both the shear layer and wall jet growth are influenced by the nozzle's axis orientation, with further analysis of the wall jets provided in subsequent sections. At an impingement distance of $x/h = 17$ (Figure 9h,k), new features emerged, including weak shocks within the main jet column, attributed to slight variations from ideal conditions due to the finite nozzle thickness and minor fluctuations in the nozzle pressure ratio during operation. The increased shear layer growth at this height was more pronounced in the minor axis compared with the major axis, aligning with observations from the Schlieren analysis. The third column presents quantitative data, such as the axial and transverse centerline velocity components. At $x/h = 5$ for the major axis (Figure 9c), the axial velocity distribution demonstrated smooth supersonic flow for most of the measured distance, with a deceleration observed around $x/h = 3.75$, which was attributed to plate shock, as inferred from the Schlieren images and qualitative PIV analysis. The transverse velocity component exhibited a minor peak near $x/h = 2$, possibly resulting from weak shock cells within the jet, with variations being relatively minor (to the order of 0.02 of the exit velocity). Changing to the minor axis orientation (Figure 9f), the axial velocity variation mirrored that of the major axis, while the transverse component displayed slightly enhanced fluctuations. At a nozzle height of $x/h = 17$ (Figure 9i), the axial component remained predominantly supersonic up to approximately $x/h = 10$–12, beyond which the velocity decreased, indicative of the potential core region and consistent with free jet studies on rectangular jets [18]. The transverse component fluctuations extended from the nozzle exit to the impingement plate with an increase in fluctuation magnitude at higher heights, a trend similarly observed on the minor axis (Figure 9l). The cessation of the core and augmented diffusion contributed to the increase in the transverse component magnitude at greater distances.

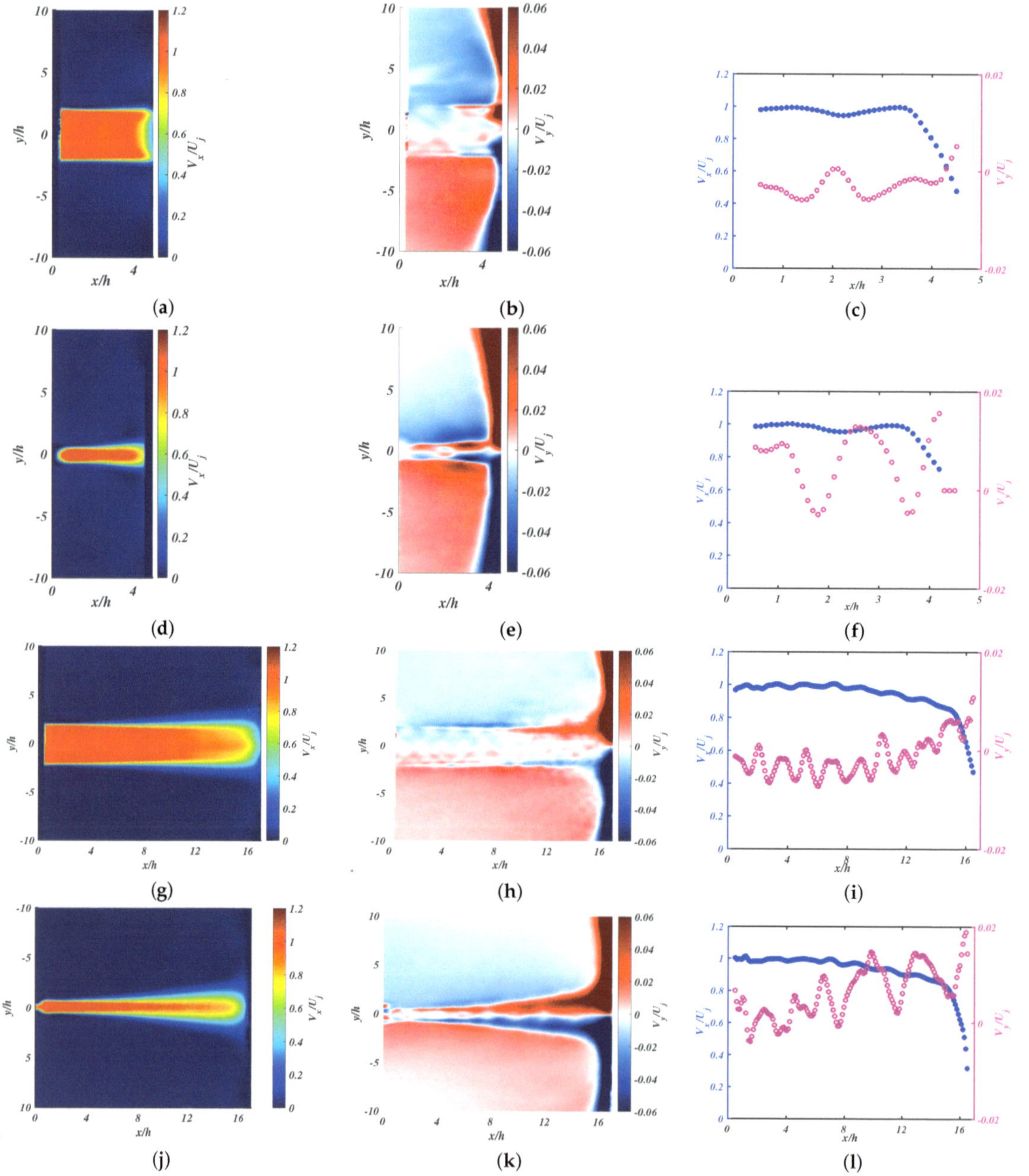

Figure 9. PIV results at different nozzle heights for major and minor axes. The first column (**a**,**d**,**g**,**j**) represents the axial velocity contours, the second column (**b**,**e**,**h**,**k**) represents the transverse velocity contours, and the third column (**c**,**f**,**i**,**l**) represents the centerline axial (blue) and transverse (pink) velocity component (flow from right to left).

Figure 10 depicts the cross-stream velocity profiles for the major and minor axes at various axial locations, with the impingement plate positioned at $x/h = 17$ from the nozzle exit. Near the nozzle exit, the velocity profiles for both orientations exhibited a classical top hat shape, with the maximum velocity at the center decreasing rapidly toward the radial directions. This trend persisted up to $x/h = 10$ for the major axis, which continued to display a top hat profile. For $x/h = 13$ and 16, the velocity profiles for both axes assumed different shapes, with peak velocities dropping below unity, indicating the end of the potential core. A notable discussion point is the variation in thickness (radial extent) of the major and minor axis profiles. A comparison at different heights reveals that the radial growth of the minor axis outpaced that of the major axis, with convergence occurring at $x/h = 13$. Extrapolating these data suggests that the minor axis thickness may surpass that of the major axis beyond $x/h = 17$, a phenomenon consistent with axis switching observed in three-dimensional jets [15,18]. Further analysis of the shear layer characteristics, including the jet half-width and shear layer spreading rates, will be provided in subsequent subsections.

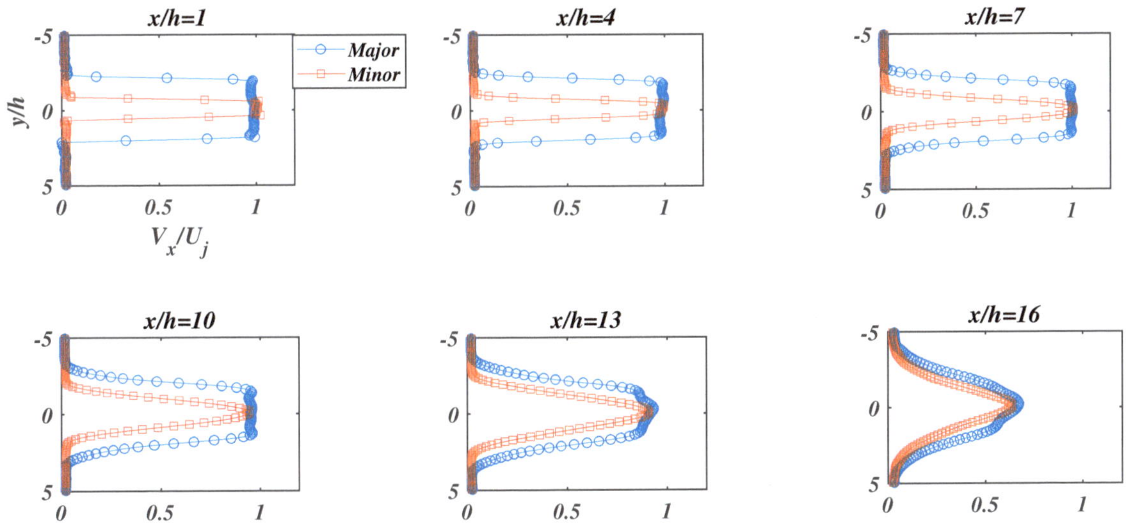

Figure 10. Cross-stream velocity profiles for different axial locations.

3.3.2. Shear Layer Properties

Figure 11a shows the results of the shear layer growth of the jet in major and minor orientations. The results are presented in terms of the jet half-width ($\delta_{0.5}$), which is non-dimensionalized with the width(y-axis) of the jet. The slope of the jet width curve is a representation of the shear layer growth. With the impingement plate fixed at a nozzle standoff distance of $x/h = 17$, the half-width along the major axis exhibited a relatively constant trend, with a minor increment at distances farther from the nozzle. Conversely, the minor axis's half-width displayed a continuous increase throughout the measured range, suggesting a more rapid shear layer expansion compared with the major axis. This variation in the growth rates of the shear layer between the axes until $x/h = 12$ implies a change in the jet's cross-section shape along the x axis. Beyond this point, the growth rates appear to converge with increasing distances. These differences in the shear layer characteristics may be ascribed to the disparate initial boundary layers at the nozzle exit for each orientation, influencing the mixing and entrainment processes and thereby affecting the jet spread. It is hypothesized that the shear layer's spread is intrinsically linked to the flow-acoustic coupling, where the jet impingement generates acoustic waves which propagate upstream, impacting the shear layer and corner vortices differently across the two axes. Given the minor axis's initially larger surface area, the coherent structures within its shear layer expanded rapidly, inducing azimuthal variations in the acoustic properties. These observations are consistent with the pronounced tones discussed in the acoustics section. The current hypothesis posits that the shear layer thickness between the free and impinging jets under identical conditions would differ, a notion supported by the findings in [32], with the authors noting discrepancies in the shear layer thickness between the free and impinging jets in circular jets operating under ideal conditions.

Figure 11b displays experimental data for a non-dimensional velocity profile in the radial direction of a supersonic rectangular impinging jet. The velocity V_x at various radial positions was normalized by a jet exit velocity. The expression $\eta* = (y - y_{0.5})/x$ was used to scale the y location from a reference position $y_{0.5}$ (where the velocity was half of the jet exit velocity U_j) normalized by x, which is the distance from the nozzle exit. The plot depicts data collected at different impingement heights corresponding to the potential core (until x/h = 12). The results are presented for both the major and minor orientations. The profile shows that the non-dimensional velocity was highest near the center of the jet and decreased toward the edges. Remarkably, all data sets, irrespective of axial distance or orientation, aligned along a single curve, indicating self-similarity in the velocity profile of the flowfield. This convergence of data onto a singular curve suggests that the flow characteristics of the rectangular jet might be encapsulated by a universal profile within the tested experimental conditions and set-up range. The span of these profiles depicted in the figure quantitatively represents the spreading rate of the shear layer in both orientations. This phenomenon, reminiscent of the data collapse observed in [33] for round jets, uniquely captures the spreading behavior of both axisymmetric and non-axisymmetric jets, as demonstrated in the current study.

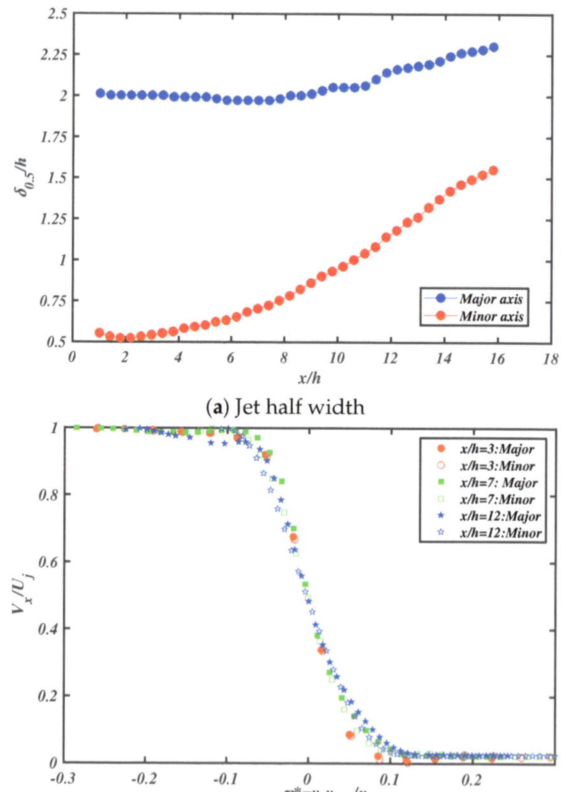

(**a**) Jet half width

(**b**) Shear layer spreading

Figure 11. Jet half-width and normalized shear layer spreading at x/h =17.

Figure 12 shows the velocity profiles of the wall jets when the plate was fixed at $x/h = 17$ for the major and minor orientations. The results were extracted roughly 0.1 h from the impingement plate, and for brevity, only one half of the wall jet velocity profile (centered at the impingement point) is presented. In general, the minor axis's velocity profile manifested a bifurcation, showcasing two distinct peaks, indicative of a complex flow structure potentially arising from interactions with the boundary layer or other flow features, such as the vortices. Beyond a peak $y/h = 4.8$, the velocity diminished, a natural consequence of jet spreading and diffusion into the ambient surroundings. Conversely, the major axis orientation exhibited a comparatively flatter and lower-magnitude profile, except within the $y/h = 1$ and 3 range. These variations in the wall jet flowfields are attributed to the distinct boundary layer effects in the two orientations, stemming from initial flow perturbations. These findings elucidate that the asymmetric nature of three-dimensional rectangular jets extends beyond the impingement zone, resulting in anisotropic wall jets.

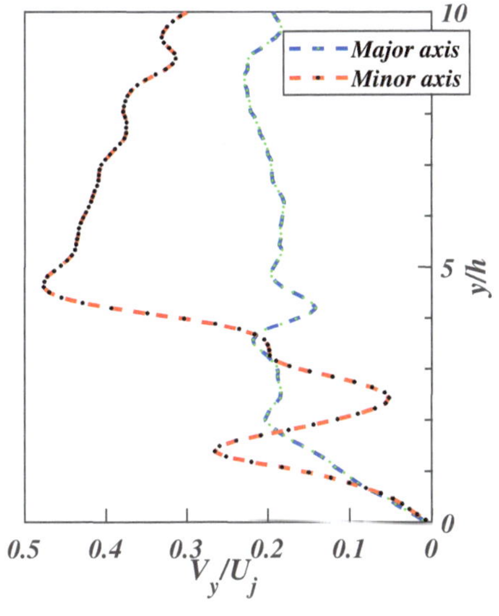

Figure 12. Wall jet velocity profile extracted 0.1 h from the plate located at $x/h = 17$.

4. Conclusions

The flowfield of supersonic impinging jets is highly oscillatory and complex in nature. This study delved into the flowfield and noise characteristics of a supersonic rectangular jet issued from a Mach 1.44 nozzle impinging on a surface. The nozzle, with an aspect ratio of 4:1, operated at a nozzle pressure ratio (NPR) of 3.37, corresponding to the design's NPR.

Flow visualization through Schlieren imaging unveiled both instantaneous and unsteady flow features, highlighting large-scale coherent structures and sharp acoustic waves, including impingement tones akin to those documented in axisymmetric supersonic impinging jet studies. These structures and the resultant impingement tones exhibited a distinct directivity, being more pronounced along one axis. This directional behavior can be attributed to the rapid shear layer growth in the minor axis compared with the major axis.

Pressure fluctuations at the impingement point and along the wall jet were measured using a high-frequency response pressure transducer. The root mean square (RMS) fluctuations indicated that unsteady pressures could reach up to 186 dB when the nozzle-to-impingement distance was approximately 5–6 h, remaining roughly constant beyond this range for the present cases.

Nearfield acoustic measurements using microphones placed in two different planes relative to the nozzle exit further elucidated the acoustic characteristics. The acoustic results revealed multiple strong impingement tones and harmonics, indicating the presence of more than one resonance mode during operation. Other observed tones were combinations of these base modes. The comparison of spectra from different microphones confirmed the hypothesis that the high-amplitude tones exhibited directional behavior. Generally, an increase in the impingement distance led to a reduction in the number of tones, displaying definitive staging behavior similar to that observed in axisymmetric impinging jets.

Additionally, particle image velocimetry (PIV) analysis provided insights into the variation in the shear layer thickness of the main jet and the resulting wall jets across different axes. These variations correlated with the observed acoustic directivity, underscoring a unique asymmetry in the resonance mechanisms of rectangular impinging jets.

Overall, this study presents a comprehensive examination of the flowfield and acoustic properties of supersonic rectangular impinging jets, highlighting the complex interplay between flow structures, pressure fluctuations, and acoustic emissions. The findings emphasize the directional nature of the acoustic tones and the impact of shear layer growth on the resonance characteristics, contributing to a deeper understanding of the behavior of supersonic impinging jets.

Author Contributions: All authors contributed equally to the paper. All authors have read and agreed to the published version of the manuscript.

Funding: This research was funded by the Federal Aviation Administration (FAA) Center of Excellence for Commercial Space Transportation (COE-CST), Co-Operative Agreement 15-C-CST-FSU-05 as Task 325.

Data Availability Statement: Raw data available on request.

Acknowledgments: The authors acknowledge the administrative and technical support of Ken Davidian and Nickolas Demidovich.

Conflicts of Interest: The authors declare no conflicts of interest.

References

1. Krothapalli, A.; Rajakuperan, E.; Alvi, F.; Lourenco, L. Flow field and noise characteristics of a supersonic impinging jet. *J. Fluid Mech.* **1999**, *392*, 155–181. [CrossRef]
2. Henderson, B. The connection between sound production and jet structure of the supersonic impinging jet. *J. Acoust. Soc. Am.* **2002**, *111*, 735–747. [CrossRef] [PubMed]
3. Alvi, F.S.; Shih, C.; Elavarasan, R.; Garg, G.; Krothapalli, A. Control of Supersonic Impinging Jet Flows Using Supersonic Microjets. *AIAA J.* **2003**, *41*, 1347–1355. [CrossRef]
4. Weightman, J.L.; Amili, O.; Honnery, D.; Edgington-Mitchell, D.; Soria, J. Nozzle external geometry as a boundary condition for the azimuthal mode selection in an impinging underexpanded jet. *J. Fluid Mech.* **2019**, *862*, 421–448. [CrossRef]
5. Liu, L.; Li, X.; Liu, N.; Hao, P.; Zhang, X.; He, F. The feedback loops of discrete tones in under-expanded impinging jets. *Phys. Fluids* **2021**, *33*, 106112. [CrossRef]
6. Gojon, R.; Bogey, C. Flow Structure Oscillations and Tone Production in Underexpanded Impinging Round Jets. *AIAA J.* **2017**, *55*, 1792–1805. [CrossRef]
7. Stahl, S.L.; Gaitonde, D.; Bhargav, V.N.; Alvi, F.S. Theoretical model for coupled dual impinging jet aeroacoustic resonance. *Phys. Rev. Fluids* **2022**, *7*, 104606. [CrossRef]
8. Mehta, Y.; Bhargav, V.N.; Kumar, R. Experimental Characterization and Control of an Impinging Jet Issued from a Rocket Nozzle. *New Space* **2021**, *9*, 187–201. [CrossRef]
9. Mehta, Y.; Natarajan, K.; Sellappan, P.; Gustavsson, J.; Kumar, R. Effect of Nozzle Spacing in Supersonic Twin Impinging Jets. *AIAA J.* **2022**, *60*, 2423–2440. [CrossRef]
10. Kumar, R.; Lazic, S.; Alvi, F.S. Control of High-Temperature Supersonic Impinging Jets Using Microjets. *AIAA J.* **2009**, *47*, 2800–2811. [CrossRef]
11. Bhargav, V.N.; Mehta, Y.; Kumar, R.; Alvi, F. Effect of Relative Momentum and Temperature on the Aeroacoustic Characteristics of Dual Impinging Jets. In Proceedings of the AIAA Aviation 2021 Forum, Online, 2–6 August 2021.
12. Powell, A. On the Edgetone. *J. Acoust. Soc. Am.* **1961**, *33*, 395–409. [CrossRef]
13. Powell, A. The sound-producing oscillations of round underexpanded jets impinging on normal plates. *J. Acoust. Soc. Am.* **1988**, *83*, 515–533. [CrossRef]
14. Husain, H.; Hussain, F. Elliptic jets. Part 3. Dynamics of preferred mode coherent structure. *J. Fluid Mech.* **1993**, *248*, 315–361. [CrossRef]

15. Zaman, K.B.M.Q. Axis switching and spreading of an asymmetric jet: The role of coherent structure dynamics. *J. Fluid Mech.* **1996**, *316*, 1–27. [CrossRef]
16. Ho, C.M.; Gutmark, E. Vortex induction and mass entrainment in a small-aspect-ratio elliptic jet. *J. Fluid Mech.* **1987**, *179*, 383–405. [CrossRef]
17. Krothapalli, A.; Baganoff, D.; Karamcheti, K. On the mixing of a rectangular jet. *J. Fluid Mech.* **1981**, *107*, 201–220. [CrossRef]
18. Valentich, G.; Upadhyay, P.; Kumar, R. Mixing characteristics of a moderate aspect ratio screeching supersonic rectangular jet. *Exp. Fluids* **2016**, *57*, 71. [CrossRef]
19. Gutmark, E.J.; Grinstein, F.F. Flow Control with Noncircular Jets1. *Annu. Rev. Fluid Mech.* **1999**, *31*, 239–272. [CrossRef]
20. Alkislar, M.B.; Krothapalli, A.; Lourenco, L.M. Structure of a screeching rectangular jet: A stereoscopic particle image velocimetry study. *J. Fluid Mech.* **2003**, *489*, 121–154. [CrossRef]
21. Upadhyay, P.; Valentich, G.; Kumar, R.; Alvi, F. Flow and acoustic characteristics of non-axisymmetric jets at subsonic conditions. *Exp. Fluids* **2017**, *58*, 52. [CrossRef]
22. Valentich, G.M.; Davis, T.B.; Fernandez, E.; Kumar, R.; Alvi, F.S.; Alphonso, M.; Harris, C. Characterization of a Supersonic Rectangular Jet over a Range of Test Conditions. In Proceedings of the 53rd AIAA Aerospace Sciences Meeting, Kissimmee, FL, USA, 5–9 January 2015. [CrossRef]
23. Wu, G.J.; Lele, S.K.; Jeun, J. Analysis of Resonance in Jet Screech with Large-Eddy Simulations. *arXiv* **2023**, arXiv:2304.00665.
24. Jun Wu, G.; Lele, S.K.; Jeun, J.; Gustavsson, J.; Sellappan, P.; Kumar, R. Unstructured Large-Eddy Simulations of Rectangular Jet Screech: Assessment and Validation. *AIAA J.* **2023**, *61*, 1224–1237. [CrossRef]
25. Krothapalli, A. Discrete tones generated by an impinging underexpanded rectangular jet. *AIAA J.* **1985**, *23*, 1910–1915. [CrossRef]
26. Norum, T.D. Supersonic rectangular jet impingement noise experiments. *AIAA J.* **1991**, *29*, 1051–1057. [CrossRef]
27. Tam, C.K.; Norum, T.D. Impingement tones of large aspect ratio supersonic rectangular jets. *AIAA J.* **1992**, *30*, 304–311. [CrossRef]
28. Thurow, B.; Samimy, M.; Lempert, W. Structure of a supersonic impinging rectangular jet via real-time optical diagnostics. In Proceedings of the 32nd AIAA Fluid Dynamics Conference and Exhibit, St Louis, MO, USA, 24–26 June 2002; p. 2865.
29. Mehta, Y.; Bhargav, V.N.; Willms, T.; Kumar, R. Flowfield and Noise Characteristics of Supersonic Rectangular Impinging Jet. In Proceedings of the AIAA AVIATION, Chicago, IL, USA, 27 June–1 July 2002; pp. 2022–3896.
30. Nageswara Rao, A.; Krishna, T.V.; Bharathwaj, M.; Sugrno, M.I. Effect of Plate Distance on Steady and Unsteady Characteristics of Impinging Rectangular Jet. *J. Spacecr. Rocket.* **2023**, *60*, 1220–1229. [CrossRef]
31. Tam, C.K.W.; Ahuja, K.K. Theoretical model of discrete tone generation by impinging jets. *J. Fluid Mech.* **1990**, *214*, 67–87. [CrossRef]
32. Davis, T.B.; Kumar, R. Shear layer characteristics of supersonic free and impinging jets. *Shock Waves* **2015**, *25*, 507–520. [CrossRef]
33. Lau, J.C.; Morris, P.J.; Fisher, M.J. Measurements in subsonic and supersonic free jets using a laser velocimeter. *J. Fluid Mech.* **1979**, *93*, 1–27. [CrossRef]

fluids

MDPI

Article

Visualization and Quantification of Facemask Leakage Flows and Interpersonal Transmission with Varying Face Coverings

Xiuhua Si [1], Jensen S. Xi [1], Mohamed Talaat [2], Jay Hoon Park [3], Ramaswamy Nagarajan [3,4], Michael Rein [5] and Jinxiang Xi [2,*]

1 Department of Mechanical Engineering, California Baptist University, Riverside, CA 92504, USA
2 Department of Biomedical Engineering, University of Massachusetts, Lowell, MA 01854, USA
3 Department of Plastics Engineering, University of Massachusetts, Lowell, MA 01854, USA
4 Fabric Discovery Center, University of Massachusetts, Lowell, MA 01854, USA
5 Advanced Functional Fabrics of America, Cambridge, MA 02139, USA
* Correspondence: jinxiang_xi@uml.edu; Tel.: +1-978-934-3259

Abstract: Although mask-wearing is now widespread, the knowledge of how to quantify or improve their performance remains surprisingly limited and is largely based on empirical evidence. The objective of this study was to visualize the expiratory airflows from facemasks and evaluate aerosol transmission between two persons. Different visualization methods were explored, including the Schlieren optical system, laser/LED-particle imaging system, thermal camera, and vapor–SarGel system. The leakage flows and escaped aerosols were quantified using a hotwire anemometer and a particle counter, respectively. The results show that mask-wearing reduces the exhaled flow velocity from 2~4 m/s (with no facemask) to around 0.1 m/s, thus decreasing droplet transmission speeds. Cloth, surgical, and KN95 masks showed varying leakage flows at the nose top, sides, and chin. The leakage rate also differed between inhalation and exhalation. The neck gaiter has low filtration efficiency and high leakage fractions, providing low protection efficiency. There was considerable deposition in the mouth–nose area, as well as the neck, chin, and jaw, which heightened the risk of self-inoculation through spontaneous face-touching. A face shield plus surgical mask greatly reduced droplets on the head, neck, and face, indicating that double face coverings can be highly effective when a single mask is insufficient. The vapor–SarGel system provided a practical approach to study interpersonal transmission under varying close contact scenarios or with different face coverings.

Keywords: face covering; mask fit; leakage flows; Schlieren optical imaging system; interpersonal transmission; self-inoculation; short-range airborne transmission; double masking

Citation: Si, X.; Xi, J.S.; Talaat, M.; Park, J.H.; Nagarajan, R.; Rein, M.; Xi, J. Visualization and Quantification of Facemask Leakage Flows and Interpersonal Transmission with Varying Face Coverings. *Fluids* **2024**, *9*, 166. https://doi.org/10.3390/fluids9070166

Academic Editors: Markus Klein, Mingming Ge, Xin-Lei Zhang and Guangjian Zhang

Received: 4 June 2024
Revised: 4 July 2024
Accepted: 13 July 2024
Published: 22 July 2024

1. Introduction

Proper fit is crucial for the safety, comfort, and functionality of face masks [1,2]. Recent advancements have greatly improved filter media, enhancing aerosol filtration efficiency while maintaining an acceptable pressure drop [3–6]. However, the effectiveness of even the most advanced filter media is compromised if the mask does not fit the wearer properly. Essentially, a mask only functions as personal protective equipment if air flows through the filter media and not around it through gaps between the mask and the skin [7]. Consequently, the protective level of the mask directly correlates with the amount of air leakage [8]. A poor fit can dramatically decrease a mask's overall protection efficacy. Notably, unlike tight-fitting respirators, disposable three-layer surgical masks often have a loose fit that can worsen during physical activities, incorrect usage, or extended wear [9–13].

Gaps between a mask and skin can significantly affect airflow and aerosol movement, depending on their size and location. However, current standardized tests cannot measure these gaps or link them to mask fit. While mask-fit testers exist for tight-fitting respirators, similar tools for loose-fitting masks, like disposable surgical masks, are lacking [14–19]. For instance, directly applying the TSI Portacount mask-fit tester to surgical masks yields unrealistically low scores (0–40) with large variability. Also, while we can measure air speed through gaps, measuring gap size is difficult. These gaps change with activity, face shape, and facial hair. Wang et al. showed that nose height and chin length notably influenced the leak site in loose-fitting masks [20]. In a similar manner, Oestenstad et al. showed that the leak site changed with the face size and shape [21]. Even gender and age matter—women often have leaks at the mask's bottom [22]. Overall, accurately measuring mask leakage is still challenging, and leakage testing should use anatomically accurate or even sub-specific face models [20,23–25].

Previous studies have explored different methods to characterize mask flows, especially leakage flows and escaping particles. Using a Schlieren optical imaging system, Tang et al. found that while surgical masks block a cough's forward jet, a poor fit can allow for air leakage from the top and sides [26]. Similarly, Su et al. reported comparable reductions in filtration efficiency among various loose-fitting masks (cloth, surgical, and KN95) against ultrafine particles [27]. Cappa et al. measured exhaled aerosols using a particle sizer and showed that imperfect sealing in surgical masks significantly reduced their ability to block expiratory particles during talking and coughing [28]. Koh et al. simulated exhalation and coughing in a manikin with different masks. Interestingly, they demonstrated that poorly fitting N95 respirators may offer less protection than a simple surgical mask [29]. Also using a manikin model, Brooks et al. demonstrated that knotting the mask band and tucking in the extra material of a surgical mask effectively improved mask fit and protection [30]. Verma et al. utilized a manikin, foot pump, smoke generator, and a laser sheet to visualize a 'cough' through different masks [31]. The team was able to identify sites of leakage in the masks and determine how far the 'cough' traveled after passing through the masks [31]. For normal breathing, the foot pump delivers 500 mL of air to the model per breath, while for coughing, the foot pump delivers 1500 mL, mimicking the average amount of air expelled during a cough [32].

Despite valuable insights obtained from previous visualization studies, they were limited to qualitative analysis and a single test participant wearing a mask. Moreover, the visualization results predominantly focused on exhalation flows. Due to ethical issues, the visualization methods with human subjects had to be safe, and many alternative optical approaches, such as laser-based methods, had to be excluded. The manikin models used in mask flow visualization were often for general purposes, with not-so-accurate facial morphologies. Even more significant, research on interpersonal droplet transmission is severely lacking, even though it is more relevant to evaluating the transmission risks of respiratory infections [33]. One exception was Xu et al., who used a Schlieren system to study the effects of ventilation on airborne aerosol transmission between two persons and suggested that only a sufficiently high personal ventilation can effectively protect against infection [34,35].

The objective of this study is to visualize mask flows and the transmission of respiratory droplets between individuals. Specific aims include the following:

1. Explore different visualization methods for expiratory and inspiratory airflows due to mask-wearing, including systems based on Schlieren, laser, LED, smoke particles, and vapor droplets.
2. Compare facemask flow dynamics under different physical activities and between different mask types.
3. Measure the particle counts and leakage flow rates from different mask types, including cloth, surgical, and KN95.
4. Investigate interpersonal droplet transmission using a vapor-based visualization system between 3D-printed head models under various interaction scenarios.

2. Methods

2.1. Mask Flow Visualization Methods

2.1.1. Schlieren Optical Imaging System

The Schlieren imaging system visualizes airflows by capturing light refraction changes between areas of differing pressure or temperature. This study used a single-mirror setup (Figure 1a) with four parts: a concave mirror (AD015, 406 mm diameter, 1.8 m focal length, Agena AstroProduct, Cerritos, CA, USA), a pinhole LED light source 3.6 m away, a razor blade, and a Canon EOS Rebel T7 camera (Canon, Tokyo, Japan). The mirror reflects light through the test area to the razor, which blocks half the light. The remainder reaches the camera, producing an image. Dimmed lighting enhances airflow visibility. Image sensitivity depends on the mirror's focal length to unobstructed object length ratio [36]. During the test, the participant sat 20 cm from the mirror, with or without face coverings, as illustrated in the rightmost panel in Figure 1a. Five activities were performed: normal breathing, coughing, soft speaking, normal breathing, and loud speaking, that represented the majority of everyday activities. For normal breathing, the effect of a bracket on mask flows was also tested. Four types of masks were evaluated in this study, including a surgical mask (JS95-01, SanJiao, KN95 (EVENTRONIC., Shenzhen, China, with a filtration efficiency of 99.1% and a breathability of 168 Pa)), cloth mask (Newmark Sports, Irvine, CA, USA), and neck gaiter (Finvizo). According to assessments by the National Personal Protective Technology (NPPTL), the JS95-01 surgical mask has a filtration efficiency of 82.4–89.0% and breathability of 71–107 Pa (test #: MTT-2020-104.1), and the EVENTRONIC KN95 has a filtration efficiency of 93.3–99.6% and a breathability of 160–206 Pa (test #: MTT-2021-57.1) [37]. The filtration efficiency and breathability of the cloth mask and neck gaiter were not measured.

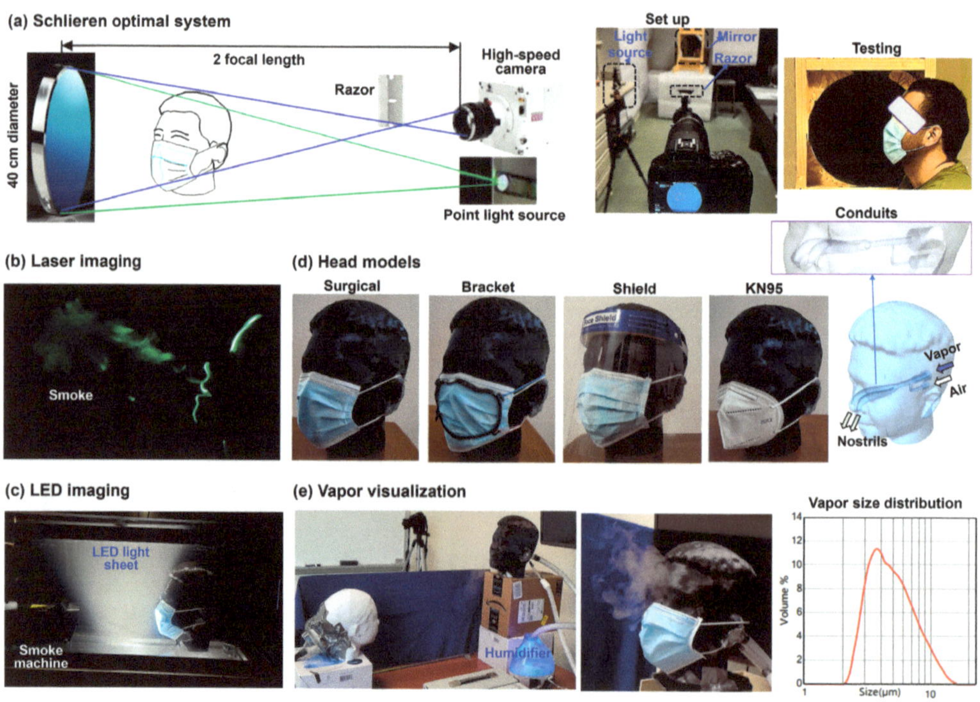

Figure 1. Experimental methods: (**a**) Schlieren optical imaging, (**b**) laser imaging, (**c**) head models with various face coverings, (**d**) LED imaging, and (**e**) vapor-based visualization.

2.1.2. Laser-Particle Imaging System for Exhalation

The laser system was created using a laser sheet, a head model, a foot pump, and a fog machine. A laser sheet (100 mW, 488 nm) was used to illuminate expiratory flows that contain fog particles (Figure 1b). The foot pump was used to pump a puff of air through the head model, simulating the way air is quickly released from the nose/mouth during coughing or breathing. A fog machine (CHAUVET DJ Hurricane 1000, Antwerpen, Belgium) was used to create tracer particles. The laser sheet was pointed toward the respiratory model to create a midsagittal cross-section of the expelled smoke. All testing was conducted in a dark room so that the illuminated particle-laden flows could be clearly visible. All testing was recorded using the same camera that was used for the Schlieren system images.

2.1.3. LED-Particle Imaging System for Inhalation and Exhalation

Considering that the foot pump could only generate expiratory flows, another visualization system was designed that allowed for the visualization of both inhalations and exhalations. To visualize the inspiratory flows that flow inwardly into the nose/mouth, seed particles need to be present outside the head model before inhalation. Ideally, these seed particles should be evenly suspended and as still as possible so that when inhalation starts, the particle trajectories converging toward the mouth/nose or mask can be sufficiently different from the surroundings. To achieve this, a 50-gallon fish tank was used, which was pre-filled with fog smoke five minutes before the test to allow for the fog particles to stabilize (Figure 1c). An LED sheet light was used to illuminate the mid-sagittal plane of the head model. A breathing machine was used to ventilate a sinusoidal pressure waveform to the head model, and the same camera that was used for the Schlieren imaging was used to record the flow dynamics around the mask (Figure 1c). Note that, even for exhalation, this LED-fish tank system is different from the laser–foot pump system described in Section 2.1.2 in that it visualizes how the ambient air is disturbed by expiratory flows, while the laser system visualizes how the exhaled particles are dispersed in the environment.

2.2. *Quantitative Measurements of Flows, Temperature, and Particles*

2.2.1. Leakage Flow Velocity and Mask Temperature

The velocity of leakage flows was measured using a TSI 9565 VelociCalc ventilation meter (Shoreview, MN, USA). Considering that the flow would decay as it moved away from the gap, the probe was positioned 2 cm from the gap sites for all tests. The sites selected for velocity measurement included two nose top ridges, two lateral sides, and the chin, based on prior observations of leakage flows of different masks. When conducting the tests, the participant wore a mask with a good fit and breathed naturally. Each test was repeated five times. Considering that the measured velocity magnitude could vary slightly even for the same leakage site depending on the probe location, the probe was positioned approximately 2 mm away from the center of the gap for all tests. Meanwhile, the mask temperature was recorded with a FLIR ONE Pro iOS thermal camera (Wilsonville, OR, USA).

2.2.2. Particle Counter Testing

A Temtop PMD 331 Particle Counter (Temtop, San Jose, CA, USA) was used to measure the number of droplets exhaled across the mask during normal breathing, as well as the size of each of the droplets. This particle counter can detect aerosol droplets from 0.3 μm to 10 μm in size and has seven channels. The SARS-CoV-2 virus is around 0.1 μm, which can escape mask filtration [38]. However, viruses typically attach to larger droplets or particles to survive and spread, ranging from 0.3 μm to large drops of several millimeters [39]. In addition to the control case (no mask), six types of masks were tested: surgical, KN95, cloth, cloth with HEPA filter, neck gaiter, and surgical with bracket. The tests were conducted in a controlled environment at 24 °C and 30% relative humidity. An air purifier was operated for one hour prior to testing to ensure that the majority of droplet particles detected by the particle counter came from breathing.

2.3. Head Models

Several head models were used in this study. One was the respiratory model (Michigan Instruments, Grand Rapids, MI, USA) that allowed for air to be expelled through the mouth only. The other two head models were developed from scans of two volunteers using the iPhone app Bellus3D (Lilburn, GA, USA). The use of human scans in this study has been approved by the Institutional Review Board (IRB) of the University of Massachusetts Lowell. The head model was further processed using SolidWorks (Dassault Systèmes, Waltham, MA, USA) to add two conduits, with one connecting to a breathing simulator (Michigan Instruments, Grand Rapids, MI, USA) and the other to a vapor source (rightmost panel, Figure 1d). The head models were manufactured using a Dimension 1200es 3D printer and ABS printing material (Stratasys, Eden Prairie, MN, USA). Due to its life-size and accurate facial topology, the head model can adequately simulate the mask's fit to the face, as demonstrated in Figure 1d.

2.4. Interpersonal Transmission Visualization System

To investigate interpersonal droplet transmission, two head models were placed at a specified distance and relative height (Figure 1e). The head orientation of the two head models could also be adjusted as needed. Each head model was ventilated to a breathing simulator with a sinusoidal waveform (Michigan Instruments, Grand Rapids, MI, USA). The source head model (black color, Figure 1e) was also connected to the vapor source to simulate exhaled droplets, while the recipient model (white color, Figure 1e) was only ventilated with tidal air flows. An ultrasonic humidifier (Pure Enrichment, Huntington Beach, CA, USA) was used to generate soft mist. A SprayLink laser diffraction spray particle size analyzer (Dickinson, TX, USA) was used to measure the size distribution of the soft mist exhaled from the head model without a mask under constant expiratory flows [40]. As shown in Figure 1e, the size distribution was 2.64, 3.92, and 8.51 μm for D10, D50, and D90, respectively. This size range is close to that of expiratory droplets from deep lungs [41–43]. During testing, the ambient temperature was 24 °C and the relative humidity was 31%.

2.5. Numerical and Statistical Methods

2.5.1. Computational Fluid Dynamics (CFD) Simulations

Complementary CFD modeling and simulations were performed. An integrated mask-wearing model was developed that included the ambient air, facemask, face geometry, and airway. Gaps at the mask–face interface were modeled using individual volumes of the mask filter medium that had the properties of air rather than a porous medium. More details of the computational model, along with the mesh generation and grid-independent study, were provided in [44,45]. The computational domain consisted of three volumes: ambient air, mask, and nose–mouth–throat airway. The ambient air and airway were defined as airflow regions, while the mask was modeled as a porous medium. Darcy's law was applied to calculate the viscous resistance (R) of the mask filter, i.e., $R = \Delta P/[(Q/A)\cdot\mu\text{L}]$, with ΔP being the breathing resistance, Q the volume flow rate during the mask test, A the test area, μ the air dynamic viscosity, and L the mask thickness. For a surgical mask, the resistance ΔP was measured to be 14.6 mmH_2O using a 8130A Automated Filter Tester (TSI, Shoreview, MN, USA), the flow rate Q was 80 L/min, the test area A was 45.6 cm^2, the flow viscosity μ was 1.825×10^{-5} kg/m·s, and the mask thickness L was 2.3 mm, leading to a viscous resistance of 3.727×10^9 1/m^2 [45]. Steady exhalations were considered, with a constant positive pressure being specified at the trachea opening and zero pressure in the far field of the ambient air. Air temperatures were set at 37.15 °C in the airway and 24 °C in the environment. The low-Reynolds number (LRN) k-ω turbulent model was utilized to simulate the multi-regime flow dynamics. ANSYS Fluent 23 (Canonsburg, PA, USA) was used to solve conservation equations for mass, momentum, and energy.

2.5.2. Statistical Analysis

Minitab 21.4 (State College, PA, USA) was applied to analyze the leakage flow velocities and cross-mask particle counts. A one-way analysis of variance (ANOVA) was used to evaluate the variability of the measurements. A p-value < 0.05 indicated statistical significance difference.

3. Results

3.1. Control Cases

3.1.1. Schlieren Optical Imaging System

Figure 2 shows the Schlieren visualization results of the exhaled flows from a participant with different masks, i.e., surgical, KN95, cloth, and neck gaiter (N-Gaiter). For each mask, five activities were tested: normal breathing, coughing, soft speaking, normal speaking, and loud speaking. The effects of wearing a mask bracket were also evaluated for all masks except the neck gaiter. Considering that a good mask should distribute airflow over the mask and seal well, the mask performance was evaluated qualitatively based on how much air is seen escaping the mask and where leakages are occurring. For all masks tested, the KN95 mask was found to have the least amount of air escaping from the mask for all activities. By comparison, the cloth, surgical, and neck gaiter exhibited large amounts of air escaping from the tops/sides of the mask, as well as through the mask filter material itself. The jet-like flows through the cloth mask and neck gaiter are due to the larger fiber pores that allow for air to pass through them more freely, while the KN95 mask has smaller filter pores. This makes it more difficult for airflows and particles to pass through and helps distribute the airflow all over the mask, thus lowering the exhaled flow speed crossing the mask and extending the effective filtration area.

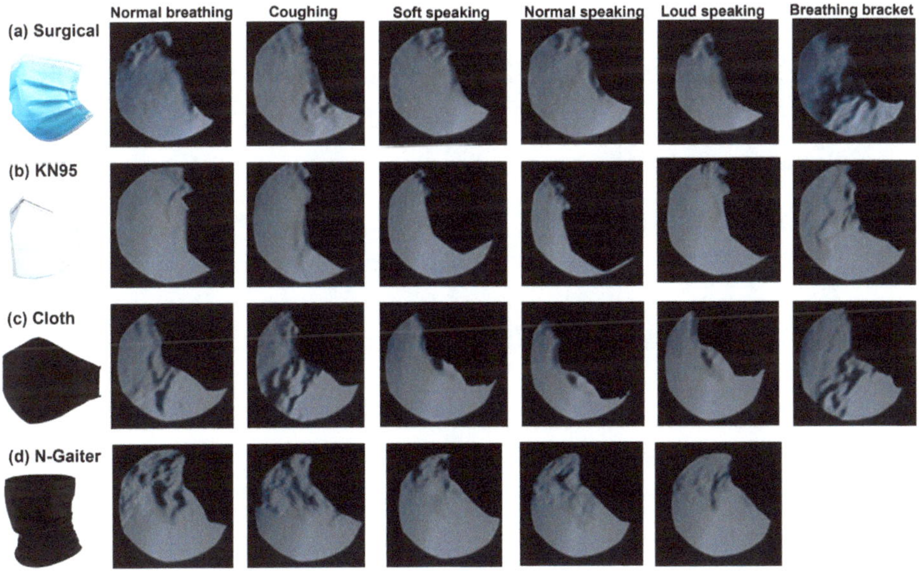

Figure 2. Representative Schlieren imaging (from multiple tests) of expiratory mask flows under various activities from a participant: (**a**) surgical mask, (**b**) KN95, (**c**) cloth mask, and (**d**) neck gaiter. Test activities include normal breathing, coughing, soft speaking, normal speaking, loud speaking, and breathing with a bracket.

Considering the three speaking scenarios (i.e., soft, normal, and loud, Figure 2), no apparent jet flows were observed through the masks, primarily because of the intermittent flow pulses during speaking, regardless of the speaking volume. When a mask bracket was used (last column, Figure 2), intensified jet flows were observed crossing the KN95 and surgical masks, as the bracket effectively confined the exhaled flows within the boundary of the bracket. The purpose of a mask bracket is to fit the mask to the face better, specifically with surgical masks that can create gaps at the sides of an individual's face.

For a given mask, we observed similar flow patterns among different activities, whether the test subject was breathing, coughing, or speaking at different volumes. This demonstrated the mask's overall consistency in effectively containing exhaled airflows under different physical activities. However, differences in flow patterns were also observed among various activities. For example, with the cloth mask (Figure 2c), coughing produced a jet flow across the mask, whereas this jet flow was absent during speaking, regardless of intensity (i.e., soft, normal, or loud). Note that the considered activities have very different dynamics, where breathing releases air constantly from the nostrils, coughing releases air from the mouth quickly at high pressure, and speaking releases air intermittently from the mouth. Moreover, the latter two activities also involve the motion of the mouth, particularly the lips, which may affect the mask fit.

3.1.2. Laser Sheet System

Figure 3 shows the laser visualization of expiratory mask flows in two head models. It is observed that a breath exhaled from the head model traveled directly forward and dispersed in the air. A commonality among all the masks was leakages out of the top of the mask. In addition, fog particulates were able to pass directly through the cloth mask without a HEPA filter and through the neck gaiter. When the HEPA filter was placed in the cloth mask, there was no passage of fog directly through the mask. This finding highlights the effectiveness of the HEPA filter in improving the effectiveness of cloth masks.

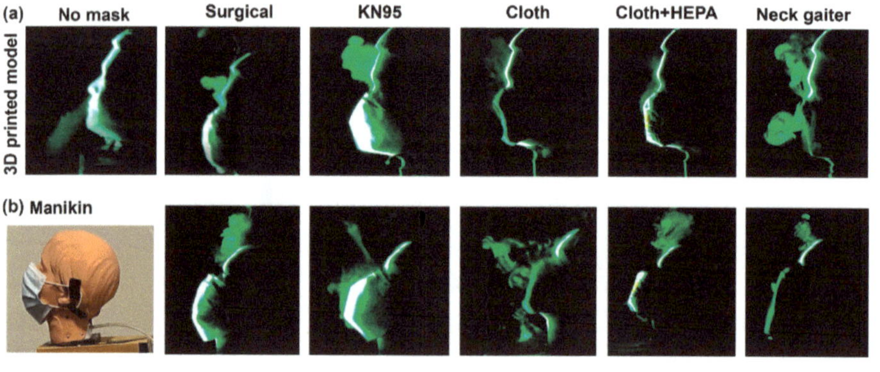

Figure 3. Representative laser visualization (from multiple tests) with different masks on two head models: (**a**) 3D-printed head model and (**b**) manikin.

Considering flows with the 3D-printed model (Figure 3a), leakages occurred at the nose top with the surgical mask, KN95, and the neck gaiter. When comparing the results between the 3D-printed head and the manikin (Figure 3a vs. Figure 3b), one can see that the cloth masks (both with and without the HEPA filter) and surgical mask all performed better with the 3D head model. This is likely due to the 3D head model being larger in size and fitting the masks better than the manikin. These findings demonstrate the importance of proper mask designs for different population groups, such as adults and children.

Based on the findings from the laser sheet experiments, the neck gaiter and the cloth mask without the HEPA filter were less effective in preventing the spread of exhaled air due to their sites of leakage and their inability to distribute the airflow evenly across the filter.

The surgical mask, the cloth mask with the HEPA filter, and the KN95 mask performed better in blocking and spreading the exhaled air despite having leakage through the top of the masks. These leakages highlight the important fact that masks are not 100% effective, necessitating social distancing and hand/face cleaning to mitigate viral transmission.

3.1.3. Particle Count across Various Masks

Figure 4 shows the exhaled particle counts across the masks in total (Figure 4a) and particle size-specific numbers (Figure 4b) for different masks. Large differences in particle counts were observed in Figure 4b between all masks and the no mask control except for the neck gaiter. As expected, the largest number of exhaled droplets was detected with no mask-wearing, followed by the neck gaiter, cloth, surgical with bracket, surgical, cloth with HEPA filter, and KN95 (Figure 4a,b). The slightly higher number of droplets from 'surgical + bracket' than from 'surgical' is because the bracket confined the exhaled flow droplets within the bracket and reduced leakages from the mask–face interface.

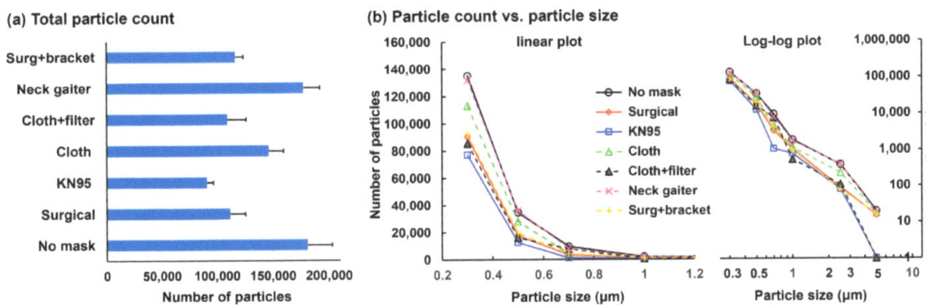

Figure 4. Exhaled particle counts across various masks from a participant: (**a**) total particle count, (**b**) particle count vs. particle size. 'Surg + bracket': Surgical plus bracket.

3.2. Leakage Flow Visualization and Quantification

3.2.1. Leakage Flow Velocity Measurement

Considering that leakages are mostly observed at the nose top and two sides during exhalation, the speeds of the leakage flows were measured using a hotwire anemometer at the right and left ridges of the nose top, two sides, and the chin (Figure 5a).

For the three masks considered (cloth, surgical, and KN95), the leakage flow speeds are much higher from the two ridges than the two lateral sides and chin. Mask-specific differences are also observed, with more leakages from the chin for the cloth and KN95 masks and more lateral leakages from the surgical mask. These observations are consistent with the peculiar mask geometry and its fitting to the facial topology. Both the cloth and KN95 masks have a flat region covering the cheeks, while the surgical mask is tucked up on both sides, forming an arch that allows for leakage flows. Among the three masks, the cloth mask had the highest leakage flow speed (1.7 ± 0.5 m/s, Figure 5b), followed by the surgical mask (1.3 ± 0.5 m/s, Figure 5c) and KN95 (0.24 ± 0.1 m/s. Figure 5d). This difference is also evident by the different ranges in the y-axis among Figure 5b–d. By comparison, the jet flow velocity exhaled from the nostrils without a mask was measured to be 2~4 m/s, while the flow velocity across the mask was 0.1~0.2 m/s.

Figure 6 shows the speeds of the inward leakage flows at different sites during inhalation for three mask types. Note that for both inhalation and exhalation, the mask was carefully pressed to conform to the participant's facial topology as much as possible, thus representing the best scenario of mask fit. The thermal images of the mask are blue in color, indicating the cooling effects from inspiratory ambient air, which contrasts with the warming effect (brown color) from the expiratory air at body temperature (Figure 6a vs. Figure 5a). The overall leakage speeds during inhalation are lower than those during exhalation (Figure 6 vs. Figure 5). This is due to the jet flow features of the expiratory flows in contrast to the converging flow features of inspiratory flows. Considering the side fitting, inward leakage is not observed in the cloth and KN95 masks (Figure 6b,d) but is observed in the surgical masks (Figure 6c), corroborating the finding that a tucked-up fold is more prone to mask–face gaps and leakage flows.

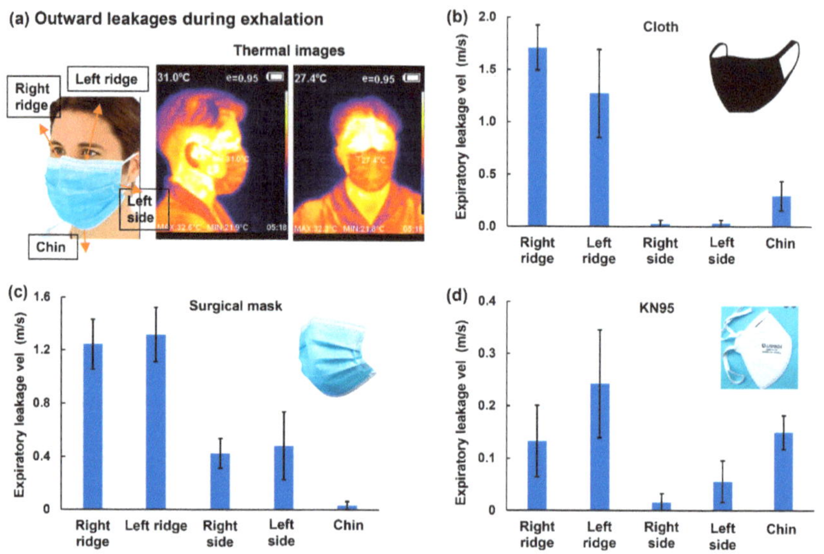

Figure 5. Leakage flow speed measurement during exhalation: (**a**) outward leakage diagram and thermal imaging, (**b**) cloth mask, (**c**) surgical mask, and (**d**) KN95.

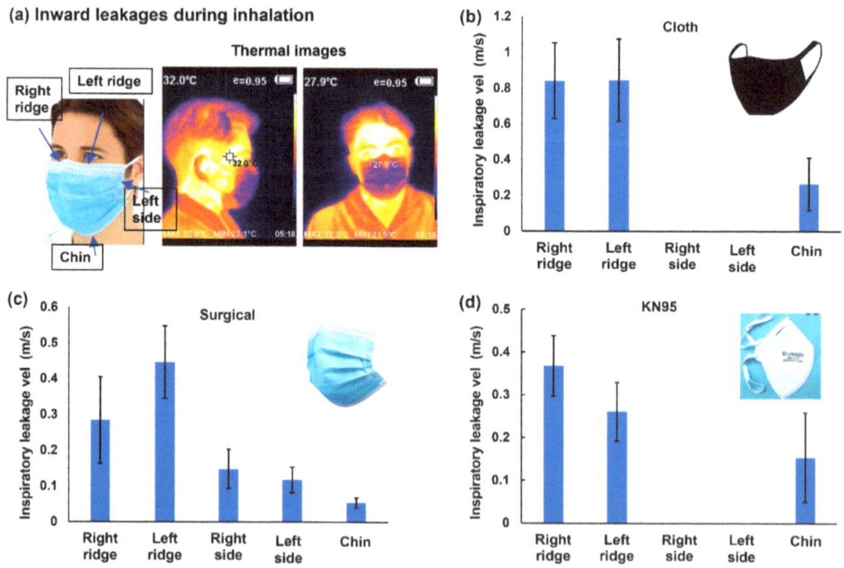

Figure 6. Leakage flow speed measurement during inhalation: (**a**) inward leakage diagram and thermal imaging, (**b**) cloth mask, (**c**) surgical mask, and (**d**) KN95.

3.2.2. In Vitro Visualization vs. CFD

The visualization of mask flows, especially underneath the mask, can provide detailed information on flow, pressure, and temperature to enhance our understanding of facemask dynamics and thermoregulation. Figure 7 compares the experimental (upper panels) and computational (lower panels) results without mask-wearing (Figure 7a), with a surgical mask and side gap (Figure 7b), and with a well-fitted surgical mask (Figure 7c). As shown in Figure 7a, the expiratory jet flows from the nostrils resembled each other between the laser-fog imaging and the predictions from computational fluid dynamics (CFD). The lower panel of Figure 7b displays the inspiratory flow streamlines, which converge from all directions but intensify considerably through the side gap [45,46]. To validate the CFD model, measurement of the flow velocities at the gap was conducted, which agreed favorably with the CFD simulations (Figure 7b).

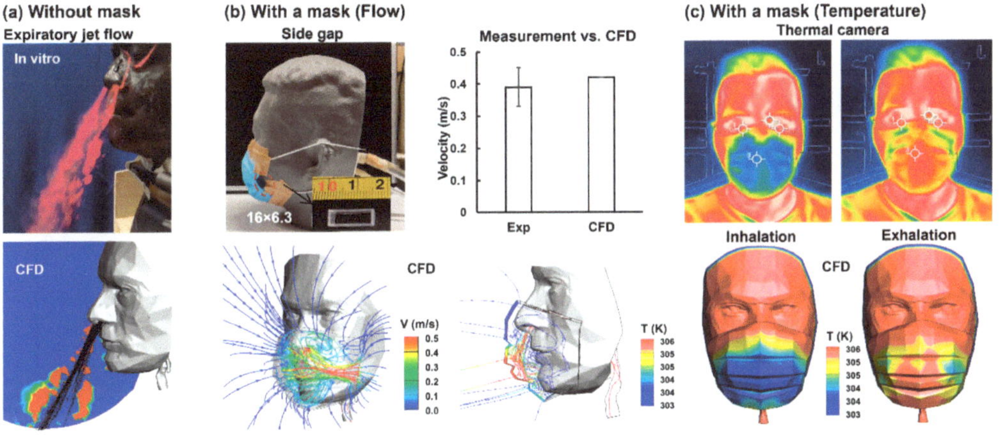

Figure 7. Experiments vs. computational fluid dynamic (CFD) simulations: (**a**) without a mask, (**b**) with a mask for flow, and (**c**) with a mask for temperature.

The lower panel of the third column (Figure 7b) shows the CFD-predicted temperature distribution during exhalation, where the warm jet flows from the nostrils are cooled down and well mixed within the mask–face space. Due to the mask resistance, strong recirculation forms under the mask, eliciting quick mixing. Note the higher temperatures of the mask where the exhaled jet flow impinges compared to other regions of the mask. Figure 7c compares the facial temperature obtained using a thermal camera and complementary CFD simulations. Again, high levels of resemblance are observed between the experiments and CFD for both inhalations and exhalations, further validating the computational model in capturing mask-wearing-associated thermoregulation, which in turn provides thermo-fluid details under the mask that are not easy to measure accurately or non-disruptively. Higher levels of similarity are also observed in comparison to the thermal images in Figures 5 and 6.

3.3. LED-Fog System to Visualize Inhalation–Exhalation Flows

Figure 8 shows the inspiratory and expiratory vector fields in two breathing cycles in a head model wearing a surgical mask. Due to leakages at the nose top, high-speed flows are observed for both inhalation and exhalation cycles. The vectors at the nose top are much longer than the ones in front of the mask. Smoke streaks are observed moving away from the mask, as indicated by the yellow arrows in Figure 8a and red arrows in Figure 8b. These streams advance forward slowly and, at the same time, oscillate due to cyclic inhalation and exhalation, forming complex flow patterns. By contrast, the flows in the immediate proximity of the mask change quickly with the breathing cycle. The slow advancement of the smoke streaks away from the mask suggests that the exhaled droplets are likely to behave similarly. Thus, the overall advancement speed can indicate whether or when virus-laden respiratory droplets from an infected person reach the recipient at a specific distance.

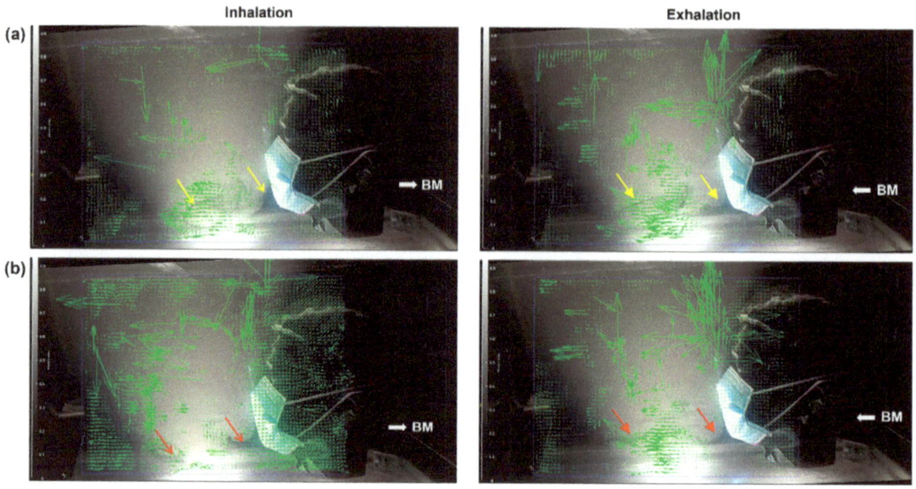

Figure 8. PIVlab analyses of the LED-illuminated smoke distribution during inhalation and exhalation in a head model with a surgical mask over two breathing cycles: (**a**) cycle A and (**b**) cycle B.

3.4. *Interpersonal Droplet Transmission*

3.4.1. Without Face Covering (1.2 m): The Control

The interpersonal transmission of exhaled droplets is shown in Figure 9 after 5 min of exposure at a distance of 1.2 m between the source and recipient. The droplet deposition is visualized using SarGel, which turns from white to pink upon contacting water, with more deposited water mass correlating with a deeper pink color [47]. The right three panels of Figure 9a,b display different views of the droplet deposition on the head model that is ventilated with a tidal breathing machine. Heterogeneous deposition distributions are observed on the face in both test cases. However, a much-enhanced deposition occurs at the lower cheek (or jaw) and neck (Figure 9b), presumably because the droplet trajectories in the second case are more aligned with the recipient's neck than in the first case.

3.4.2. With Face Covering (1.5 m)

Figure 10 shows the effects of various types of face coverings on interpersonal droplet transmission at a distance of 1.5 m. In both cases, the source wore a surgical mask and a face shield. Thus, the exhaled droplets mainly escaped from the bottom, and a smaller fraction escaped from the two lateral sides of the face shield. Note that none of these three sites pointed to the recipient. The exposure lasted for twenty minutes, and photos of the droplet deposition were taken every five minutes with the mask on and removed. The deposition after 5 min of exposure was negligible in both cases and is thus not presented here.

Two observations are noteworthy regarding the droplet deposition from 10–20 min. First, the progression of deposition is nonlinear with time, which is unnoticeable during 0–5 min, becomes discernable during 5–10 min, but increases quickly between 10 and 20 min (Figure 10a). This suggests that the exhaled droplet plume needs time to propagate from the source to the recipient. However, once the plume reaches the recipient, the deposition will be continuous and become noticeable after a short time, as illustrated by the quick color change from 10 to 15 min. Second, more deposition was observed in the mouth–nose region and the jaw–neck region. The deposition on the hairs of the frontal head is also notable.

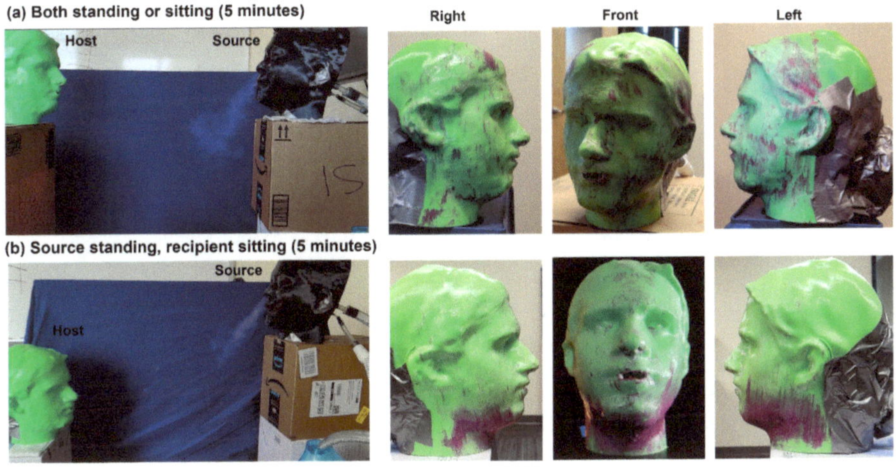

Figure 9. Visualization of interpersonal transmission of respiratory droplets after 5 min exposure when both the source and recipient do not wear a mask: (**a**) both standing or sitting and (**b**) source standing and recipient sitting.

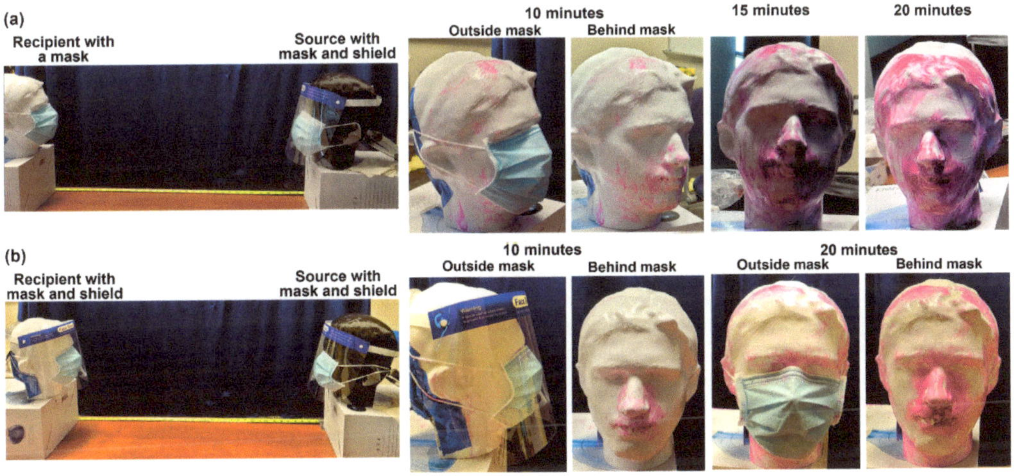

Figure 10. Effects of a face shield on inter-person transmission for varying exposure durations with the source person wearing both a surgical mask and face shield: (**a**) the recipient wearing a surgical mask and (**b**) the recipient wearing a surgical mask and face shield.

Figure 10b illustrates the effects of double protection with the recipient adding a face shield on top of a surgical mask. Compared to wearing a surgical mask only, the face deposition is significantly lower after 10 min of exposure, indicating the effectiveness of extra face-shield protection, which indeed reduced the number of droplets reaching the face underneath the surgical mask. In this case, the droplets need to avoid impacting the face shield by moving downward in front of the shield and subsequently moving upward to reach the mask. This would decrease the leakage flows from the nose top and make more airflow cross the mask filter medium. Wearing a face shield might also have pressed down the tucked-up arch at two sides, reducing side leakages. Slight deposition in the philtrum region is observed after a 10 min exposure, with no discernable deposition in other regions of the head model (Figure 10b). An additional 10 min of exposure remarkably increased the deposition in the philtrum region, as well as in the nasofacial sulcus, and to a lesser degree, on the nasal ridge and neck (Figure 10b).

4. Discussion

Even though mask-wearing has become ubiquitous, knowledge to quantify its performance or improve its performance is surprisingly limited and empirical at best. This study explored multiple methods to visualize mask-related flows and droplet transmission to better understand mask performance, including the Schlieren optical system, laser imaging system, thermal camera, and vapor–SarGel system. Interesting observations and further thoughts on mask-wearing are discussed below.

4.1. Leakage Flow Characterization

In this study, for all masks considered, leakages from the nose top were observed to be notably higher than those from any other sites (i.e., two lateral sides and chin). Therefore, efforts to improve the mask fit at the nose top are both compelling and rewarding. For instance, using a wider and stronger nasal strap in the surgical mask will not only improve the mask fit to the nose ridge but also help the mask maintain its good fit during various physical activities [48]. A 'Knot-and-tuck' method also promises to achieve a better fit [49]. By knotting the ear loops of a surgical mask and tucking excess material under the edges, a 3D cup-shaped mask forms that better conforms to the facial topology, thus improving the mask–face fit and reducing sideway leakages. It is noted that standardized methods or devices to quantify leakages or fit for disposable masks are lacking, even though commercial devices for tight-fitting respirators are available. One example is the PortaCount Respirator Fit Tester 8038 (TSI, Shoreview, MN, USA); a fit score of no less than 200 under all tested activities is considered a pass. However, when used for loose-fitting disposable masks, such as surgical masks, the fit scores based on the PortaCount Respirator Fit Tester are very low and exhibit large variability (e.g., 0–40). These erroneous readings are mainly due to the device's working principle, which estimates the fit score based on the aerosol concentrations (generated from NaCl-solution) inside and outside the respirator.

For a loosely fit disposable mask, the aerosols will be easily inhaled across the filter medium or through the inward leakage, equalizing the aerosol concentrations inside and outside the mask, losing the good foundation for fit estimation [27]. Furthermore, while wearing a mask diffuses the expiratory flow compared to the jet-like flow without a facemask, the resulting flow is still not isotropic. The variations in leakage velocity among different leaking sites (Figures 5 and 6) reflect both this flow anisotropy and the size differences of the leakage sites. Additional research is needed to quantify the relative contributions of these two factors. A more practical method is to leverage the temperature variations that are dependent on the leakage flow volume [46,50]. Moreover, the periodic temperature variation within one breathing cycle is expected to provide further information on leakage flows, including the leakage volume (gap size) and site, because the cooling effect during inhalation and warming effect during exhalation are highly sensitive to the leakage flow patterns [51–53].

4.2. Droplet Deposition Visualization and Implications

Interpersonal transmission testing with vapor and SarGel provides a practical approach to visualize the transport and deposition of respiratory droplets that are often invisible in life conditions. It is observed that the relative orientation and distance can notably affect the transport and deposition of respiratory droplets (Figures 9 and 10), making it necessary to systematically study these factors to gain a comprehensive understanding of viral transmission and mask protection. Zhang et al. reviewed the close-contact (or short-range) transmission of aerosols between two individuals and highlighted influencing physical parameters, such as their distance/orientation, body/head motion, exposure duration, and breathing intensity, among others [54]. Zhang et al. also suggested that an exhaled droplet would need at least 0.6 s to reach the receiver at a distance of 1.5 m, based on an expiratory flow velocity of 2.4 m/s (i.e., 1.5 m over 2.4 m/s equals 0.6 s). Note that it can take much longer than 0.6 s for an exhaled droplet to travel from the source to the recipient because the exhaled droplets will quickly slow down within the ambient air. In this study, we observed that it took even longer (i.e., five minutes, Figure 9) for the SarGel on the head model to turn pink under no-mask conditions and ten minutes and more when both the source and recipient were wearing masks (Figure 10). It is reminded that the color change of SarGel with vapor deposition is a gradual process and becomes a darker pink with accumulating deposition. To reach the color depth on the face in Figure 10a, a deposition intensity of 0.24 mg/cm^2 of vapor is needed [55].

One salient feature in vapor droplet deposition with no mask-wearing is the elevated deposition in the neck, chin, and jaw, as displayed in Figure 10b, where the source is standing and the recipient is sitting. This preferential deposition can cause secondary self-inoculation of the recipient's mucous membrane because people spontaneously touch their eyes, cheeks, chin, and mouth, unknowingly [56–58]. According to a study involving 26 students in 2015, a person touched his or her own face 23 times per hour, with around 44% touching the mucous membranes (mouth, nose, and eyes) and 56% touching the non-mucosal areas (chin, cheek, neck, ears, forehead, and hair), with insignificant differences between right and left hands [59]. In a more recent systematic review that considered ten observational studies, facial self-touching per hour was counted as 50.06 (±47) times [60]. Considering the high deposition in the neck, chin, jaw, and mouth–nose region, hand/face/neck washing is highly recommended to reduce the chance of self-inoculation via spontaneous face-touching.

4.3. *Effect of Double Masking*

Double masking has been recommended as an extra measure during the COVID-19 pandemic to further curb viral transmission more effectively than wearing a single mask [61]. It can significantly enhance protection against respiratory droplets that may contain viruses, including COVID-19 [62]. This practice involves wearing one mask over another—typically a cloth mask over a surgical mask—to improve the overall fit and filtration capability of the masks [63]. The CDC has noted that this method reduces exposure by 95% when both masks are worn properly [30]. In this research, we evaluated the effect of double masking on interpersonal droplet transmission by comparing the recipient wearing a surgical mask vs. a surgical mask plus a face shield. Adding a face shield led to improvements: (a) prolonged the time when the droplet deposition became noticeable and (b) reduced the overall deposition on the face and neck of the recipient. Adding a face shield resulted in significant improvements in reducing deposition on the recipient, especially on the cheek and neck, even though deposition in the nose–mouth region was still observed (Figure 10b vs. Figure 10a). The time for deposition onset was also prolonged.

There might be two reasons behind these improvements: modified flow-particle dynamics and enhanced surgical mask fit. The presence of the face shield causes airflow and aerosols that approach the wearer to either collide with the shield or divert upwards from beneath the shield. This not only filters out a fraction of aerosols but also lowers their speeds, making it more difficult for these aerosols to be inhaled across the surgical mask. Furthermore, the face shield can also enhance the fit of the surgical mask to the face by pressing down the tucked-up folds, thus minimizing gaps around the edges of the mask. The seal of the face shield at the forehead also reduces the influx of unfiltered air through the nose top. Thus, adding a face shield to a surgical mask can effectively increase the number of barriers against potentially infectious aerosols while introducing relatively minimal extra resistance due to its side opening to the environment. It can be particularly useful in scenarios where higher protection is needed, such as crowded indoor spaces or when interacting closely with others. Also note that double masking is not recommended with two disposable masks or with two N95 masks, as these combinations do not enhance fit or filtration effectively and can make breathing more difficult [64–66].

4.4. Limitations and Future Studies

The limitations of this study included an unbalanced focus on exhalation flows, the qualitative nature of results, and a limited number of interpersonal and mask-wearing scenarios. Overall, it is easier to visualize the expiratory flows than inspiratory flows both with and without a mask, due to the jet-like exhalation flows vs. the converging inhalation flows (Figure 7a vs. Figure 7b). The latter often have extremely low velocities (i.e., proportional to $1/r^3$), except in the proximity of the nostrils and face–mask gaps (Figures 7b and 8). By contrast, the exhalation core flow has a much higher velocity, penetrates a longer distance, and maintains a large temperature difference relative to the ambient air (Figure 7b). It is unclear, for a mask with given gaps, whether the outward and inward leakage flow rates are the same or not. An accurate answer to this question will affect the estimation of the mask-fit effect on viral source control or receiver protection.

The major limitation of the visualization methods in this study is that they are mostly qualitative. The Schlieren optical system can also be used to visualize interpersonal droplet transport [26,34]. This, however, requires a large concave mirror, whose cost increases drastically with the mirror diameter [67]. By comparison, the vapor–SarGel system developed in this study is low-cost and directly relevant to interpersonal viral transmission. The vapor droplets are measured to be 3.92 μm on average, which is consistent with the respiratory droplets generated in the alveolar region from liquid film ruptures [41–43]. Note that the alveolar region is the major site of SARS-CoV-2 viral infections [68]. Droplets larger than 10 μm were not considered in this study. Note that exhaled aerosols or droplets can be internally generated at different regions of the respiratory system and the droplet size can also vary significantly depending on the source regions (i.e., 1–145 μm) [69].

Only four scenarios of interpersonal short-range airborne transmission were considered in this study. However, there are countless scenarios that can markedly influence the transmission risks. Edmunds et al. summarized such behaviors into four categories, with conversation accounting for 95% of all possible close contact events, i.e., conversation without physical contact (61%) and with physical contact (34%) [70]. Zhang et al. [54] proposed quantitative physical parameters to describe these close contact transmissions, including the two persons' locations and head orientations (1), head/body movement (2), close contact frequency and duration (3), and breathing patterns/intensities (4). So far, the parameters most frequently studied are the interpersonal distance and exposure duration, which have been demonstrated to play an important role in viral transmission [54]. In this study, in addition to the distance and duration, we also observed that the head relative height (i.e., recipient standing and recipient sitting) and face covering type can notably alter the aerosol transmission, including the exposure time for SarGel color change, the variation in deposition with time, and the deposition distribution (Figures 9 and 10). Furthermore, real-life scenarios are often much more complex. An individual changes his/her activity and position throughout the day and may be exposed to varying levels of viral sources. Consequently, the total viral load received by an individual would be an integration of transient exposures over a prescribed period. Future studies are needed to examine how other close-contact parameters influence the transmission risk. Such datasets are vital for creating models of viral spread and planning future outbreak responses.

5. Conclusions

In summary, various visualization methods were attempted to better understand mask flows and interpersonal droplet transmission. Leakage flow velocities at the nose top, two sides, and chin were measured for the cloth, surgical, and KN95 masks, with significant differences among the three masks and between exhalation and inhalation for each mask. Particle counts were measured for five mask types (including neck gaiter and cloth + filter) with a good fit, which were consistent with their filtration efficiency in comparison to that with no mask. The vapor–SarGel visualization system for interpersonal droplet transmission provided a practical method to simulate the temporospatial transport and deposition of respiratory droplets under varying scenarios or with different face coverings. Significant deposition occurred in the mouth–nose region, as well as in the neck, chin, and jaw, increasing the risk of self-inoculation due to spontaneous face-touching, urging post-exposure hand/face washing. Adding a face shield to a surgical mask notably reduced total droplet deposition on the head and regional deposition in the neck and face, supporting that double face covering can be highly effective when one face covering is deemed inadequate.

Author Contributions: Conceptualization, J.H.P., R.N., X.S. and J.X.; methodology, J.S.X., M.T. and X.S.; software, M.T. and J.X.; validation, J.S.X., J.H.P., R.N., X.S., J.X. and M.R.; formal analysis, J.S.X., M.T. and X.S.; investigation, J.S.X., M.T. and X.S.; Resources: X.S., J.H.P., R.N. and J.X. data curation, M.T.; writing—original draft preparation, J.X.; writing—review and editing, X.S.; visualization, X.S.; supervision, J.X. and M.R. All authors have read and agreed to the published version of the manuscript.

Funding: This study was partially sponsored by the Advanced Functional Fabrics of America and U.S. Army ARDEC under Agreement number W15QKN-16-3-0001.

Institutional Review Board Statement: The study was conducted in accordance with the Declaration of Helsinki, and the protocol was approved by the Ethics Committee of the University of Massachusetts Lowell (protocol code 22-194 and date of approval: 10/02/2022).

Informed Consent Statement: All subjects gave their informed consent for inclusion before they participated in the study.

Data Availability Statement: The data presented in this study are available upon request from the corresponding author.

Acknowledgments: We thank Amr Seifelnasr at UMass Lowell Biomedical Engineering for editing and proofreading this manuscript. Claire Lepont is gratefully acknowledged for the project management.

Conflicts of Interest: The authors declare that they have no known competing financial interests or personal relationships that could have appeared to influence the work reported in this paper. Michael Rein is employee of Advanced Functional Fabrics of America, who has no conflict of interest to declare in this study, and this study received funding from Advanced Functional Fabrics of America and U.S. Army ARDEC. The funder had no role in the design of the study; in the collection, analysis, or interpretation of data, in the writing of the manuscript, or in the decision to publish the results.

References

1. O'Kelly, E.; Arora, A.; Pirog, S.; Ward, J.; Clarkson, P.J. Comparing the fit of N95, KN95, surgical, and cloth face masks and assessing the accuracy of fit checking. *PLoS ONE* **2021**, *16*, e0245688. [CrossRef] [PubMed]
2. Kähler, C.J.; Hain, R. Fundamental protective mechanisms of face masks against droplet infections. *J. Aerosol Sci.* **2020**, *148*, 105617. [CrossRef] [PubMed]
3. Essa, W.K.; Yasin, S.A.; Saeed, I.A.; Ali, G.A.M. Nanofiber-based face masks and respirators as COVID-19 protection: A review. *Membranes* **2021**, *11*, 250. [CrossRef] [PubMed]
4. Yu, B.; Chen, J.; Chen, D.; Chen, R.; Wang, Y.; Tang, X.; Wang, H.-L.; Deng, W. Visualization of the interaction of water aerosol and nanofiber mesh. *Phys. Fluids* **2021**, *33*, 092106. [CrossRef] [PubMed]
5. Shen, H.; Zhou, Z.; Wang, H.; Zhang, M.; Han, M.; Durkin, D.P.; Shuai, D.; Shen, Y. Development of electrospun nanofibrous filters for controlling coronavirus aerosols. *Environ. Sci. Technol. Lett.* **2021**, *8*, 545–550. [CrossRef] [PubMed]
6. Hao, J.; Passos de Oliveira Santos, R.; Rutledge, G.C. Examination of nanoparticle filtration by filtering facepiece respirators during the COVID-19 pandemic. *ACS Appl. Nano Mater.* **2021**, *4*, 3675–3685. [CrossRef]
7. Frankfort, M.G.H.; Lauwers, I.; Pruijn, E.M.C.; Dijkstra, S.F.; Boormans, L.H.G.; Schouten, N.A.; van Donkelaar, C.C.; Janssens, H.M. Minimizing aerosol leakage from facemasks in the COVID-19 pandemic. *J. Aerosol Med. Pulm. Drug Deliv.* **2023**, *36*, 101–111. [CrossRef] [PubMed]
8. Solano, T.; Ni, C.; Mittal, R.; Shoele, K. Perimeter leakage of face masks and its effect on the mask's efficacy. *Phys. Fluids* **2022**, *34*, 051902. [CrossRef]
9. Bradford Smith, P.; Agostini, G.; Mitchell, J.C. A scoping review of surgical masks and N95 filtering facepiece respirators: Learning from the past to guide the future of dentistry. *Saf. Sci.* **2020**, *131*, 104920. [CrossRef]

10. Smith, J.D.; MacDougall, C.C.; Johnstone, J.; Copes, R.A.; Schwartz, B.; Garber, G.E. Effectiveness of N95 respirators versus surgical masks in protecting health care workers from acute respiratory infection: A systematic review and meta-analysis. *Can. Med. Assoc. J.* **2016**, *188*, 567–574. [CrossRef]
11. Lai, A.C.K.; Poon, C.K.M.; Cheung, A.C.T. Effectiveness of facemasks to reduce exposure hazards for airborne infections among general populations. *J. R. Soc. Interface* **2012**, *9*, 938–948. [CrossRef] [PubMed]
12. Lee, L.Y.-K.; Lam, E.P.-W.; Chan, C.-K.; Chan, S.-Y.; Chiu, M.-K.; Chong, W.-H.; Chu, K.-W.; Hon, M.-S.; Kwan, L.-K.; Tsang, K.-L.; et al. Practice and technique of using face mask amongst adults in the community: A cross-sectional descriptive study. *BMC Public Health* **2020**, *20*, 948. [CrossRef] [PubMed]
13. Barnawi, G.M.; Barnawi, A.M.; Samarkandy, S. The association of the prolonged use of personal protective equipment and face mask during COVID-19 pandemic with various dermatologic disease manifestations: A systematic review. *Cureus* **2021**, *13*, e16544. [CrossRef]
14. Ruhle, K.H.; Randerath, W. Measurement of mask leakage during CPAP in patients with obstructive sleep apnea. *Pneumologie* **2000**, *54*, 422–424. [PubMed]
15. Persson, B.N.J. Side-leakage of face mask. *Eur. Phys. J. E. Soft Matter.* **2021**, *44*, 75. [CrossRef]
16. Leidag, M.; Hader, C.; Keller, T.; Meyer, Y.; Rasche, K. Mask leakage in continuous positive airway pressure and C-Flex. *J. Physiol. Pharmacol.* **2008**, *59* (Suppl. 6), 401–406.
17. Mueller, J.T.; Karimi, S.; Poterack, K.A.; Seville, M.T.A.; Tipton, S.M. Surgical mask covering of N95 filtering facepiece respirators: The risk of increased leakage. *Infect. Control Hosp. Epidemiol.* **2021**, *42*, 627–628. [CrossRef]
18. He, X.; Grinshpun, S.A.; Reponen, T.; McKay, R.; Bergman, M.S.; Zhuang, Z. Effects of breathing frequency and flow rate on the total inward leakage of an elastomeric half-mask donned on an advanced manikin headform. *Ann. Occup. Hyg.* **2014**, *58*, 182–194.
19. Crutchfield, C.D.; Park, D.L. Effect of leak location on measured respirator fit. *Am. Ind. Hyg. Assoc. J.* **1997**, *58*, 413–417. [CrossRef]
20. Wang, T.K.; Solano, T.; Shoele, K. Bridge the gap: Correlate face mask leakage and facial features with 3D morphable face models. *J. Expo. Sci. Environ. Epidemiol.* **2022**, *32*, 735–743. [CrossRef]
21. Oestenstad, R.K.; Bartolucci, A.A. Factors affecting the location and shape of face seal leak sites on half-mask respirators. *J. Occup. Environ. Hyg.* **2010**, *7*, 332–341. [CrossRef]
22. Solano, T.; Mittal, R.; Shoele, K. One size fits all?: A simulation framework for face-mask fit on population-based faces. *PLoS ONE* **2021**, *16*, e0252143. [CrossRef] [PubMed]
23. Schmitt, J.; Wang, J. A critical review on the role of leakages in the facemask protection against SARS-CoV-2 infection with consideration of vaccination and virus variants. *Indoor Air* **2022**, *32*, e13127. [CrossRef] [PubMed]
24. Pushpawela, B.; Chea, P.; Ward, R.; Flagan, R.C. Quantification of face seal leakage using parallel resistance model. *Phys. Fluids* **2023**, *35*, 127127. [CrossRef]
25. Larsen, P.S.; Heebøll, J.; Meyer, K.E. Measured air flow leakage in facemask usage. *Int. J. Environ. Res. Public Health* **2023**, *20*, 2363. [CrossRef] [PubMed]
26. Tang, J.W.; Liebner, T.J.; Craven, B.A.; Settles, G.S. A schlieren optical study of the human cough with and without wearing masks for aerosol infection control. *J. R. Soc. Interface* **2009**, *6* (Suppl. 6), S727–S736. [CrossRef] [PubMed]
27. Su, W.C.; Lee, J.; Xi, J.; Zhang, K. Investigation of mask efficiency for loose-fitting masks against ultrafine particles and effect on airway deposition efficiency. *Aerosol Air Qual. Res.* **2022**, *22*, 210228. [CrossRef]
28. Cappa, C.D.; Asadi, S.; Barreda, S.; Wexler, A.S.; Bouvier, N.M.; Ristenpart, W.D. Expiratory aerosol particle escape from surgical masks due to imperfect sealing. *Sci. Rep.* **2021**, *11*, 12110. [CrossRef] [PubMed]
29. Koh, X.Q.; Sng, A.; Chee, J.Y.; Sadovoy, A.; Luo, P.; Daniel, D. Outward and inward protection efficiencies of different mask designs for different respiratory activities. *J. Aerosol Sci.* **2022**, *160*, 105905. [CrossRef]
30. Brooks, J.T.; Beezhold, D.H.; Noti, J.D.; Coyle, J.P.; Derk, R.C.; Blachere, F.M.; Lindsley, W.G. Maximizing fit for cloth and medical procedure masks to improve performance and reduce SARS-CoV-2 transmission and exposure, 2021. *MMWR Morb. Mortal Wkly. Rep.* **2021**, *70*, 254–257. [CrossRef]
31. Verma, S.; Dhanak, M.; Frankenfield, J. Visualizing the effectiveness of face masks in obstructing respiratory jets. *Phys. Fluids* **2020**, *32*, 061708. [CrossRef] [PubMed]
32. Eni, M.; Mordoh, V.; Zigel, Y. Cough detection using a non-contact microphone: A nocturnal cough study. *PLoS ONE* **2022**, *17*, e0262240. [CrossRef] [PubMed]
33. Crawford, F.W.; Jones, S.A.; Cartter, M.; Dean, S.G.; Warren, J.L.; Li, Z.R.; Barbieri, J.; Campbell, J.; Kenney, P.; Valleau, T.; et al. Impact of close interpersonal contact on COVID-19 incidence: Evidence from 1 year of mobile device data. *Sci. Adv.* **2022**, *8*, eabi5499. [CrossRef] [PubMed]
34. Xu, C.; Wei, X.; Liu, L.; Su, L.; Liu, W.; Wang, Y.; Nielsen, P.V. Effects of personalized ventilation interventions on airborne infection risk and transmission between occupants. *Build. Environ.* **2020**, *180*, 107008. [CrossRef] [PubMed]
35. Xu, C.; Nielsen, P.V.; Liu, L.; Jensen, R.L.; Gong, G. Human exhalation characterization with the aid of schlieren imaging technique. *Build. Environ.* **2017**, *112*, 190–199. [CrossRef] [PubMed]
36. Talaat, M.; Barari, K.; Si, X.A.; Xi, J. Schlieren imaging and video classification of alphabet pronunciations: Exploiting phonetic flows for speech recognition and speech therapy. *Vis. Comput. Ind. Biomed. Art* **2024**, *7*, 12. [CrossRef] [PubMed]

37. CDC, NPPTL Respirator Assessments to Support the COVID-19 Response. Available online: https://www.cdc.gov/niosh/npptl/respirators/testing/default.html#:~:text=Workplaces%20should%20make%20every%20effort,Support%20the%20COVID-19%20Response (accessed on 22 June 2024).
38. Forouzandeh, P.; O'Dowd, K.; Pillai, S.C. Face masks and respirators in the fight against the COVID-19 pandemic: An overview of the standards and testing methods. *Saf. Sci.* **2021**, *133*, 104995. [CrossRef] [PubMed]
39. Ranga, U. SARS-CoV-2 aerosol and droplets: An overview. *Virusdisease* **2021**, *32*, 190–197. [CrossRef] [PubMed]
40. Talaat, M.; Si, X.; Xi, J. Breathe out the secret of the lung: Video classification of exhaled flows from normal and asthmatic lung models using CNN-long short-term memory networks. *J. Respir.* **2023**, *3*, 237–257. [CrossRef]
41. April Si, X.; Talaat, M.; Xi, J. SARS-CoV-2 virus-laden droplets coughed from deep lungs: Numerical quantification in a single-path whole respiratory tract geometry. *Phys. Fluids* **2021**, *33*, 023306. [CrossRef]
42. Li, H.; Leong, F.Y.; Xu, G.; Kang, C.W.; Lim, K.H.; Tan, B.H.; Loo, C.M. Airborne dispersion of droplets during coughing: A physical model of viral transmission. *Sci. Rep.* **2021**, *11*, 4617. [CrossRef] [PubMed]
43. Bake, B.; Larsson, P.; Ljungkvist, G.; Ljungström, E.; Olin, A.C. Exhaled particles and small airways. *Respir. Res.* **2019**, *20*, 8. [CrossRef] [PubMed]
44. Xi, J.; Kim, J.; Si, X.A.; Mckee, E.; Corley, R.A.; Kabilan, S.; Wang, S. CFD modeling and image analysis of exhaled aerosols due to a growing bronchial tumor: Towards non-invasive diagnosis and treatment of respiratory obstructive diseases. *Theranostics* **2015**, *5*, 443–455. [CrossRef] [PubMed]
45. Xi, J.; Barari, K.; Si, X.A.; Abdollahzadeh Jamalabadi, M.Y.; Park, J.H.; Rein, M. Inspiratory leakage flow fraction for surgical masks with varying gaps and filter materials. *Phys. Fluids* **2022**, *34*, 041908. [CrossRef]
46. Barari, K.; Si, X.; Xi, J. Impacts of mask wearing and leakages on cyclic respiratory flows and facial thermoregulation. *Fluids* **2024**, *9*, 9. [CrossRef]
47. Xi, J.; Lei, L.R.; Zouzas, W.; April Si, X. Nasally inhaled therapeutics and vaccination for COVID-19: Developments and challenges. *MedComm* **2021**, *2*, 569–586. [CrossRef] [PubMed]
48. Roberge, R.J.; Palmiero, A.J.; Liu, Y.; Kim, J.H.; Zhuang, Z. Effect of upper strap downward displacement on n95 filtering facepiece respirator fit factors: A pilot study. *J. Occup. Environ. Hyg.* **2014**, *11*, 338–341. [CrossRef] [PubMed]
49. South-Shore_Health. How the 'Knot-and-Tuck' Method Achieves a Better Fitting Face Mask. Available online: https://www.southshorehealth.org/wellness/blog/how-knot-and-tuck-method-achieves-better-fitting-face-mask (accessed on 27 May 2024).
50. Cherrie, J.W.; Wang, S.; Mueller, W.; Wendelboe-Nelson, C.; Loh, M. In-mask temperature and humidity can validate respirator wear-time and indicate lung health status. *J. Expo. Sci. Environ. Epidemiol.* **2019**, *29*, 578–583. [CrossRef] [PubMed]
51. Zheng, Z.; Dai, K.; Zhou, X.; Liu, J.; Liu, W.; Lu, J.; Fang, Z. Field investigation of thermal comfort with face masks in outdoor spaces in South China: A case study. *Urban Clim.* **2023**, *51*, 101632. [CrossRef]
52. Hu, R.; Liu, J.; Xie, Y.; Jiao, J.; Fang, Z.; Lin, B. Effects of mask wearing duration and relative humidity on thermal perception in the summer outdoor built environment. *Build. Simul.* **2023**, *16*, 1601–1616. [CrossRef]
53. Oner, E.; Seçkin, A.Ç.; Egeli, D.; Seçkin, M. Investigation of the Thermal Comfort Properties of Masks Used during the COVID-19 Pandemic. *Int. J. Environ. Res. Public Health* **2022**, *19*, 11275. [CrossRef] [PubMed]
54. Zhang, N.; Chen, W.; Chan, P.T.; Yen, H.L.; Tang, J.W.; Li, Y. Close contact behavior in indoor environment and transmission of respiratory infection. *Indoor Air* **2020**, *30*, 645–661. [CrossRef]
55. Xi, J.; Yuan, J.E.; Zhang, Y.; Nevorski, D.; Wang, Z.; Zhou, Y. Visualization and quantification of nasal and olfactory deposition in a sectional adult nasal airway cast. *Pharm. Res.* **2016**, *33*, 1527–1541. [CrossRef] [PubMed]
56. Kwok, Y.L.; Gralton, J.; McLaws, M.L. Face touching: A frequent habit that has implications for hand hygiene. *Am. J. Infect. Control* **2015**, *43*, 112–114. [CrossRef]
57. Chen, Y.J.; Qin, G.; Chen, J.; Xu, J.L.; Feng, D.Y.; Wu, X.Y.; Li, X. Comparison of face-touching behaviors before and during the coronavirus disease 2019 pandemic. *JAMA Netw. Open* **2020**, *3*, e2016924. [CrossRef]
58. Wiener, R.C.; Trickett Shockey, A.K.; Waters, C.; Bhandari, R. Face-touching behavior during the COVID-19 pandemic: Self-inoculation and transmission potentials. *J. Dent. Hyg.* **2021**, *95*, 41–46.
59. Mueller, S.M.; Martin, S.; Grunwald, M. Self-touch: Contact durations and point of touch of spontaneous facial self-touches differ depending on cognitive and emotional load. *PLoS ONE* **2019**, *14*, e0213677. [CrossRef]
60. Rahman, J.; Mumin, J.; Fakhruddin, B. How frequently do we touch facial T-zone: A systematic review. *Ann. Glob. Health* **2020**, *86*, 75. [CrossRef]
61. Nalunkuma, R.; Abila, D.B.; Ssewante, N.; Kiyimba, B.; Kigozi, E.; Kisuza, R.K.; Kasekende, F.; Nkalubo, J.; Kalungi, S.; Muttamba, W.; et al. Double face mask use for COVID-19 infection prevention and control among medical students at Makerere University: A cross-section survey. *Risk Manag. Healthc. Policy* **2022**, *15*, 111–120. [CrossRef]
62. Sickbert-Bennett, E.E.; Samet, J.M.; Prince, S.E.; Chen, H.; Zeman, K.L.; Tong, H.; Bennett, W.D. Fitted filtration efficiency of double masking during the COVID-19 pandemic. *JAMA Intern. Med.* **2021**, *181*, 1126–1128. [CrossRef]
63. Blachere, F.M.; Lemons, A.R.; Coyle, J.P.; Derk, R.C.; Lindsley, W.G.; Beezhold, D.H.; Woodfork, K.; Duling, M.G.; Boutin, B.; Boots, T.; et al. Face mask fit modifications that improve source control performance. *Am. J. Infect. Control* **2022**, *50*, 133–140. [CrossRef] [PubMed]
64. NewYork-Presbyterian. Should You Be Double Masking? Available online: https://healthmatters.nyp.org/should-you-be-double-masking/ (accessed on 27 May 2024).

65. Htwe, Y.Z.N.; Mamat, H.; Osman, B.; Mahmud, H. Performance comparison of single and double masks: Filtration efficiencies, breathing resistance and CO_2 content. *Arab J. Sci. Eng.* **2023**, *48*, 8349–8357. [CrossRef] [PubMed]
66. Patra, S.S.; Nath, J.; Panda, S.; Das, T.; Ramasamy, B. Evaluating the filtration efficiency of commercial facemasks' materials against respiratory aerosol droplets. *J. Air. Waste Manag. Assoc.* **2022**, *72*, 3–9. [CrossRef] [PubMed]
67. Gena, A.W.; Voelker, C.; Settles, G.S. Qualitative and quantitative schlieren optical measurement of the human thermal plume. *Indoor Air* **2020**, *30*, 757–766. [CrossRef] [PubMed]
68. Bridges, J.P.; Vladar, E.K.; Huang, H.; Mason, R.J. Respiratory epithelial cell responses to SARS-CoV-2 in COVID-19. *Thorax* **2022**, *77*, 203–209. [CrossRef] [PubMed]
69. Johnson, G.R.; Morawska, L.; Ristovski, Z.D.; Hargreaves, M.; Mengersen, K.; Chao, C.Y.H.; Wan, M.P.; Li, Y.; Xie, X.; Katoshevski, D.; et al. Modality of human expired aerosol size distributions. *J. Aerosol Sci.* **2011**, *42*, 839–851. [CrossRef]
70. Edmunds, W.J.; Kafatos, G.; Wallinga, J.; Mossong, J.R. Mixing patterns and the spread of close-contact infectious diseases. *Emerg. Themes Epidemiol.* **2006**, *3*, 10. [CrossRef]

fluids

MDPI

Article

Experimental Study of Oil–Water Flow Downstream of a Restriction in a Horizontal Pipe

Denghong Zhou [1], Kanat Karatayev [1], Yilin Fan [1,*], Benjamin Straiton [2] and Qussai Marashdeh [2]

1 Petroleum Engineering Department, Colorado School of Mines, Golden, CO 80401, USA
2 Tech4Imaging LLC, Columbus, OH 43235, USA
* Correspondence: yilinfan@mines.edu

Abstract: This work presents an experimental study on oil–water flow downstream of a restriction. The flow pattern, volumetric phase distribution, and their impacts on pressure drop are discussed. We employed two techniques to visualize the oil-water flow patterns, a high-speed camera and an Electrical Capacitance Volume Tomography (ECVT) system. The ECVT system is a non-intrusive device that measures the volumetric phase distribution at the pipe cross-section with time, which plays a critical role in determining the continuous phase in the oil–water flow, and therefore the oil-water flow pattern. In this study, we delved into the oil–water flow pattern and volumetric phase distribution for different valve openings, flow rates, and water cuts, and how they impact the pressure drop. The experimental results have demonstrated a strong relationship between the oil–water flow pattern and the pressure gradient, while the oil–water flow pattern is significantly influenced by the flowing conditions and the valve openings. The impacts of water cuts on the oil–water flow pattern are more obvious for smaller valve openings. For large valve openings, the oil and water phases tend to be more separated. This results in a moderate variation in the pressure gradient as a function of water cuts. However, it becomes more complex as the valve opening decreases. The pressure gradient generally increases with decreasing valve openings until the flow pattern becomes an oil-in-water dispersed flow. The impact of the valve on the pressure gradient is more pronounced in water-dominated flow when the water cut is above the inversion point, while it seems to be most obvious for medium water cut conditions.

Keywords: oil–water flow; flow pattern; electrical capacitance volume tomography; flow through restrictions; volumetric phase distribution; pressure gradient

Citation: Zhou, D.; Karatayev, K.; Fan, Y.; Straiton, B.; Marashdeh, Q. Experimental Study of Oil–Water Flow Downstream of a Restriction in a Horizontal Pipe. *Fluids* **2024**, *9*, 146. https://doi.org/10.3390/fluids9060146

Academic Editor: Kambiz Vafai

Received: 16 April 2024
Revised: 27 May 2024
Accepted: 12 June 2024
Published: 20 June 2024

1. Introduction

Oil–water two-phase flow widely exists in various industries. In the oil and gas industry, water production is becoming inevitable as the field matures. The oil–water flow pattern in pipes can significantly influence the pressure drop, which determines pipeline and wellbore design, pump selection and operation, and flow assurance. The behaviors of oil–water two-phase flow are less predictable compared to those of gas–liquid flow due to the similar physical properties and low interfacial shear stress between the two phases. It can be influenced by many factors, such as flow rates, pipe geometry, piping components such as elbows, reducers, multiphase flow meters, chokes or valves, etc.

Restrictions have ubiquitous applications in multi-phase flow pipelines in the oil and gas industry. Chokes are essential in adjusting a reasonable production rate for oil and gas wells to avoid an early water breakthrough into the wellbore [1]. They also help protect downstream equipment from high pressures to ensure safety in field operations [1]. However, the presence of valves can also cause other flow-assurance-related problems, such as the formation of stable emulsions in oil–water flow. The excessive mixing by the valves can disperse one phase into the other phase, and natural surfactants such as asphaltenes can stabilize the dispersions for a very long time, even forming tight emulsions. This can cause extra demulsification costs to the fields. There are several studies in the literature investigating valve effects on droplet breakup mechanisms and downstream droplet sizes (Van der Zande et al., 1999; Malot et al., 2003; Dalmazzone et al., 2005; Fossen et al., 2006; Fossen and Schümann, 2016; Paolinelli and Yao, 2018; Silva et al., 2019) [2–8]. Yet, little attention has been paid to the effects of choke on downstream fluid flow behaviors such as flow patterns and their impacts on the pressure drop. Table A1 in the Appendix A summarizes the previous studies on the effects of restriction on oil–water flow for reference.

Choke can impact the downstream flow patterns and, consequently, the pressure gradient. Shmueli et al. (2019) observed that the oil–water flow pattern changed from stratified flow to dispersed flow under conditions of 0.2 bar pressure drop choking implemented near the inlet of their test section [9]. This alternation was verified by measuring water cut profiles at the pipe center in the vertical direction using a traversing narrow-beam gamma densitometer. A homogeneous water cut profile was observed for the 0.2 bar inlet mixing, while stratification was shown in the water volumetric fraction profile for the case without inlet choking. Schümann et al. (2016) [10] conducted experiments in a 25 m long horizontal pipe using a static mixer near the inlet of the test section. They compared the measured pressure drop at a distance of 200 pipe diameters downstream of the inlet mixer for cases with and without the mixer and concluded that there existed a substantial pressure increase for the water-dominated flow with the presence of an inlet mixer. By contrast, the pressure drop almost stayed constant for the oil-dominated flow with and without the mixer. However, none of these studies systematically examined how various choke sizes or openings impact the downstream flow behaviors.

The objectives of this study are to examine how various choke openings would impact the downstream flow pattern and volumetric phase distribution, and how they influence the pressure gradient. We employed two visualization methods to study the flow pattern: a high-speed camera that visualizes the flow pattern from the side and an Electrical Capacitance Volume Tomography (ECVT) system that sheds light on the phase volumetric distribution within the pipe. Their impacts on the pressure gradient are discussed afterward.

2. Materials and Methods

The experimental work was conducted in a three-phase flow loop at the Colorado School of Mines. This section introduces the experimental facilities, equipment, fluid properties, and the testing matrix used in this study.

2.1. Experimental Facilities

The flow loop in this study consists of a horizontal pipe with a length of 13.72 m and an inner diameter of 5.25 cm, as shown in Figure 1. The system was designed to introduce water and oil simultaneously through wye connections at the inlet. The experimental setup utilized a progressive cavity pump for the oil phase to minimize the shear effects and avoid the formation of tight emulsions. The oil flow rate was controlled by a frequency converter. A submersible pump was deployed to introduce water into the system, and the flow rate was manually controlled by a needle valve at the inlet. The oil and water flow rates were metered by two Coriolis flow meters (Emerson Micro Motion R100S) separately. The facility can also be used for three-phase flow study, and air can be introduced right after the wye-shaped oil connection. This study mainly focuses on oil–water two-phase flow.

To examine the effect of different choke openings on downstream flow behaviors, a 2 in. ball valve was installed immediately after the introduction of all phases. Pressure and temperature transducers were installed in the test section to monitor the pressures and temperatures. Two differential pressure transducers (Emerson Rosemount 3051S differential pressure transducer) were used to measure the pressure drops. One was installed across the ball valve and the other was set in the test section around 120 L/D (length/pipe inner diameter) downstream of the valve over a span of 5.5 m. The differential pressure sensors have a pressure range of −250 to 250 in H_2O with an accuracy of 0.025% of span. All the sensors were wired to a data acquisition system controlled by a LabView (2020) project, which was used to automate the data acquisition and valve control to achieve the desired flow rates. The operational data were recorded at a time interval of 100 milliseconds.

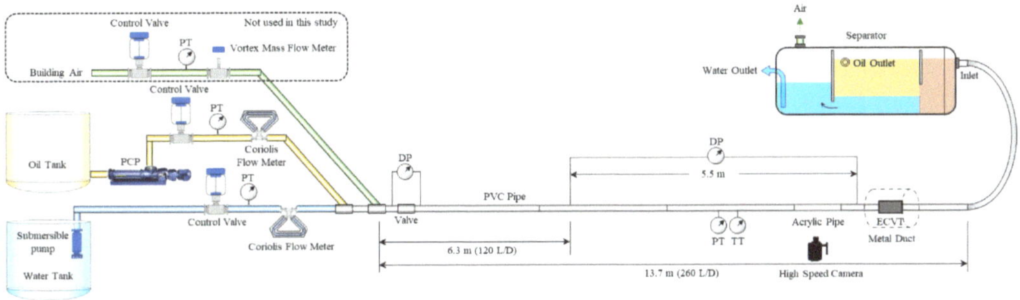

Figure 1. The schematic of the flow loop.

2.2. Equipment for Flow Visualization

We employed two pieces of equipment to study the oil–water flow pattern in the pipe, a high-speed camera and an Electrical Capacitance Volume Tomography (ECVT) system (Tech4Imaging LLC, Columbus, OH, USA). Figure 2 shows a picture of the high-speed camera while running the experiments. It was placed in front of a clear acrylic pipe as shown in Figure 1.

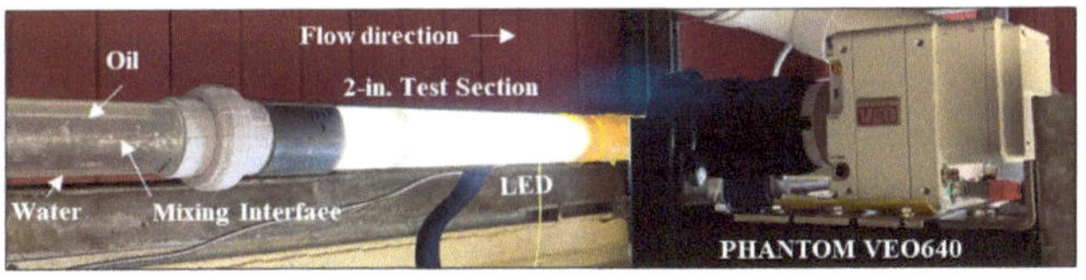

Figure 2. A picture of the facility and the high-speed camera while running experiments.

The ECVT system was deployed downstream of the pressure drop measurement section to monitor the in situ volumetric phase distribution inside the pipe. It was clamped around the pipe and the system was calibrated with pure water and oil before actual measurements were taken. Pictures of the system are shown in Figure 3. The ECVT system was used in several previous studies, such as [11–13].

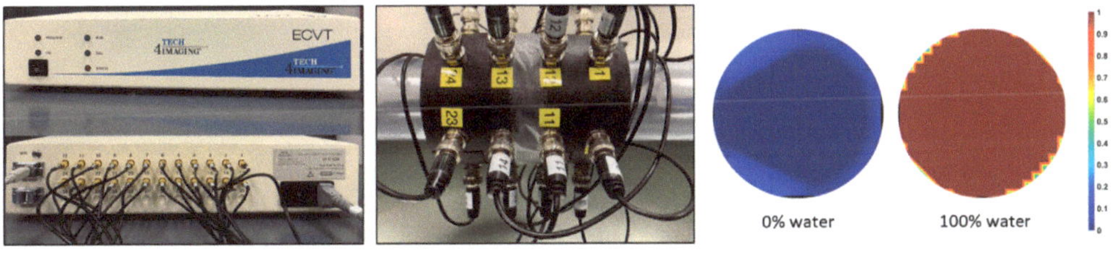

Figure 3. Pictures of the ECVT system and volumetric phase distribution at the pipe cross-section for 0% and 100% water cuts.

ECVT is a soft-field imaging modality in which a plurality of electrically conductive sensing plates are placed around a region of interest (RoI) [14]. Traditional ECVT utilizes 12, 24, or 36 rectangular electrodes configured around a cylindrical RoI. However, different numbers and shapes of plates can be configured around differently shaped RoIs including elbow, square, and spherical plates [15,16]. Twenty-four rectangular plates are used in this study with 4 rows of 6 plates in an offset configuration between rows.

The mutual capacitance is measured between all plate pairs surrounding the region such that, if there are a number of plates, N, there are n mutual capacitance measurements created by n plate pairs as given by Equation (1) [14]. This excludes plates paired with themselves and duplicate plate pairs in which only the excite and detect functions of the plates are switched.

$$n = \frac{N(N-1)}{2}, \tag{1}$$

There are two methods of measuring the mutual capacitance of each plate pair—alternating current (AC) and direct current (DC) [17]. In both cases, the measured signal is modulated by the permittivity of the mixture in between the plates. The AC version is used in this study. Lower permittivity corresponds to a lower signal and higher permittivity corresponds to a higher signal. Generally the capacitance, C, between a parallel plate capacitor can be modeled according to Equation (2), where ε_0 is the permittivity of free space, ε_r is the relative permittivity of the dielectric, A is the area of the plate, and d is the distance between the two plates. While most of the plate pairs will not fall into the parallel plate category, the general principle is useful.

$$C = \frac{\varepsilon_0 \varepsilon_r A}{d}, \tag{2}$$

Typically, two phases are introduced into the sensing region and each capacitance measurement is normalized between the signal when the RoI is completely filled with each phase. The lower dielectric phase may be designated as the "empty" phase and normalized to 0 whereas the higher dielectric phase may be designated as the "full" phase and normalized to 1. Normalization is critical because each plate pair has its own biases due to the sensor geometry and variations in the data acquisition system channels. In this case, oil has $\varepsilon_r \sim 2$ and water has $\varepsilon_r \sim 80$ at room temperature.

Once the capacitance measurements are obtained, an image reconstruction algorithm must be used to generate the image from the data. A sensitivity matrix is used to map the capacitance data to the voxels (volumetric pixels) of the image. The number of voxels and size of the image can be determined by the user. In this study, a $20 \times 20 \times 20$ voxel image is used. These voxels are vectorized into an 8000×1 vector, g. The vector, c, contains the capacitance measurements and is 276×1. The sensitivity matrix, S, is therefore 276×8000 and contains the sensitivity or weight of each capacitance measurement to each voxel in the image. The forward problem is, therefore,

$$c = Sg \tag{3}$$

This problem is both ill-conditioned and ill-posed and there are many proposed solutions. In this study, we used LBP. S is typically generated in a multi-physics simulation, c is measured, and g is solved for through solving the inverse problem. For more details on related topics, please see [18,19].

Our initial tests with pure oil or water alone flowing in the pipe showed reasonable results, as indicated in the right plots in Figure 3. The pure oil is indicated in blue, whereas the pure water is shown in red. In dispersed flow, the red color should denote a water-continuous state while the blue or yellow color indicates an oil-continuous state. In the experimental results, we will show the volumetric phase distribution variation with time inside the pipe in the axial direction at the center of the pipe, and at the pipe cross-section at different times.

2.3. Fluid Properties and Text Matrix

Tap water was used as the aqueous phase and mineral oil (Isopar V) as the oil phase. The oil density is 810 kg/m^3 and viscosity is 14.6 mPa·s at 20 °C and atmospheric pressure. The interfacial tension (IFT) between the oil and tap water was measured by a Krüss tensiometer DSA 100 using the pendant drop method. The measured IFT demonstrated a transient behavior and stabilized at 40 mN/m after 5 min. More data on the IFT between Isopar V and tap water can be found in [20]. The oil density and viscosity were measured by an Anton Paar SVM 3001 Stabinger Viscometer at various temperatures. A linear relationship was established for density as a function of temperature, and a logarithmic relationship was obtained for viscosity. Those equations were used for estimating the in situ oil density and viscosity at the different temperatures experienced in the test section during the experiments.

Table 1 lists the testing matrix of this study. Five valve openings, seven water cuts, and two mixture velocities were studied. Each flowing condition was stabilized for at least 10 min before measurement started.

Table 1. Testing matrix.

Mixture Velocity [1], v_M	Water Cut [2], WC	Valve Opening
0.2 m/s	20%, 60%, 80%	100%, 75%, 50%, 30%, 20%
0.5 m/s	60%	100%, 75%, 50%, 30%, 20%
0.2 m/s	0%, 20%, 30% [3], 40%, 60%, 80%, 100%	100%, 50%, 30%
0.5 m/s	0%, 20%, 30% [3], 40%, 60%, 80%, 100%	100%, 75%, 50%, 30%

[1] Mixture velocity is the total velocity of oil and water. Mathematically, it is the total volumetric flow rate of oil and water divided by the pipe's cross-sectional area. [2] Water cut is defined as the volumetric fraction of water in the injected liquid stream. Mathematically, it is the superficial water velocity (water volumetric flow rate divided by the pipe's cross-sectional area) divided by the mixture velocity. [3] The 30% water cut was only investigated for certain valve openings.

3. Results

In this section, we first discuss the flow pattern generally observed in this study, followed by two subsections that discuss the impacts of valve opening and water cuts, respectively, on the flow pattern, volumetric phase distribution, and pressure gradient.

3.1. Flow Patterns

Figure 4 shows the flow patterns observed in the current study. The pictures on the left were taken from the high-speed camera videos. The axial plots in the middle are the volumetric phase distribution with time at the center of the pipe obtained from the ECVT system, while the cross-sectional plots on the right show the volumetric phase distribution in the pipe at different time frames also obtained from the ECVT system. The red color represents water, and the blue is oil. The ECVT data were recorded at a frequency of 41.73 Hz.

Overall, seven major flow patterns were observed in the current study, including stratified flow with a mixing interface (ST&MI), a free oil layer above a water-in-oil dispersed layer (O&W/O), a water-in-oil dispersed layer above a free water layer (W/O&W), a water-in-oil dispersed layer above an oil-in-water dispersed layer (W/O&O/W), an oil-in-water dispersed layer above a free water layer (O/W&W), oil-in-water dispersion (O/W), and water-in-oil dispersion (W/O).

The ST&MI flow pattern is one of the most observed flow patterns in this study. It mainly occurs when there are low flow rates, large valve openings, and medium water cut conditions. O&W/O flow mainly occurs under conditions of high liquid flow or when there are small valve openings with low water cut conditions. Occasionally, there is a sub-flow pattern in O&W/O, in which rolling waves made of large water droplets are present above the interface. A picture of it is shown in Figure 4. W/O&W flow normally occurs at high water cut conditions. Under certain rare flow conditions, small oil slugs were observed at the top part of the pipe. This type of flow pattern should be around the transition but is classified as W/O&W considering the relatively small size of oil pockets observed in this study. W/O&O/W, which is also referred to as dual dispersion or dual continuous flow by some previous studies [21–26], occurs at medium water cut and high flow rate or small valve opening conditions. O/W&W flow occurs at high water cut, small valve opening, or high liquid flow rate conditions. Occasionally, random W/O slugs appear within O/W&W flow at the top port of the pipe under some flowing conditions, which is clearly shown in the ECVT images. O/W flow occurs at small valve openings at medium or high water cut conditions. W/O flow occurs at low water cut, small valve opening, and high flow rate conditions. The next section discusses in detail the flow pattern transition with different valve openings, liquid mixture velocities, and water cuts. Flow pattern maps are presented together with the pressure gradient plots to enhance the understanding of their correlations. It is also worth noting that the flow pattern transitions in a gradual manner, resulting in some fluctuation or intermittency, which is shown clearly in the ECVT images. Understanding this phenomenon aids in comprehending its impacts on the pressure gradient, which are discussed in the next section.

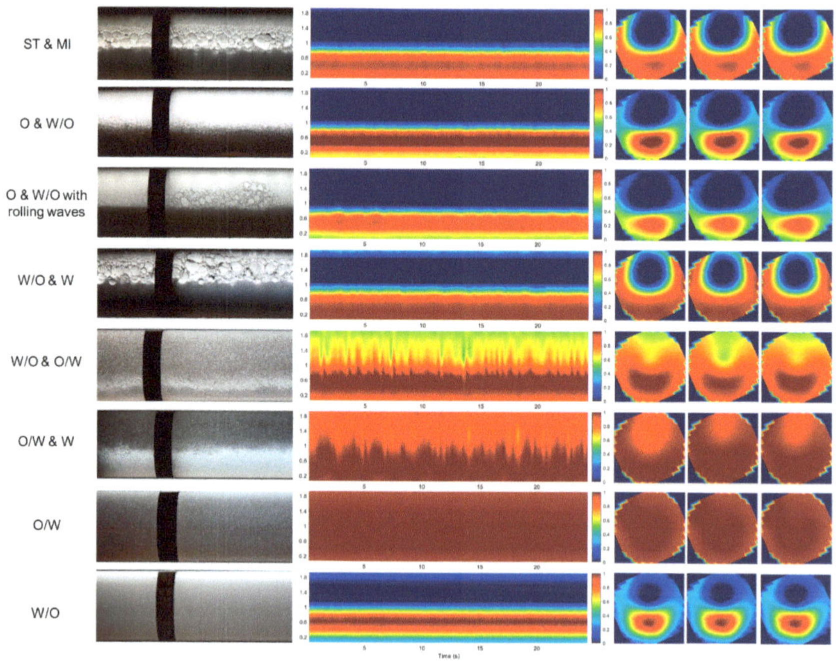

Figure 4. Oil–water two-phase flow patterns observed in this study. ((**Left**): high-speed camera video pictures; (**middle**): volumetric phase distribution at the center of the pipe in the axial direction obtained from ECVT; (**right**): volumetric phase distribution at the cross-section of the pipe at three different times obtained from ECVT).

3.2. *Effect of Valve Opening on Flow Pattern and Pressure Gradient*

The changes in downstream flow behavior with valve openings are illustrated by a series of experiments with a mixture velocity of 0.2 m/s at 20%, 60%, and 80% water cuts, and a mixture velocity of 0.5 m/s at 60% water cut. As anticipated, the valve plays a critical role in determining the oil–water flow patterns. Our experimental data also demonstrated that the oil–water flow pattern impacts the corresponding pressure gradient dramatically.

Figure 5 shows the flow patterns at different valve openings when the input water cut is 20%. The pictures on the left were captured by the high-speed camera. The images in the middle demonstrate the volumetric phase distribution in the vertical direction at the pipe center as a function of time, obtained from the ECVT system. The square images on the right are the volumetric phase distributions at the cross-section of the pipe at three different time frames, also obtained from the ECVT system. The percentages shown on the left are the valve openings. As shown by Figure 5, the flow pattern for 100% to 50% valve openings is ST&MI based on the high-speed camera videos, with an increase in mixing at the interface as the valve opening is reduced. Transition to O&W/O flow occurs when the valve opening decreases to 30%. Some rolling waves made of large water droplets were observed occasionally above the interface. They grow as the valve opening is further reduced, indicating transitioning to W/O.

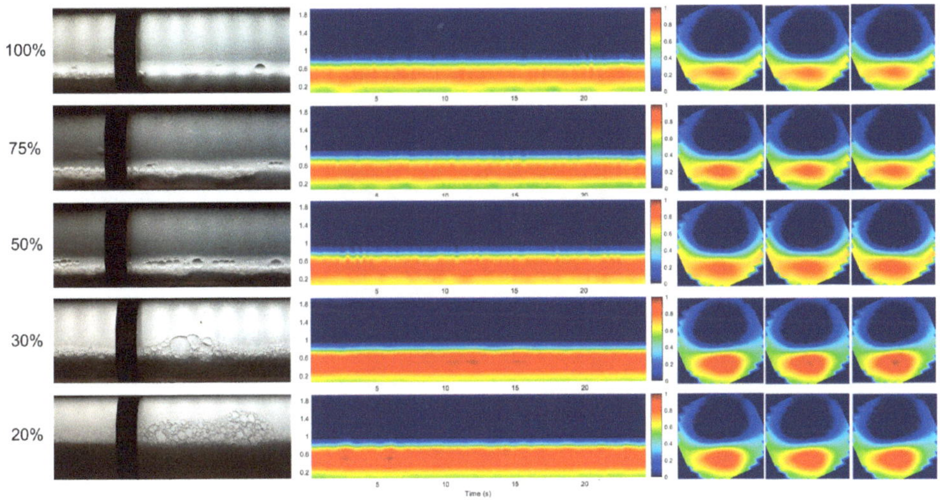

Figure 5. Flow pattern observed from the high-speed camera (**left**) and variation of volumetric phase distribution at the pipe center in the vertical direction (**middle**) and volumetric phase distribution at the pipe's cross-section at three different time frames (**right**). v_{SL} = 0.2 m/s with a 20% water cut (the percentages in the picture indicate the opening of the valve).

The corresponding pressure gradient measurement is shown as a green line in Figure 6, with a flow pattern indicated for each testing point. For a 20% water cut, the pressure gradient shows a little increase as the valve opening decreases from 100% to 50%, when the flow pattern remains ST&MI. This increase could be due to the increasing mixing at the oil and water interface that leads to a higher frictional pressure drop. As the flow pattern transitions to O&W/O when the valve opening reduces to 30% and 20%, the pressure gradient starts to show a noticeable increase with decreasing valve opening, which is due to the increase in mixture viscosity in the bottom dispersed layer. For a better understanding of the flow pattern transition, Figure 7 provides a flow pattern map using coordinates of water cut and valve opening.

Figure 8 shows the flow pattern and the volumetric phase fraction in the axial and cross-sectional planes for a 60% water cut at the same liquid mixture velocity (0.2 m/s). It can be seen that the flow pattern remains ST&MI for valve openings $\leq$ 50%, but with more mixing at the interface and a thicker water layer compared to that for the 20% water cut. The flow pattern transitions to O/W&W as the valve opening is reduced to 30%. Some small W/O slugs appear intermittently at the top part of the pipe, which is shown more clearly in the ECVT images. We anticipate that this point could be near the transition boundaries. The upper part of the fluid flow could generate high frictional pressure losses due to the densely packed dispersed O/W and W/O flows [10], leading to a peak in the pressure gradient plot shown in Figure 6 (black line). As the valve opening is reduced to 20%, the flow pattern starts to transition to O/W flow, where W/O slugs are occasionally present at the pipe top as illustrated from the ECVT axial plot. This leads to a decrease in the pressure drop when the valve opening is reduced from 30% to 20%.

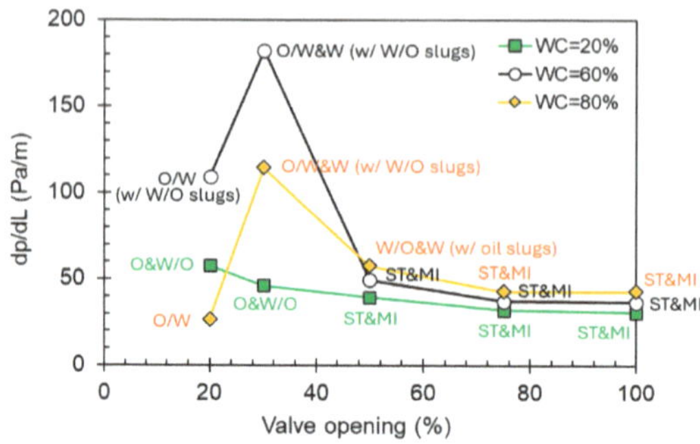

Figure 6. Pressure gradient in the test section as a function of valve opening at 0.2 m/s liquid mixture velocity and different water cuts.

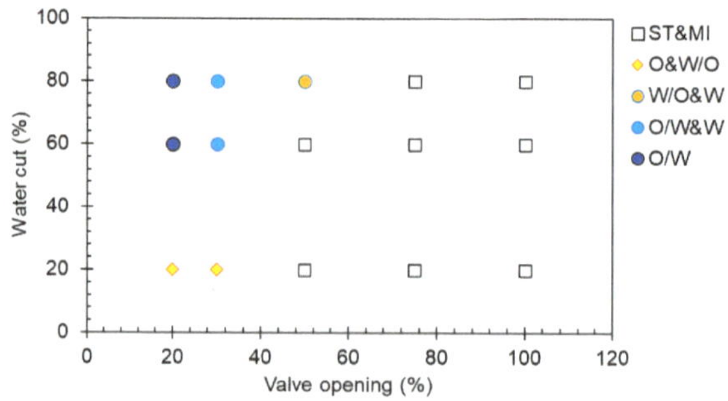

Figure 7. Flow pattern map for 0.2 m/s total mixture velocity as a function of valve opening in the x-axis and water cut in the y-axis.

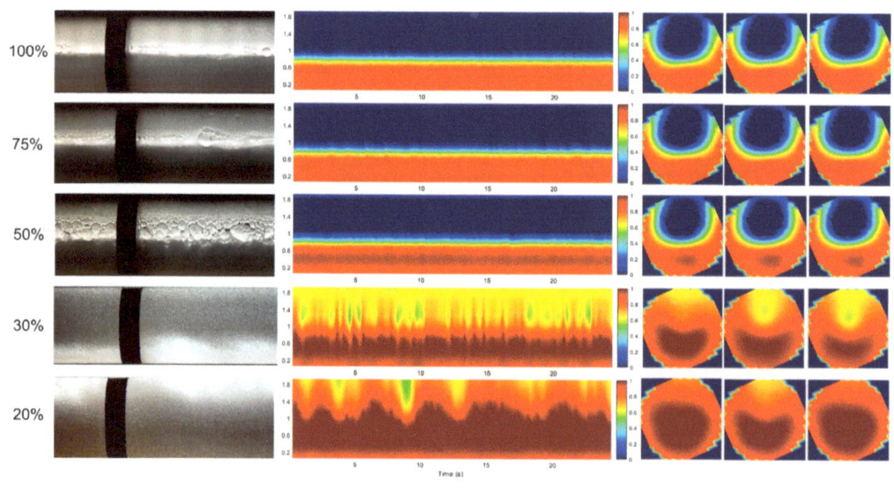

Figure 8. Flow pattern observed from the high-speed camera (**left**) and variation of volumetric phase distribution at the pipe center in the vertical direction (**middle**) and volumetric phase distribution at the pipe's cross-section at three different time frames (**right**). v_{SL} = 0.2 m/s with a 60% water cut (the percentages in the picture indicate the opening of the valve).

A similar flow pattern transition occurs for cases with an 80% water cut as for a 60% water cut, but with a much thicker mixing layer. Figure 9 shows the flow pattern and the volumetric phase fraction in the axial and cross-sectional planes for an 80% water cut at the same liquid mixture velocity (0.2 m/s). ST&MI flow was observed for valve openings of 100% and 75%, although the oil layer at the top of the pipe was very thin. Transition to W/O&W starts at a valve opening of 50%, earlier than for the 60% water cut where the transition occurred at a valve opening of around 30%. The flow pattern starts to change to O/W flow when the valve opening is further decreased. For the 30% valve opening, we suspect that the top layer contains intermittent W/O and O/W dispersed flows as illustrated in the ECVT images in the axial direction. The corresponding pressure gradient measurement is shown in Figure 6 (orange line). Similar trends were observed for 80% and 60% water cuts, i.e., the pressure gradient increases as the valve opening is reduced until the flow becomes dispersed oil in water.

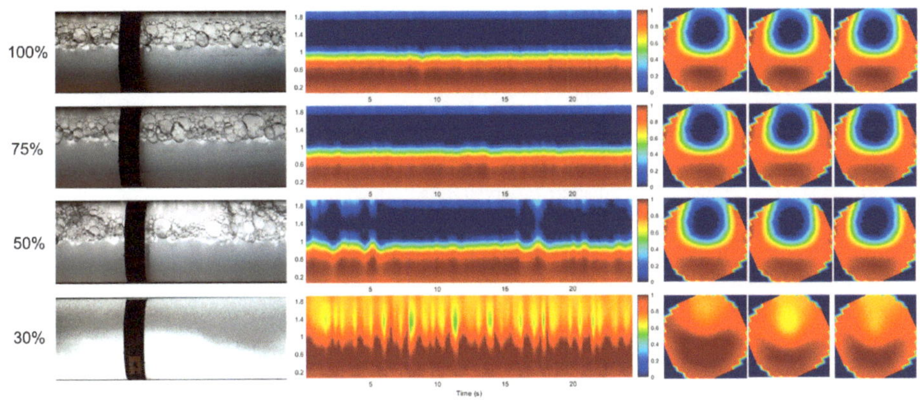

Figure 9. Flow pattern observed from the high-speed camera (**left**) and variation of volumetric phase distribution at the pipe center in the vertical direction (**middle**) and volumetric phase distribution at the pipe's cross-section at three different time frames (**right**). v_{SL} = 0.2 m/s with an 80% water cut (the percentages in the picture indicate the opening of the valve).

Another interesting observation from Figure 6 is that the valve's impact on the pressure gradient is more obvious for water-dominated flow than oil-dominated flow (note that the inversion point is between a water cut of 20% to 30% for our testing fluids). In our experiments, the valve impact is most obvious for a 60% water cut, while it is less noticeable for a 20% water cut.

The effect of choking on flow pattern and pressure gradient was also investigated at a higher liquid mixture velocity. Figure 10 shows the corresponding pictures from the high-speed camera and the images from the ECVT system for a mixture velocity of 0.5 m/s. When compared with Figure 8, one can notice that the transition from ST&MI flow to the other flow patterns starts earlier at a valve opening of 50%, whereas this transition occurred at a valve opening of 30% when the mixture velocity was 0.2 m/s. The elevated mixture velocity shifts the flow pattern transition to a larger valve opening.

For the 50% valve opening, we think the flow pattern was O/W&W, but with some small W/O pockets in the upper parts based on the ECVT axial image. The upper part should still be dominated by O/W. The viscosity of this upper densely packed dispersed flow could be high even though it is water-continuous, which eventually leads to the peak in the pressure gradient plots shown in Figure 11. This is consistent with Schümann et al. (2016)'s results, who observed similar phenomena for a liquid mixture velocity of 0.5 m/s at water cuts of 50% and 80%, and first described the top layer as a densely packed water-continuous layer [27]. They used quick-closing valves to observe the separation of the top layer, from which they concluded that the top layer should be oil droplets in the water phase.

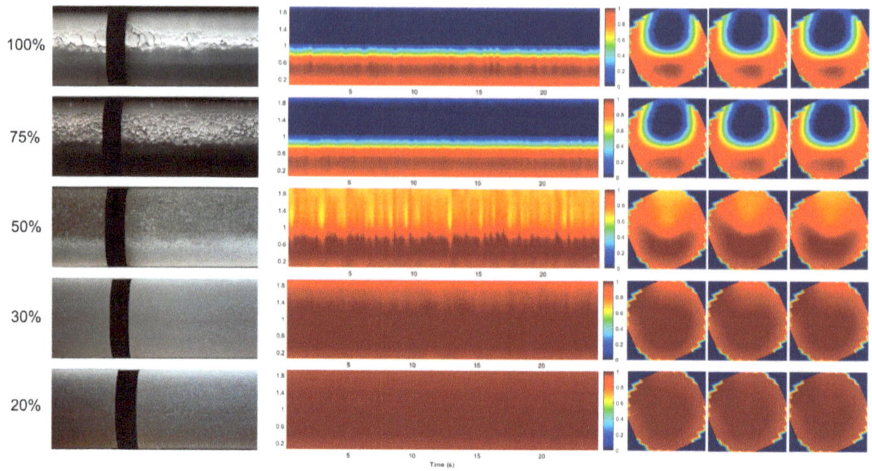

Figure 10. Flow pattern observed from the high-speed camera (**left**) and variation of volumetric phase distribution at the pipe center in the vertical direction (**middle**) and volumetric phase distribution at the pipe's cross-section at three different time frames (**right**). v_{SL} = 0.5 m/s with a 60% water cut (the percentages in the picture indicate the opening of the valve).

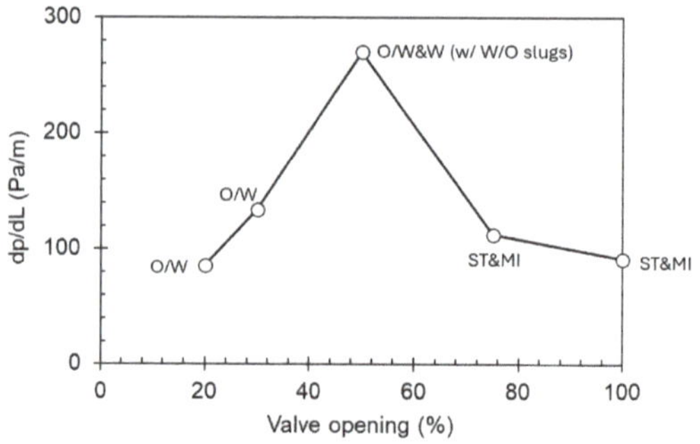

Figure 11. Pressure gradient in the test section as a function of valve opening at 0.5 m/s liquid mixture velocity and 60% water cut.

When the valve opening is reduced to 30%, the flow pattern becomes O/W, demonstrated by both the high-speed camera videos and the ECVT images. The pressure gradient drops abruptly as the valve opening decreases from 50% to 30%, consistent with our flow pattern observations. However, the oil droplets are still a little bit concentrated towards the top of the pipe due to buoyancy for the 30% valve opening, as illustrated from the ECVT images. The flow becomes more homogeneous as the valve opening is reduced to 20%, resulting in an even smaller pressure gradient.

3.3. Effect of Water Cut on Flow Pattern and Pressure Gradient

As indicated in the previous section, the oil–water flow pattern, volumetric phase distribution, and pressure gradient are also closely related to the water cut. In this section, we emphasize the water cut's effect on the flow pattern, phase distribution, and pressure gradient in the test section downstream of the inlet valve. The effect of the water cut is investigated at two different mixture velocities and four different valve openings.

Figure 12 shows the flow pattern and phase distribution for different water cuts at 0.2 m/s mixture velocity and a valve opening of 100%. ST&MI was generally observed for all water cuts from 20% to 80%. An increase in the water cut leads to a thicker mixing layer at the interface and a thicker free water layer as expected. The increased mixing layer thickness could be induced by the increased superficial water velocity at a higher water cut, which enhances the dispersion of water droplets into the upper oil layer.

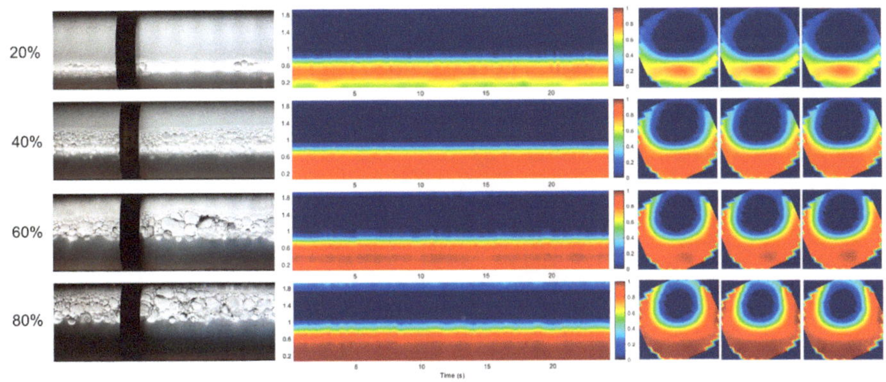

Figure 12. Flow pattern observed from the high-speed camera (**left**) and variation of volumetric phase distribution at the pipe center in the vertical direction (**middle**) and volumetric phase distribution at the pipe's cross-section at three different time frames (**right**). v_{SL} = 0.2 m/s with a 100% valve opening (the percentages in the picture indicate the water cut).

The black line in Figure 13 shows the corresponding pressure drop as a function of the water cut, with the flow pattern noted on the side. Figure 14 shows the corresponding flow pattern map in terms of water cut on the *x*-axis and valve opening on the *y*-axis. The pressure gradient first decreases slightly when the water cut changes from 0% to 20% due to the addition of the free water layer that has a lower viscosity than the oil phase. Then it increases until the water cut reaches 80%. This is mainly due to the increase in the thickness of the mixing layer, which has a higher viscosity. Although the free water layer also grows as the water cut increases, their negative impact on the pressure gradient is outweighed by the positive and larger impact from the mixing layer.

The impacts of the water cut on the oil–water flow pattern are more obvious for smaller valve openings. Figure 15 shows the flow pattern and phase distribution for a valve opening of 50% at the same mixture velocity. The corresponding pressure gradient measurement is shown as a green line in Figure 13. The flow pattern is ST&MI for water cuts $\leq$ 60%, and transitions to W/O&W at an 80% water cut, at which the pressure gradient reaches the maximum. Oil slugs are occasionally present at an 80% water cut, which are captured by the high-speed video camera. The variation in pressure gradient with varying water cuts is consistent with the flow pattern observations.

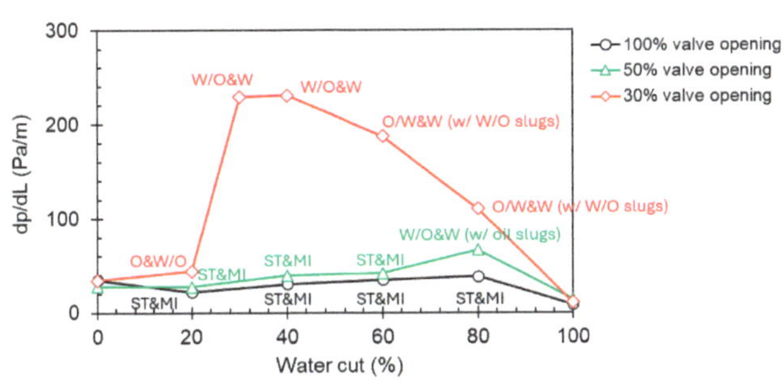

Figure 13. Pressure gradient in the test section as a function of the water cut at 0.2 m/s liquid mixture velocity and different valve openings.

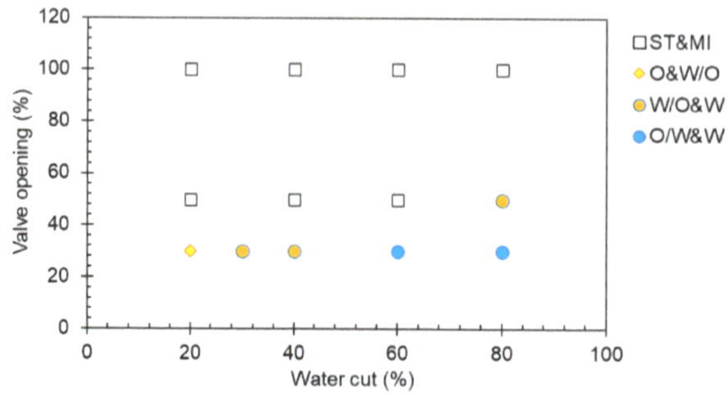

Figure 14. Flow pattern map for 0.2 m/s total mixture velocity as a function of the water cut on the x-axis and valve opening on the y-axis.

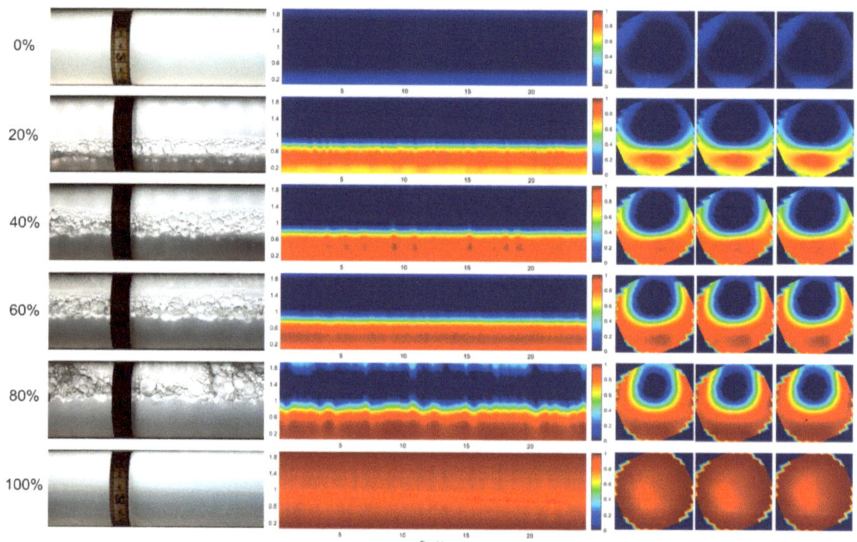

Figure 15. Flow pattern observed from the high-speed camera (**left**) and variation of volumetric phase distribution at the pipe center in the vertical direction (**middle**) and volumetric phase distribution at the pipe's cross-section at three different time frames (**right**). v_{SL} = 0.2 m/s with a 50% valve opening (the percentages in the picture indicate the water cut).

Figure 16 shows the fluid flow behavior for a valve opening of 30%. O&W/O was observed under 20% water cut conditions, W/O&W for 30% and 40% water cut conditions, and O/W&W for 60% and 80% water cut conditions, which is also indicated in Figures 13 and 14. Since W/O dispersion has a higher mixture viscosity than O/W dispersion for the same type of fluids, the pressure gradient is at its maximum at around the 30% and 40% water cuts (red line in Figure 13). Figure 13 clearly shows that the pressure gradient of oil–water two-phase flow is strongly impacted by water cut and flow pattern, and the latter could be influenced significantly by pipe restrictions.

Similar behavior was observed for a higher mixture velocity of 0.5 m/s. Figures 17 and 18 show the flow patterns and phase distributions for valve openings of 100% and 50%, respectively. The corresponding pressure gradients as functions of the water cuts are shown in Figure 19, and the flow pattern is shown in Figure 20. It can be noticed that the flow tends to be more mixed at the higher mixture velocity, resulting in flow pattern transitions occurring at smaller water cuts. The relationship between the pressure gradient and the water cut for a valve opening of 50% at 0.5 m/s is similar to that of a valve opening of 30% at 0.2 m/s. At 0.5 m/s, when the valve opening is 30%, the fluid flow is dispersed flow for all water cuts, with an inversion point between the 20% and 30% water cuts. This phenomenon is reflected in the ECVT images (Figure 21) and the pressure gradient measurements shown in Figure 19, which show a peak at around the 30% water cut.

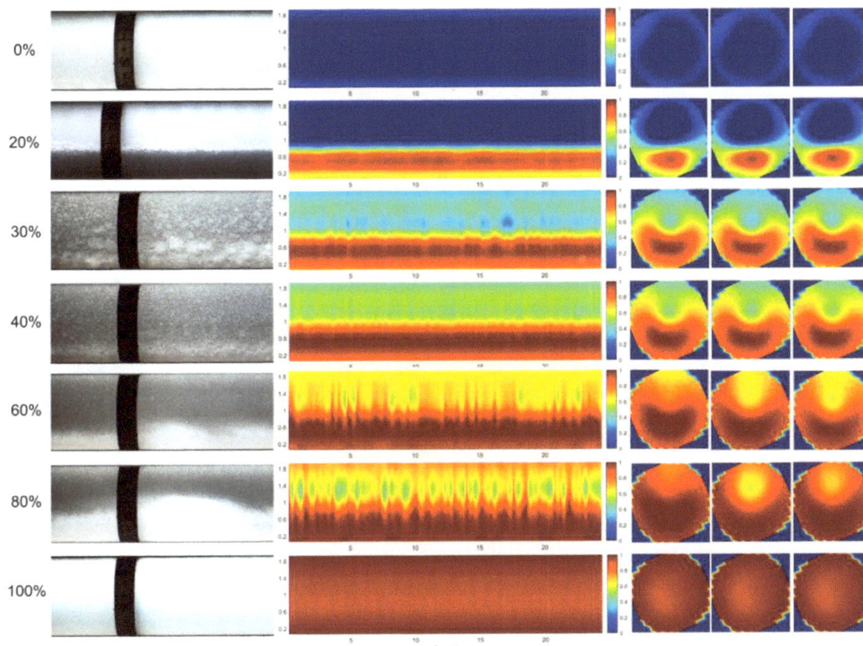

Figure 16. Flow pattern observed from the high-speed camera (**left**) and variation of volumetric phase distribution at the pipe center in the vertical direction (**middle**) and volumetric phase distribution at the pipe's cross-section at three different time frames (**right**). v_{SL} = 0.2 m/s with a 30% valve opening (the percentages in the picture indicate the water cut).

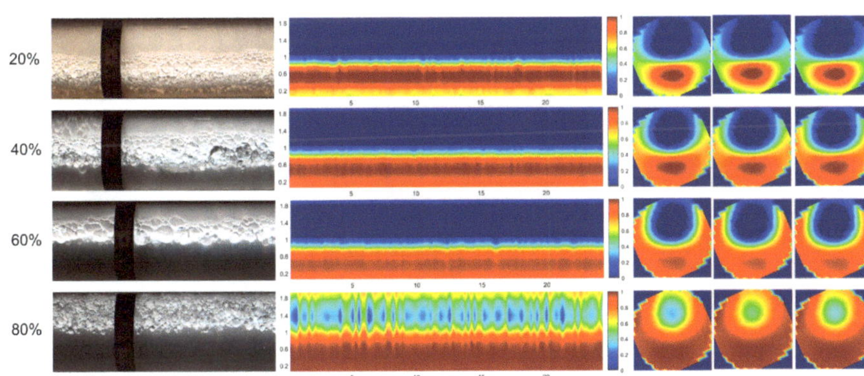

Figure 17. Flow pattern observed from the high-speed camera (**left**) and variation of volumetric phase distribution at the pipe center in the vertical direction (**middle**) and volumetric phase distribution at the pipe's cross-section at three different time frames (**right**). v_{SL} = 0.5 m/s with a 100% valve opening (the percentages in the picture indicate the water cut).

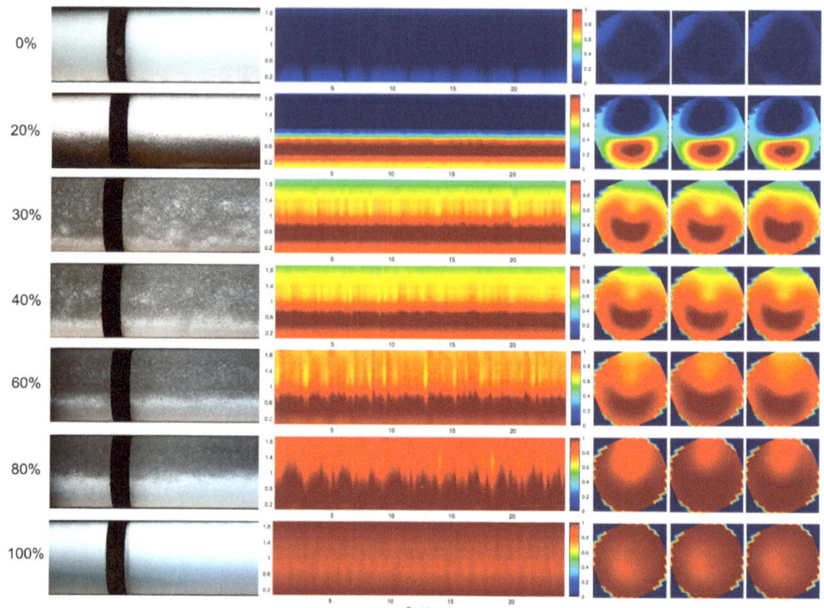

Figure 18. Flow pattern observed from the high-speed camera (**left**) and variation of volumetric phase distribution at the pipe center in the vertical direction (**middle**) and volumetric phase distribution at the pipe's cross-section at three different time frames (**right**). v_{SL} = 0.5 m/s with a 50% valve opening (the percentages in the picture indicate the water cut).

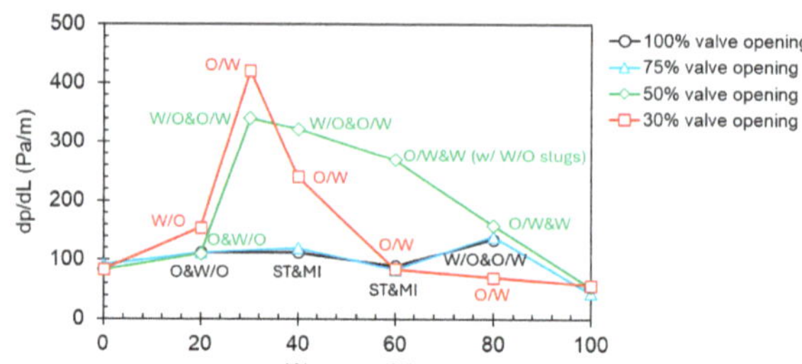

Figure 19. Pressure gradient in the test section as a function of water cut at 0.5 m/s liquid mixture velocity and different valve openings. (Note: the flow patterns in black color are for both 100% and 75% valve openings).

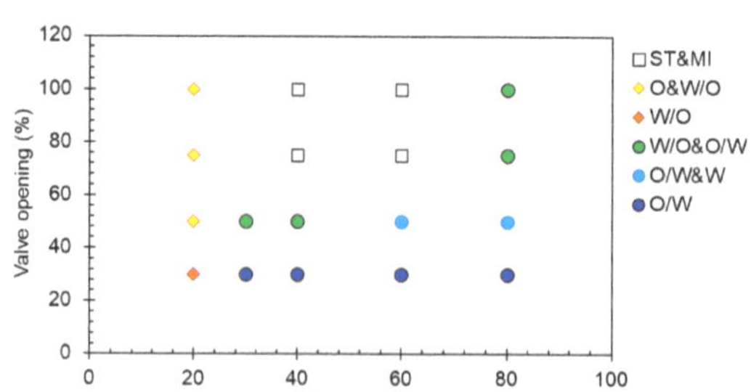

Figure 20. Flow pattern map for 0.5 m/s total mixture velocity as a function of the water cut on the x-axis and valve opening on the y-axis.

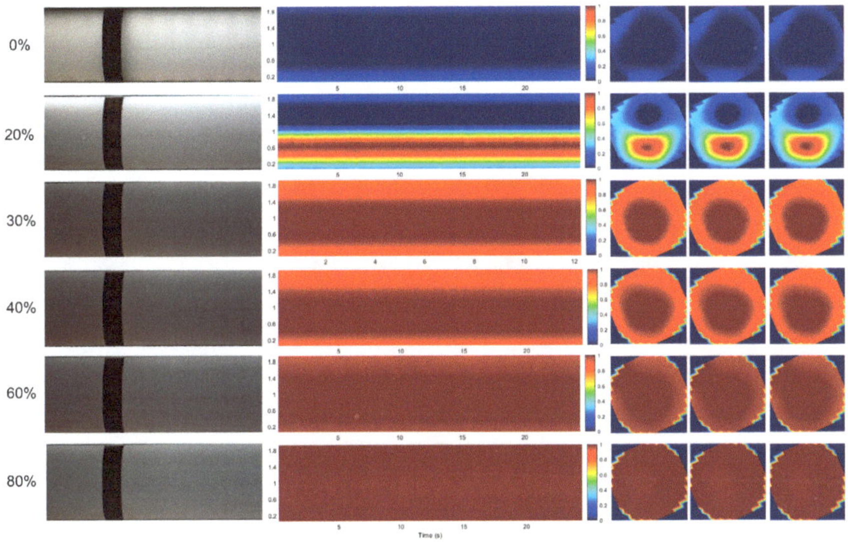

Figure 21. Flow pattern observed from the high-speed camera (**left**) and variation of volumetric phase distribution at the pipe center in the vertical direction (**middle**) and volumetric phase distribution at the pipe's cross-section at three different time frames (**right**). v_{SL} = 0.5 m/s with a 30% valve opening (the percentages in the picture indicate the water cut).

4. Discussion

Oil–water flow is widely encountered in various industries. In this study, we emphasized the oil–water flow pattern downstream of restrictions, and their impacts on the pressure drop, which is a crucial factor in production system design and optimization, flow assurance, etc. [1]. For fluids containing natural surfactants, the restriction effect is more pronounced because these surfactants stabilize the formed droplets and prevent their coalescence. In other words, restriction effects on multiphase flow can persist over very long distances when natural surfactants are present. Additionally, understanding the distribution of oil and water phases downstream of a restriction is crucial for gas hydrate prediction and management. Besides pressure and temperature, gas hydrate formation depends on the interfacial area between different phases, which is a function of the flow pattern and phase distribution. Restrictions induce significant turbulence to the fluid flow, enhancing phase mixing and consequently affecting the interfacial area and gas hydrate formation. Furthermore, accurately predicting the pressure drop in surface flow lines, which involve numerous flowline components and chokes, is crucial for flow allocation assessment using hydraulic models. For the first time, this study provides a systematic study of the effects of restrictions on oil–water flow patterns, volumetric phase distribution in the axial and cross-sectional directions, and the pressure gradient.

We employed two visualization methods, a high-speed camera and an ECVT system, to better understand the behavior of oil–water fluid flow. The high-speed camera videos were recorded on the side of the horizontal pipe, which provided limited information on the fluid flow behavior within the pipe. By combining these videos with the ECVT system, we have a better understanding of the phase distributions, especially the continuous phase in dispersed flows such as O/W&W, W/O&W, O/W, and W/O. However, ECVT also has some limitations. It has low accuracy at the phase interface near the pipe wall; low detectability of large water droplets in the oil phase, such as the ones in ST&MI and W/O&W flows; and low detectability of the oil phase volumetric fraction in water-continuous dispersed flow. Therefore, we recommend using both a high-speed camera and the ECVT system for a better understanding of the flow behavior for future studies.

Although the hydraulic modeling work on oil–water flow has been ongoing for decades, the restriction effects are still not well-captured. Our next goal is to develop a mechanistic model for oil–water flow in horizontal pipes that considers the effects of restriction, water cut, and flow rate on the pressure gradient.

5. Conclusions

An experimental study was conducted in this study to investigate the oil–water flow behavior downstream of restrictions in a horizontal pipe, and their impact on the pressure gradient. We employed two visualization methods to study the flow patterns, a high-speed camera that visualizes the flow pattern from the side and an Electrical Capacitance Volume Tomography (ECVT) system that sheds light on the volumetric phase distribution within the pipe.

Seven oil–water flow patterns were observed in this study, namely stratified flow with a mixing interface (ST&MI), a free oil layer with a dispersed water-in-oil layer (O&W/O), a dispersed water-in-oil layer with a free water layer (W/O&W), a dispersed water-in-oil layer with a dispersed oil-in-water layer (W/O&O/W), a dispersed oil-in-water layer with a free water layer (O/W&W), dispersed water-in-oil (W/O), and dispersed oil-in-water (O/W).

The experimental results have demonstrated a strong relationship between the oil–water flow pattern and the pressure gradient, while the oil–water flow pattern is significantly influenced by the flowing conditions and the valve openings. The impacts of water cuts on the oil–water flow pattern are more obvious for smaller valve openings. For large valve openings, the oil and water phases tend to be more separated and the flow pattern tends to be ST&MI at medium water cuts or low mixture velocities, and the droplets are generally large. This results in a moderate variation in the pressure gradient as a function of water cut. However, it becomes more complex as the valve opening decreases. The pressure gradient generally reaches the maximum when W/O&O/W flow occurs. On the other hand, the pressure gradient increases with decreasing valve openings until the flow pattern becomes O/W flow. The impact of the valve on the pressure gradient is more pronounced in water-dominated flow when the water cut is above the inversion point, while it seems to be most obvious for medium water cut conditions.

Author Contributions: Conceptualization, D.Z. and Y.F.; methodology, D.Z., K.K. and Y.F.; software, D.Z., Y.F., B.S. and Q.M.; validation, D.Z. and Y.F.; formal analysis, D.Z. and Y.F.; investigation, D.Z., K.K. and Y.F.; resources, D.Z. and Y.F.; data curation, D.Z. and K.K.; writing—original draft preparation, D.Z. and Y.F.; writing—review and editing, K.K., B.S. and Q.M.; visualization, Y.F.; supervision, Y.F.; project administration, Y.F.; funding acquisition, Y.F. All authors have read and agreed to the published version of the manuscript.

Funding: This research received no external funding.

Data Availability Statement: The raw data supporting the conclusions of this article may be made available by the authors on request.

Conflicts of Interest: Authors Benjamin Straiton and Qussai Marashdeh were employed by the company Tech4Imaging. The remaining authors declare that the research was conducted in the absence of any commercial or financial relationships that could be construed as a potential conflict of interest.

Appendix A

Table A1 in this appendix summarizes the previous experimental studies of oil–water flows downstream of restrictions.

Table A1. Experimental studies of oil–water flows downstream of restrictions.

Authors	Restriction Type	ID (mm)	Θ * (°)	Flow Rate Range (m/s)	Fluids	Properties	Measurements
van der Zande et al. (1999) [7]	Orifice	4.5, 15.25	-	0.5–5.5	n-heptane, Vitrea 9, 46, 68; tap water and demineralized water	μ_o = 0.4–410 cP; ρ_o = 684–882 kg/m^3	Drop breakup
Malot et al. (2003) [8]	Orifice	4.8	90	0.46–0.92	crude oil; brine	μ_o = 10.6 cP; ρ_o = 847 kg/m^3	Drop breakup, droplet size
Dalmazzone et al. (2005) [2]	Orifice	30	90	0.007	Heptane; tap water	μ_o = 0.45 cP; ρ_o = 684 kg/m^3	Drop breakup, droplet size
Fossen et al. (2006) [3]	Needle valve	6.35	0	0.263	Exxsol D60; 3.5 wt.% NaCl	ρ_o = 780 kg/m^3	droplet size
Schümann et al. (2016) [10]	Static mixer	100	0	0.5–1	Primol 352 + Exxsol D80; tap water	μ_o = 35–120 cP; ρ_o = 853–866 kg/m^3	Flow pattern, phase distribution, pressure drop, droplet size
Fossen and Schümann (2017) [4]	Butterfly valve	100	0	0.18–0.71	Primol 352 + Exxsol D60; tap water + NaOH	μ_o = 4 cP; ρ_o = 800 kg/m^3	Droplet size
Paolinelli and Yao (2018) [5]	Globe valve	100	0	1.1–1.6	Isopar V; 0.1 wt.% NaCl	μ_o = 9 cP; ρ_o = 810 kg/m^3	Droplet size
Shmueli et al. (2019) [9]	Ball valve	69	0	0.5–2	Exxsol D80; tap water	μ_o = 1.8 cP; ρ_o = 803 kg/m^3	Flow pattern, water distribution, pressure drop
Silva et al. (2019) [28]	Gate valve	12.7	-	4.52–4.49	Mineral oil; tap water	μ_o = 12, 26 cP; ρ_o = 865.6 kg/m^3	Droplet size
Skjefstad et al. (2020) [29]	Ball valve	67.8	0	1.38–2.31	Exxsol D60; 3.2 wt.% NaCl	μ_o = 1.6 cP; ρ_o = 795.74 kg/m^3	Separation performance

* Inclination angle from horizontal, 90° is upward vertical, 0° is horizontal, - is not available.

References

1. Al-Safran, E.M.; Brill, J.P. *Applied Multiphase Flow in Pipes and Flow Assurance: Oil and Gas Production*; Society of Petroleum Engineers: Richardson, TX, USA, 2017.
2. Dalmazzone, C.; Noik, C.; Galinat, S.; Masbernat, O.; Guiraud, P. Flow of a Concentrated Oil-Water Dispersion through a Restriction. In Proceedings of the 12th International Conference on Multiphase Production Technology, Barcelona, Spain, 25–27 May 2005.
3. Fossen, M.; Arntzen, R.; Hemmingsen, P.V.; Sjöblom, J.; Jakobsson, J. A Laboratory-Scale Vertical Gravity Separator for Emulsion Characterization. *J. Dispers. Sci. Technol.* **2006**, *27*, 453–461. [CrossRef]
4. Fossen, M.; Schümann, H. Experimental study of the relative effect of pressure drop and flow rate on the droplet size downstream of a pipe restriction. *J. Dispers. Sci. Technol.* **2017**, *38*, 826–831. [CrossRef]
5. Paolinelli, L.D.; Rashedi, A.; Yao, J. Characterization of droplet sizes in large scale oil–water flow downstream from a globe valve. *Int. J. Multiph. Flow* **2018**, *99*, 132–150. [CrossRef]
6. Silva, F.S.; Medronho, R.A.; Barca, L.F. Experimental study of water droplet break up in water in oil dispersions using an apparatus that produces localized pressure drops. *Oil Gas Sci.Technol.–Rev. d'IFP Energ. Nouv.* **2019**, *74*, 1. [CrossRef]
7. Van der Zande, M.J.; Van Heuven, K.R.; Muntinga, J.H.; Van den Broek, W. Effect of flow through a choke valve on emulsion stability. In Proceedings of the SPE Annual Technical Conference and Exhibition, Houston, TX, USA, 3–6 October 1999.
8. Malot, H.; Noïk, C.; Dalmazzone, C. Experimental investigation on water-in-crude oil emulsion formation through a model choke-valve: Droplet break-up and phase inversion. In Proceedings of the Multiphase Conference, San-Remo, Italy, 11–13 June 2003.

9. Shmueli, A.; Unander, T.E.; Schümann, H. Oil/Water Pipe-flow dispersions: From traditional flow loops to real industrial-transport conditions. *SPE Prod. Oper.* **2019**, *34*, 260–269. [CrossRef]
10. Schümann, H.; Tutkun, M.; Yang, Z.; Nydal, O.J. Experimental study of dispersed oil-water flow in a horizontal pipe with enhanced inlet mixing, Part 1: Flow patterns, phase distributions and pressure gradients. *J. Pet. Sci. Eng.* **2016**, *145*, 742–752. [CrossRef]
11. Wang, A.; Marashdeh, Q.; Fan, L.-S. ECVT imaging and model analysis of the liquid distribution inside a horizontally installed passive cyclonic gas–liquid separator. *Chem. Eng. Sci.* **2016**, *141*, 231–239. [CrossRef]
12. Haryono, D.; Harjanto, S.; Rahmandani, H.F.; Oediyani, S.; Nugraha, H.; Huda, M.A. A Preliminary study of three-phase column flotation process monitoring using electrical capacitance volume tomography. *IOP Conf. Ser. Mater. Sci. Eng.* **2019**, *673*, 012130. [CrossRef]
13. Park, H.C.; Choi, H.S. Visualization of flow structures inside a conical spouted bed by electrical capacitance volume tomography. *Particuology* **2019**, *42*, 15–25. [CrossRef]
14. Warsito, W.; Marashdeh, Q.; Fan, L.-S. Electrical Capacitance Volume Tomography. *IEEE Sens. J.* **2007**, *7*, 525–535. [CrossRef]
15. Wang, F.; Marashdeh, Q.; Fan, L.-S.; Warsito, W. Electrical capacitance volume tomography: Design and applications. *Sensors* **2010**, *10*, 1890–1917. [CrossRef]
16. Chowdhury, S.M.; Charleston, M.A.; Marashdeh, Q.M.; Teixeira, F.L. Propellant Mass Gauging in a Spherical Tank under Micro-Gravity Conditions Using Capacitance Plate Arrays and Machine Learning. *Sensors* **2023**, *23*, 8516. [CrossRef]
17. Yang, W.Q. Hardware design of electrical capacitance tomography systems. *Meas. Sci. Technol.* **1996**, *7*, 225. [CrossRef]
18. Wang, A.; Marashdeh, Q.; Teixeira, F.; Fan, L. Electrical capacitance volume tomography: A comparison between 12- and 24-channels sensor systems. *Prog. Electromagn. Res. M* **2015**, *41*, 73–84. [CrossRef]
19. Warsito, W.; Fan, L.-S. *Development of 3-Dimensional Electrical Capacitance Tomography Based on Neural Network Multi-Criterion Optimization Image Reconstruction*; International Society for Industrial Process Tomography: Łódź, Poland, 2003.
20. Zhou, D.; Karatayev, K.; Fan, Y. Water-in-oil emulsion temporal and spatial evolution at static conditions and its impact on pipeline restart. *SPE J.* **2024**. [CrossRef]
21. Lovick, J.; Angeli, P. Experimental studies on the dual continuous flow pattern in oil–water flows. *Int. J. Multiph. Flow* **2004**, *30*, 139–157. [CrossRef]
22. Trallero, J.L. Oil-Water Flow Patterns in Horizontal Pipes. Ph.D. Thesis, The University of Tulsa, Tulsa, OK, USA, 1995.
23. Nädler, M.; Mewes, D. Flow induced emulsification in the flow of two immiscible liquids in horizontal pipes. *Int. J. Multiph. Flow* **1997**, *23*, 55–68. [CrossRef]
24. Ahmed, S.A.; John, B. Liquid–Liquid horizontal pipe flow—A review. *J. Pet. Sci. Eng.* **2018**, *168*, 426–447. [CrossRef]
25. Al-Wahaibi, T.; Yusuf, N.; Al-Wahaibi, Y.; Al-Ajmi, A. Experimental study on the transition between stratified and non-stratified horizontal oil–water flow. *Int. J. Multiph. Flow* **2012**, *38*, 126–135. [CrossRef]
26. Vielma, M.; Atmaca, S.; Sarica, C.; Zhang, H. Characterization of oil/water flows in horizontal pipes. *SPE Proj. Facil. Constr.* **2008**, *3*, 1–21. [CrossRef]
27. Schümann, H.; Tutkun, M.; Nydal, O.J. Experimental study of dispersed oil-water flow in a horizontal pipe with enhanced inlet mixing, Part 2: In-situ droplet measurements. *J. Pet. Sci. Eng.* **2016**, *145*, 753–762. [CrossRef]
28. Silva, C.A.; Filho, D.R.; Nunes, G.M.; Bassani, G.S.; Almeida, N.L.; Panossian, Z. Corrosion in Multiphase Slug Flow Loop in Deep-Water Oil and Gas Exploitation. In Proceedings of the Offshore Technology Conference, Brazil, Rio de Janeiro, Brazil, 29–31 October 2019.
29. Skjefstad, H.S.; Dudek, M.; Øye, G.; Stanko, M. The effect of upstream inlet choking and surfactant addition on the performance of a novel parallel pipe oil–water separator. *J. Pet. Sci. Eng.* **2020**, *189*, 106971. [CrossRef]

fluids

MDPI

Article

Wind Tunnel Experiments on Parallel Blade–Vortex Interaction with Static and Oscillating Airfoil

Andrea Colli *, Alex Zanotti and Giuseppe Gibertini

Dipartimento di Scienze e Tecnologie Aerospaziali, Politecnico di Milano, Via La Masa 34, 20156 Milan, Italy; alex.zanotti@polimi.it (A.Z.); giuseppe.gibertini@polimi.it (G.G.)
* Correspondence: andrea.colli@polimi.it

Abstract: This study aims to experimentally investigate the effects of parallel blade–vortex interaction (BVI) on the aerodynamic performances of an airfoil, in particular as a possible cause of blade stall, since similar effects have been observed in literature in the case of perpendicular BVI. A wind tunnel test campaign was conducted reproducing parallel BVI on a NACA 23012 blade model at a Reynolds number of 300,000. The vortex was generated by impulsively pitching a second airfoil model, placed upstream. Measurements of the aerodynamic loads acting on the blade were performed by means of unsteady Kulite pressure transducers, while particle image velocimetry (PIV) techniques were employed to study the flow field over the blade model. After a first phase of vortex characterisation, different test cases were investigated with the blade model both kept fixed at different incidences and oscillating sinusoidally in pitch, with the latter case, a novelty in available research on parallel BVI, representing the pitching motion of a helicopter main rotor blade. The results show that parallel BVI produces a thickening of the boundary layer and can induce local flow separation at incidences close to the stall condition of the airfoil. The aerodynamic loads, both lift and drag, suffer important impulsive variations, in agreement with literature on BVI, the effects of which are extended in time. In the case of the oscillating airfoil, BVI introduces hysteresis cycles in the loads, which are generally reduced. In conclusion, parallel BVI can have a detrimental impact on the aerodynamic performances of the blade and even cause flow separation, which, while not being as catastrophic as in the case of dynamic stall, has relatively long-lasting effects.

Keywords: blade–vortex interaction; BVI; fluid–structure interaction; rotorcraft; wind tunnel; PIV; oscillating airfoil

Citation: Colli, A.; Zanotti, A.; Gibertini, G. Wind Tunnel Experiments on Parallel Blade–Vortex Interaction with Static and Oscillating Airfoil. *Fluids* **2024**, *9*, 111. https://doi.org/10.3390/fluids9050111

Academic Editors: Guangjian Zhang and Stéphane Le Dizès

Received: 28 March 2024
Revised: 28 April 2024
Accepted: 7 May 2024
Published: 10 May 2024

1. Introduction

Blade–vortex interaction (BVI), which can broadly be defined as the interaction between a rotor blade and a coherent vortical structure, is an unsteady phenomenon characteristic of the complex aerodynamic environment of rotary-wing aircraft [1]. While many of the early works regarded helicopter flight—and in particular, powered descent with the main rotor essentially passing through its own wake—as the main example of BVI occurrence, nowadays, the range of interest has greatly increased given the wide variety of vehicle configurations featuring multiple rotors, from tilt-rotors to compound helicopters, drones, micro aerial vehicles (MAVs), and multi-copter configurations typical of electric vertical take-off and landing (eVTOL) machines. Part of this interest in BVI has classically been associated with its relevance concerning noise production [2,3] (e.g., blade slap sound), but in the literature, the effect of the interaction on the blade aerodynamic loads, in terms of unsteady impulsive variations, often very significant, has always been recognised [4].

Typically, different typologies of BVI are identified based on the relative orientation between the impacting vortex and the blade, resulting in three main categories: parallel BVI, perpendicular BVI, and normal (i.e., out-of-plane) BVI, each with its own peculiar behaviour. An overview of the associated phenomena can be found in [5]. Naturally, in a real-case rotor environment, any interaction will generally be a combination of the three; nonetheless, this subdivision is useful for a more fundamental study of the behaviour of BVI.

In recent times, particular attention has been brought to the topic of perpendicular BVI, as several studies, both experimental and numerical [6–9], on oscillating blades reported that an interaction of this kind can be the triggering source of dynamic stall of the blade, with a significant detrimental effect on the performances of the airfoil. On the other hand, the research on parallel BVI has mostly been limited to static airfoil models, and highlighted issues with flow separation only in the case of an airfoil at a very high incidence [10–13]; studies of vortex interaction with oscillating bodies have been focused on leading edge impingement or on the behaviour of periodic vortex patterns [5], rather than on the loads generated.

In the last few years, moreover, research in parallel BVI has been performed also outside of the field of rotary-wing aerodynamics, for example, investigating the improvement effect of streamwise vortices on airfoil efficiency [14] or, concerning parallel BVI, studying the interaction of an airfoil or wing with a vortical gust. In the latter case, the literature includes high-fidelity CFD simulations [15,16] and wind tunnel experiments [17,18], albeit at low Reynolds number, and reports on the impulsive changes in the aerodynamic loads during BVI. In particular, the work by [15] studied the viscous effects associated with the vortex–airfoil interaction, showing boundary layer instability and separation.

The aim of the present study is to assess the behaviour of parallel BVI in terms of the influence on the airfoil aerodynamic performances, with a particular focus on macroscopic features of flow separation and stall, to investigate whether effects similar to those observed for perpendicular BVI can occur. By considering a static airfoil at different incidences, the dependence on the blade loading of the BVI effects can be characterised. Finally, the interaction is also tested on a periodically oscillating blade, the study of which extends the literature on the topic of parallel BVI and is of key importance to the application to helicopter aerodynamics.

For these purposes, an experimental test rig was developed for the generation of a suitable vortex and the following study of its interaction with an airfoil model, and an extensive wind tunnel test campaign was conducted including both unsteady pressure measurements to evaluate the loads on the blade and two-dimensional, two-component particle image velocimetry (2D-2C PIV) techniques to investigate the flow field resulting from the interaction. For the vortex generation, the strategy of having an impulsively pitched upstream airfoil was employed, which is common in the literature on parallel BVI [19–21]: the generated vortex is essentially the starting vortex which is released at the trailing edge of the airfoil as a consequence of this sharp motion, resulting from the rolling up of the wake; the vortex is then convected downstream by the wind tunnel flow.

This relatively simple, two-dimensional, test case was chosen to focus on the fundamental behaviour of the vortex–airfoil interaction; taking into account the limitations of the available literature to the interaction with low-incidence airfoils, for the present work, relatively high-load conditions were chosen, which could be representative of those encountered by a helicopter rotor retreating blade, particularly susceptible to stall behaviour.

In the following, the experimental setup will be described together with the data analysis and reduction techniques; then, the results of the campaign will be presented and discussed. For the sake of brevity, throughout the text, the generic term "vortex" will be used in place of the appropriated scientific term "vortical structure", when referring to the main structure generated to produce the interaction.

2. Materials and Methods

The tests were conducted in the "S. De Ponte" subsonic wind tunnel at Politecnico di Milano, a closed-loop wind tunnel with a 1 by 1.6 m test chamber, a maximum speed of 55 m/s, and a maximum turbulence level of 0.1%.

As previously mentioned, the setup consisted of two airfoil models in a tandem configuration, with one (the vortex generator) placed upstream of the other (the blade model) in the wind tunnel test chamber, as shown in Figure 1.

(**a**) (**b**)

Figure 1. Schematic view (**a**) and picture (**b**) of the test rig assembly in the wind tunnel; in the latter, the blade model is visible in the foreground, while the vortex generator is in the background.

The blade model is a pre-existing aluminum machined assembly composed of two lateral sections and a narrower middle section, attached to an internal metallic frame of four airfoil ribs connected by three wing boxes. Two different middle sections can be interchanged: a plain one, and one with a series of holes spanning the section chord, on both the upper and lower surfaces, to be used as taps for pressure measurements. The model chord is $c = 0.3$ m with a uniform span-wise distribution, while the span is $b = 0.9$ m. The NACA 23012 airfoil section was used, as is typical of helicopter blades. Tubular shafts, connected to the ribs at each extremity in correspondence of the quarter-chord axis, allow the model to be mounted on self-aligning bearings and pivoted around that axis. More details on the blade model characteristics, including locations of the pressure taps and structural analysis, are available in [22].

The blade model is placed horizontally in the wind tunnel test section by means of an external supporting structure of aluminum profiles. In order to control the rotation of the model around the pitch axis, a Parker SME115 brushless electrical motor (Parker Hannifin Corp., Cleveland, OH, USA) is connected to one of the shafts through a double-cardanic steel laminae coupling, to account for angular and axial displacements, and through a 12:1 planetary gear. The control of the pitching motion was exploited to easily set the geometrical incidence of the model during the static interaction study, and then employed to realise the oscillating motion of the model.

The vortex generator model is composed of a steel spar, with a square section of 3 cm side, around which a carbon fibre skin was laminated. To allow for the presence of the spar, a custom airfoil section shape was employed, joining the leading edge portion of a NACA 0018 to the trailing edge section of a NACA 0016, with the spar in the quarter-chord position. A styrofoam filler was shaped accordingly and used as a reference for the lamination process. The resulting model has a chord $c_g = 0.2$ m and a span $b_g = 0.9$ m. An assembly of tubular shafts, bearings, laminae coupling, and supporting structure, similar to that used for the blade model, allows one to mount the vortex generator in the wind tunnel test section, 1.2 m upstream of the blade model quarter-chord axis; the vertical position was manually adjustable by sliding the support braces along vertical struts. The vortex generator is attached on one end to a Parker SME170 brushless electrical motor (Parker Hannifin Corp., Cleveland, OH, USA), with a nominal torque of 35 Nm and a peak torque of 100 Nm, which is used to control its pitching motion.

A programmable logic controller (PLC), commanded via a LabView (v. 2018) graphical interface, is used to control both motors. To measure their angular position, the signal from the internal Heidenhain SinCos EnDat absolute encoders (Dr. Johannes Heidenhain GmbH, Traunreut, Germany), with an accuracy of $\pm 400''$, is acquired; any mechanical backlash or model deformation is neglected in the following study. The initial angular positioning in pitch with respect to the wind tunnel centreline was measured by a digital inclinometer, with an accuracy of 0.01°. The motion of the vortex generator was of fundamental importance: in order for a single, well-defined, concentrated vortex to be released, the pitching motion has to be fast compared to a time scale relative to the free-stream velocity and the generator chord; in particular, the rotation should be completed in a time interval smaller than c_g/V_∞ [20]. This relation poses an upper limit to the free-stream velocity, and therefore to the flow Reynolds number, depending on the speed achievable by the motion system. The amplitude of the rotation was chosen to be of $\Delta\alpha_g = 10^{\circ}$ and a trial-and-error procedure was conducted to fine-tune the parameters of the motor drive internal PID controller and obtain the fastest possible motion without overshoot or oscillations; the result was a time interval to complete the motion of $\Delta t = 11$ ms. Based on this result, a wind tunnel free-stream velocity of $V_\infty = 15\,\mathrm{m\,s^{-1}}$ was selected, in compliance with the aforementioned constraint, corresponding to a Reynolds number of $Re = (cV_\infty)/\nu$ = 300,000, relative to the airfoil chord; this value of the Reynolds number is in the range of the ones found in the literature for similar studies [19,20].

The pitching motion of the generator was realised by feeding the PLC a sawtooth-like input signal, with a comparatively slow ramp to set the airfoil at the incidence of $\alpha_g = 0^{\circ}$ followed by an impulsive rotation to $\alpha_g = 10^{\circ}$, which causes the release of the vortex. These values were chosen in order to have a high value of lift, and thus circulation, of the vortex generator and to have a counter-clockwise vortex. Both before and after the ramp, i.e., before and after the vortex interaction, a waiting period is added to allow the flow to reach steady conditions. The final input signal has a total period of 2 s. More details on the mechanism of vortex generation can be found in [23].

The same motion of the generator was employed also for the tests with the oscillating blade model, while the latter was pitched according to a sine wave with amplitude 10° given by the law $\alpha = \alpha_0 + 10^{\circ}\sin 2\pi f t$, where α_0 is the starting airfoil incidence and f is the frequency of the motion; for the former, different values were chosen, ranging from 5° to 7°, to study the differences in BVI as the blade incidence increases, while the latter was chosen as to achieve a reduced frequency $k = \pi f c/V_\infty = 0.1$, resulting in a motion with a period of 0.628 s. These values are typical of "light" dynamic stall tests. The motion of the generator was adjusted in order to have a vortex interaction every three periods of the blade model, i.e., every 1.884 s, while keeping the duration of the impulsive pitching equal to that of the previous static tests so that the two kinds of interactions could be compared. This condition is different from that experienced by a rotor blade in a real-case scenario, where the BVI would occur at each cycle, assuming periodic conditions, but the results showed that the effects of the interaction exhaust themselves well before a full cycle, which leads to the hypothesis that the BVI behaviour would be closely similar in the two cases. Finally, the commanded time histories of the motion of the two models were shifted with respect to each other in order to have the interaction happening at two different positions during the airfoil oscillation, both in the descending portion, one right after the peak incidence is reached ("high incidence interaction", HII) and one when the airfoil is at an incidence 5° lower than the peak one, approximately ("low incidence interaction" LII); the relative time shift between the HII and LII oscillation time histories is of 30 ms.

Concerning the vortex generator, particular care is needed in taking into account its presence in the wind tunnel in order to correctly interpret the actual test conditions and the subsequent results. As previously mentioned, after the vortex release, the vortex generator is at an incidence of $\alpha_g = 10^{\circ}$; therefore, its induction on the blade model has to be included in the reference condition ("baseline") against which to evaluate the effects of the interaction; on the other hand, the conditions before the vortex generation will be different from the baseline: in fact, the process of vortex generation can be interpreted by substituting the generator model with the vortex itself, held fixed in the same position, and then allowed to be convected by the free-stream flow.

2.1. PIV Measurements

The PIV setup is composed of a dual Nd-YAG Evergreen laser (LUMIBIRD SA, Lannion, France), with an energy of 200 mJ per pulse, and two ILA sCMOS cameras (ILA_5150 GmbH, Aachen, Germany), with a resolution of 2560 × 2160 pixel. The laser was placed below the floor of the wind tunnel test section, in which a mid-span optic window allowed the light to illuminate the flow; because of this positioning, the blade model was installed with the suction side facing the floor when $\alpha = 0^\circ$. Oil droplets, with diameter in the range of 1–2 µm, generated by a Laskin-nozzle particle generator, were employed as tracer particles. An ILA synchroniser (ILA_5150 GmbH, Aachen, Germany) was used to control the acquisition timing, managing the laser and the cameras; an Hall-effect sensor was mounted on one of the vortex generator support shafts, and its signal, corresponding to $\alpha_g = 0^\circ$, was used as a trigger. The PIV images were processed with a multi-pass method over a 32 × 32 pixel grid, with no previous filtering but with outlier detection and interpolation, and finally phase-averaged; the PIVTEC PIVview2C [24] software (v. 3.9) was used for the task. Considering a 0.2 px sub-pixel interpolation accuracy [25], the error on the velocity field can be estimated to be below 2% of the free-stream velocity.

A first phase of PIV measurement was conducted in order to characterise the generated vortex velocity profile; for this phase, the blade model was not installed in the wind tunnel.

Secondly, the blade model was inserted, and its vertical position was adjusted in such a way that the vortex would impinge approximately at the leading edge of the airfoil, with an upward trajectory going over the suction side. This procedure was executed for a geometrical incidence of the blade model of 14° and was not repeated for different angles as it was seen that the change in position due to the rotation was offset by the change in circulation around the airfoil, resulting in a similar vortex relative trajectory over a range of incidences 10°–16°.

For the study of the interaction, PIV measurements were taken at different time instants by adding a time delay to the triggering signal; a range of times starting from the vortex approaching the blade model leading edge and up to the vortex having travelled downstream of the trailing edge was considered. Measurements with no vortex interaction were also performed, with the vortex generator fixed at $\alpha_g = 0^\circ$ and 10°; in those cases, the acquisition sequence was manually triggered.

2.2. Pressure Measurements

For the pressure measurement phase, the mid-span section of the blade model was swapped with the appropriate one, instrumented with 21 pressure taps distributed chord-wise in the middle of the section; details on the positions of the taps can be found in [22]. The taps were fitted with nylon pipes in which Kulite XCS-093-2D (Kulite Semiconductor Products, Inc., Leonia, NJ, USA) unsteady pressure transducers were inserted, with sealing rubber O-rings; the combined non-linearity, hysteresis, and repeatability error is indicated to be typically within 0.1% of the full scale output. The signals from the transducers were acquired by a NI cDAQ-9172 with NI 9237 bridge modules (National Instruments Corporation, Austin, TX, USA) at a sampling rate of 10 kHz; simultaneously, the angular positions of the vortex generator and of the blade model were acquired by a NI 9215 module (National Instruments Corporation, Austin, TX, USA) on the same cDAQ. It is to be noticed that during the test campaign, five transducers suffered a failure, which therefore reduced the number of available data points.

The pressure measurements during the vortex interaction study were acquired for 20 s, starting from the same trigger signal used for the PIV acquisition. Given the periodicity of the motion, 10 cycles were included in this period; each acquisition was repeated 10 times, for a total of 100 cycles of vortex generation and interaction acquired for each test condition; the time histories were finally phase-averaged, and a moving average filter with a ten-point kernel was applied to reduce noise. Measurements with no vortex interaction were also acquired at different incidences in order to characterise the airfoil polar, with the vortex generator fixed at $\alpha_g = 0°$ and $10°$; in those cases, the acquisition sequence was manually triggered and the data were time-averaged and filtered to obtain the steady values.

The data from the pressure measurements were postprocessed by a custom Python (v. 3.10) script, performing the computation of the pressure coefficient and the integration to obtain the lift and drag acting on the blade model; in particular, the airfoil section was divided into straight panels, over which the pressure was obtained from the values measured at the taps through linear interpolation based on the curvilinear abscissa. The lift C_l and drag C_d coefficients were then computed as

$$C_l = \frac{L}{\frac{1}{2}\rho V_\infty^2 c} \qquad C_d = \frac{D}{\frac{1}{2}\rho V_\infty^2 c}, \tag{1}$$

where L and D are, respectively, the lift and drag per unit span obtained by the integration of the pressure data and ρ is the air density. Finally, the two-dimensional boundary corrections from [26] were applied to the data, in particular concerning wake blockage and solid blockage, especially relevant for the higher incidences tested. It is to be noticed that no attempt was made to extend these corrections to the trajectory of the vortex itself, that is, to determine the influence of the test being conducted in the wind tunnel rather than in free air on the convection and evolution of the vortex.

In the case of the unsteady measurements during vortex interaction, the processing method was applied to the time history of the pressure distribution, obtaining the corresponding time histories of the aerodynamic coefficients. The same corrective factors computed for the stationary case were used for the whole time history.

The same procedure was also applied to the analysis of the data recorded from the interaction on the oscillating airfoil; in this case, however, no corrective factors were applied. Measurements were taken of the oscillating airfoil with no vortex interaction to serve as the baseline reference.

3. Results and Discussion

In the analysis of the following results, a reference frame placed at the quarter-chord point of the blade model and aligned with the free-stream velocity is employed, as shown in Figure 2.

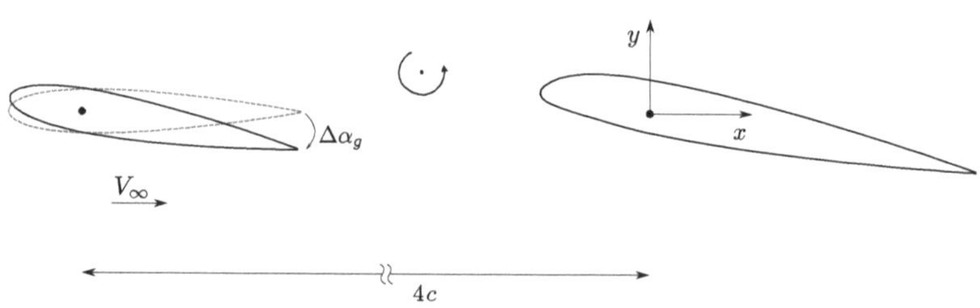

Figure 2. Schematics of the setup and reference frame employed. The vortex generator is on the left, and the blade model on the right.

The non-dimensional time $\tau = tV_\infty/c$ is also introduced, where t is the time in seconds and $t = 0$ is chosen as the instant at which the impulsive motion of the vortex generator ends. One time unit, therefore, corresponds to the time it takes for the vortex, assumed to be convected by the free-stream velocity, to travel over the blade model chord.

3.1. Vortex Characterisation

The averaged PIV measurements of the isolated vortex (Figure 3) show that the vortex core, defined as the flow region delimited by the local extrema in the velocity profile, presents a circular shape, with a radius of $r_{\text{core}}/c = 0.0328$. The circulation, computed by integrating along the core perimeter, is $\Gamma/(cV_\infty) = 0.066$, corresponding to a vortex

Reynolds number $Re_\Gamma = \Gamma/\nu = \Gamma/(cV_\infty)Re$ = 19,800. By comparing with literature data from experiments on tip vortices released by helicopter rotor models [27], the size of the present vortex fits very well in the typical range for tip vortices, while its core circulation is consistent with newly released vortices. Concerning the vortex Reynolds number, which is related to turbulent diffusion and vortex growth rate [28], the present value falls within the range of measurements for model-scale rotors [29]. The induced tangential velocity u_θ profile is also reported in Figure 3 and compared to a Vatistas model [30]:

$$u_\theta = \frac{r}{\left(1 + r^{2n}\right)^{1/n}}, \tag{2}$$

where r is the distance from the vortex centre, normalised to have a core size of $r = 1$; it can be seen that a close fit to the experimental data is obtained by setting $n = 1$.

To characterise the dispersion associated with vortex meandering and the repeatability of the generation process, a statistics analysis over all the PIV image couples was performed by considering each couple separately and identifying the vortex centre position through the λ_2 criterion [31]: the results showed that the data from the averaged PIV measurements overestimate the vortex size by less than 1%, with respect to the size determined by statistical analysis on the separate image couples. Therefore, no specific procedure was adopted to align the position of the vortex across all the images prior to the averaging process; such a procedure, moreover, would not have been applicable to the following measurements with the blade model.

In conclusion, the vortex generation process was deemed suitable for the parallel BVI study, being able to reliably produce an isolated, coherent vortex with adequate size and strength.

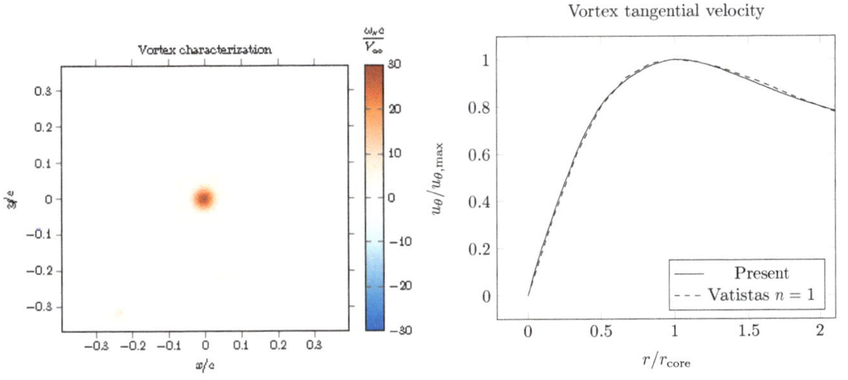

(a) (b)

Figure 3. Vorticity contour (**a**) and tangential velocity u_θ profile (**b**) for the averaged vortex, from PIV measurements.

3.2. Airfoil Polar

The aerodynamic coefficients computed from the pressure measurements on the stationary airfoil (Figure 4), considering the case with $\alpha_g = 0°$, show a behaviour of the lift curve that is consistent with the literature work on NACA 23012 at a similarly low Reynolds number [32–34]. In particular, the maximum lift coefficient has a value of 1.163, which is found in correspondence of $\alpha = 14.6°$. It is to be noticed that no evident laminar separation bubble, which is a common feature in airfoils of this type, was visible from the experimental data; while this might be ascribed to the relatively low number of pressure taps in the relevant portion of the airfoil, it could also hint to three-dimensional flow behaviour influencing boundary layer transition and separation. Wind tunnel data from [35] also show a limited presence, if any, of a laminar separation bubble, which could confirm a sensibility of the phenomenon to experimental conditions. The lift curve in the condition with $\alpha_g = 10°$, i.e., the baseline condition, shows a less abrupt stall behaviour with a slightly lower maximum lift coefficient of 1.133 for $\alpha = 15.6°$; this increase in the stall geometric incidence is expected due to the effect of the upstream vortex generator, which decreases the effective incidence of the blade model by about 1°. During the tests, a hysteresis of the stall behaviour was observed. It should be noticed that the reported drag values include only pressure drag, being obtained from the pressure measurements.

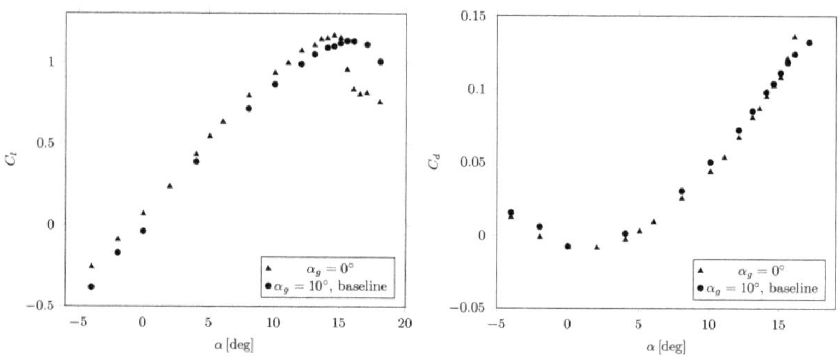

(**a**) (**b**)

Figure 4. Lift (**a**) and drag (**b**) coefficients from integration of pressure measurements for the baseline condition $\alpha_g = 10°$ and for $\alpha_g = 0°$.

3.3. Vortex Interaction—Static Airfoil

The measurements of the interaction between the vortex and the blade model show the vortex approaching with a trajectory impinging on the leading edge of the airfoil, and then moving towards the suction side and along it. During the interaction, the vortex remains coherent, with a core region which is clearly identifiable, although distorted into a more oval shape; the dimensions of the core also increase slightly.

In Figure 5 the sequence of the interaction with the airfoil at incidence $\alpha = 14°$ shows how the main effect of the vortex passage is a very noticeable thickening of the boundary layer along the suction side of the airfoil, as indicated by the vorticity contours. At first, the thickening is more severe at the chord-wise positions corresponding to and immediately downstream of the vortex position, while it reduces drastically immediately upstream: this behaviour is expected given the counter-clockwise rotation of the vortex, as the induced velocity field tends to displace the flow away from the airfoil surface ahead of the vortex, and towards the surface behind it. Once at approximately the mid-chord position, the upstream edge of the increased thickness region appears to lag behind the vortex, while its downstream edge continues to correspond to the vortex position: this results in the widening of the region and, at later times, in the formation of two almost separate "bubbles" of increased vorticity; in this region, moreover, the vorticity has opposite sign to that of the airfoil and is comparable in magnitude. As the vortex gets closer to the trailing edge, a portion of recirculating flow can be identified as associated with the vorticity bubble. This viscous behaviour is consistent with the observations by [36] and the analysis of [37,38], concerning the vortex interaction with a wall: these authors report the formation of a similar vorticity bubble and indicate its cause in the adverse pressure gradient produced by the presence of the vortex, which induces a suction peak beneath the core. The fact that no "eruption" of this bubble can be seen may be attributed to the different flow conditions of the present case, with an airfoil at incidence, with respect to a plane wall.

Similarities can also be found with the results of the LES computations of [15], which report thickening of the boundary layer, with laminar separation and the formation of a counter-rotating vortical disturbance. While detailed comparisons are difficult to make, given the lower incidence of the airfoil and the much larger interacting vortex that they used in the work, the mechanism of flow separation presently observed is likely to be similar.

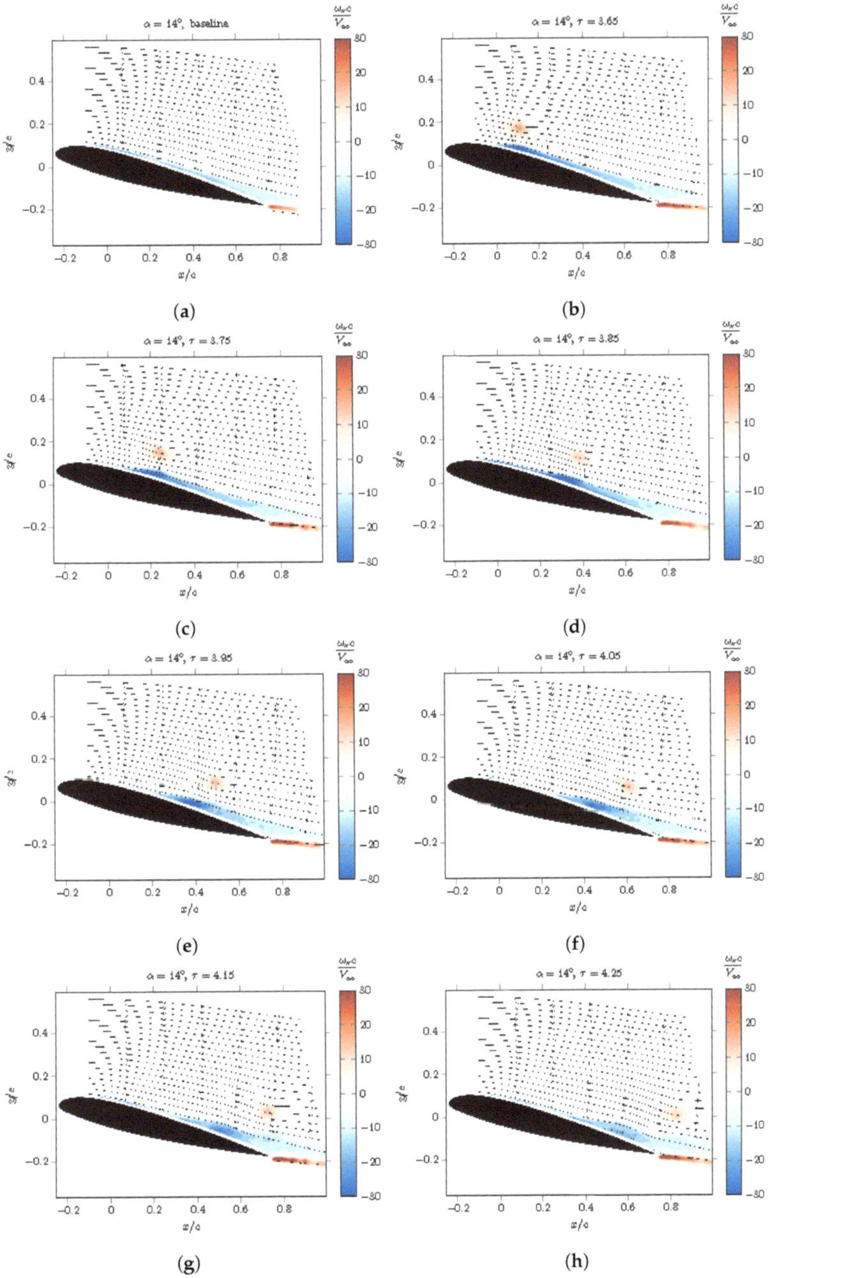

Figure 5. Vorticity contours from PIV measurements of the flow field during the vortex interaction with the static airfoil at 14° and comparison with baseline: (**a**) baseline, (**b**) $\tau = 3.65$, (**c**) $\tau = 3.75$, (**d**) $\tau = 3.85$, (**e**) $\tau = 3.95$, (**f**) $\tau = 4.05$, (**g**) $\tau = 4.15$, (**h**) $\tau = 4.25$.

The description just presented can be applied to the behaviour of the interaction with the airfoil at $\alpha = 15°$, as shown in Figure 6, with the main differences being that in the latter case, the thickness increase does not reduce downstream, thus not forming a bubble as definite as in the former case; and that recirculating flow regions are clearly visible, starting from earlier interaction times. In particular, the boundary layer is more severely displaced and the separated region extends from around 50% of the chord, persisting well after the vortex has passed the trailing edge of the blade. This behaviour could be expected given that the incidence is closer to the stall condition and the boundary layer would be more prone to separating under disturbances. While the appearance of separated flow regions, both for $\alpha = 14°$ and $\alpha = 15°$, could be explained in terms of the essentially inviscid effect of the vortex induced velocity increasing the effective incidence of the airfoil, viscous effects also play a significant role, as shown by the vorticity distribution in the boundary layer being heavily influenced by the passage of the vortex, with the appearance of secondary structures and their subsequent evolution.

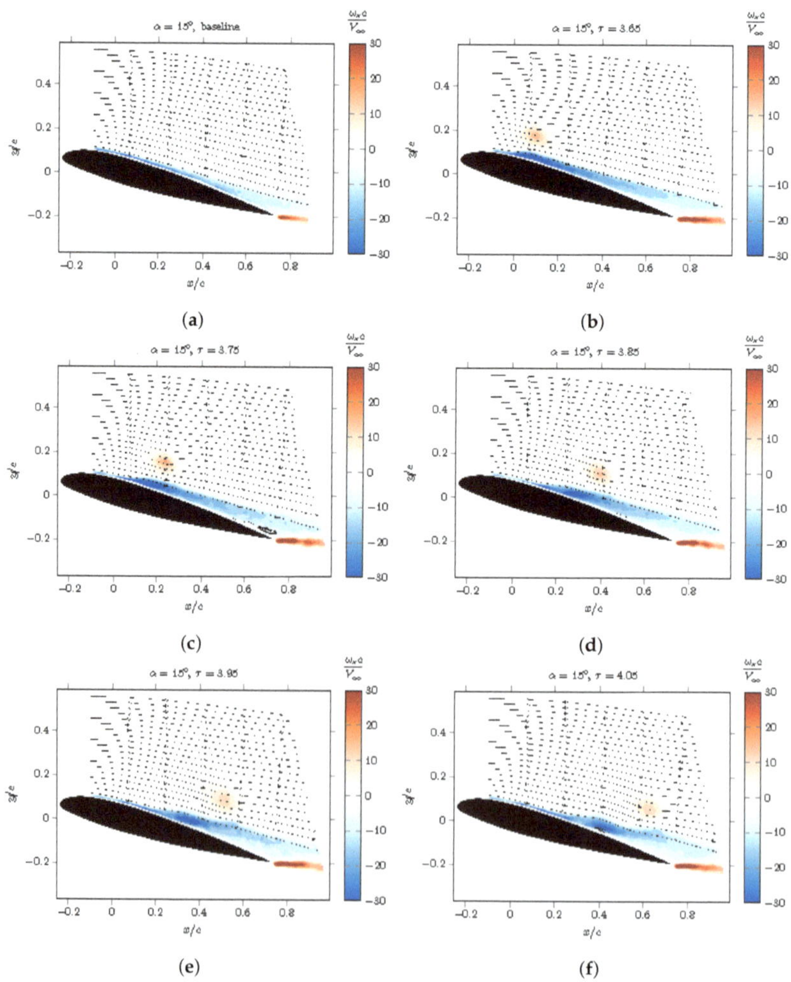

Figure 6. *Cont.*

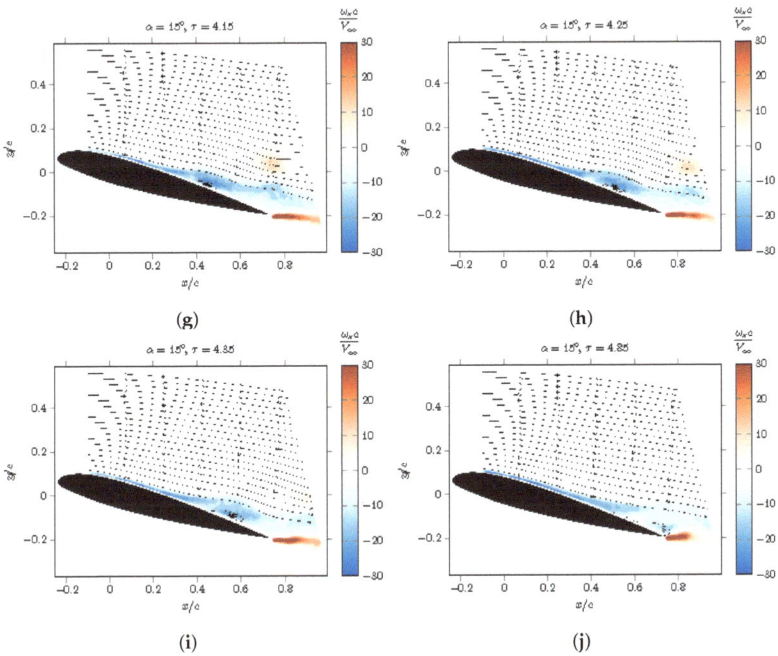

Figure 6. Vorticity contours from PIV measurements of the flow field during the vortex interaction with the static airfoil at 15° and comparison with baseline: (**a**) baseline, (**b**) $\tau = 3.65$, (**c**) $\tau = 3.75$, (**d**) $\tau = 3.85$, (**e**) $\tau = 3.95$, (**f**) $\tau = 4.05$, (**g**) $\tau = 4.15$, (**h**) $\tau = 4.25$, (**i**) $\tau = 4.35$, (**j**) $\tau = 4.85$.

The measurements of the interaction in the case with the airfoil at $\alpha = 10°$, presented in Figure 7, show that the above reasoning can still be applied, but the effect of the vortex is much reduced, with the vorticity bubble being smaller and more localised, and no separation of the flow being evident.

From the PIV measurements, therefore, it can be concluded that the interaction produces a thickening of the airfoil boundary layer, with the development of a vorticity bubble; this effect is local, limited to a region which moves downstream approximately following the vortex chordwise position. The magnitude of this effect, both in terms of the increase in thickness and the size of the affected region, is greater for the higher incidences of the airfoil, particularly for those close to the stall condition. The transiency of the observed phenomenon—that is, the fact that upstream of the vortex, the flow field tends to return to the undisturbed conditions—is to be expected by the nature of the interaction, as explained above. In conclusion, trailing-edge stall-like separation is observed as a result of the parallel blade–vortex interaction for high incidences of the airfoil, close to the maximum lift conditions. This is consistent with the interpretation of the effect of the vortex as inducing an increase in incidence, given also the comparatively smooth stall behaviour observed from the airfoil polar. This reasoning, however, is too simplistic since the behaviour cannot be reduced to a mere variation of incidence in unperturbed conditions, but the unsteadiness of the phenomenon must be taken into account. By projecting the results shown here, moreover, it could be argued that a vortex of higher circulation could disrupt more severely the boundary layer downstream of it in such a way that even the restoring influence of the upstream induced velocity is not able to reattach the flow, which would produce a separated condition all over the airfoil chord. This hypothesis could not be tested with the present test rig, as the strength of the vortex is limited by the pitching motion of the vortex generator.

Other works in the literature [10,11,13] presented flow separation as a result of parallel BVI, also in the case of highly loaded airfoil only, already close to the stall conditions.

The behaviour of the interaction in those cases, however, differed from the present as it triggered a separation bubble in the leading edge region, which then propagated downstream. This difference could be explained by the overall blade model behaviour, which, as already mentioned, did not feature a definite laminar separation bubble, suggesting a more turbulent flow; three-dimensional effects could also modify the flow conditions.

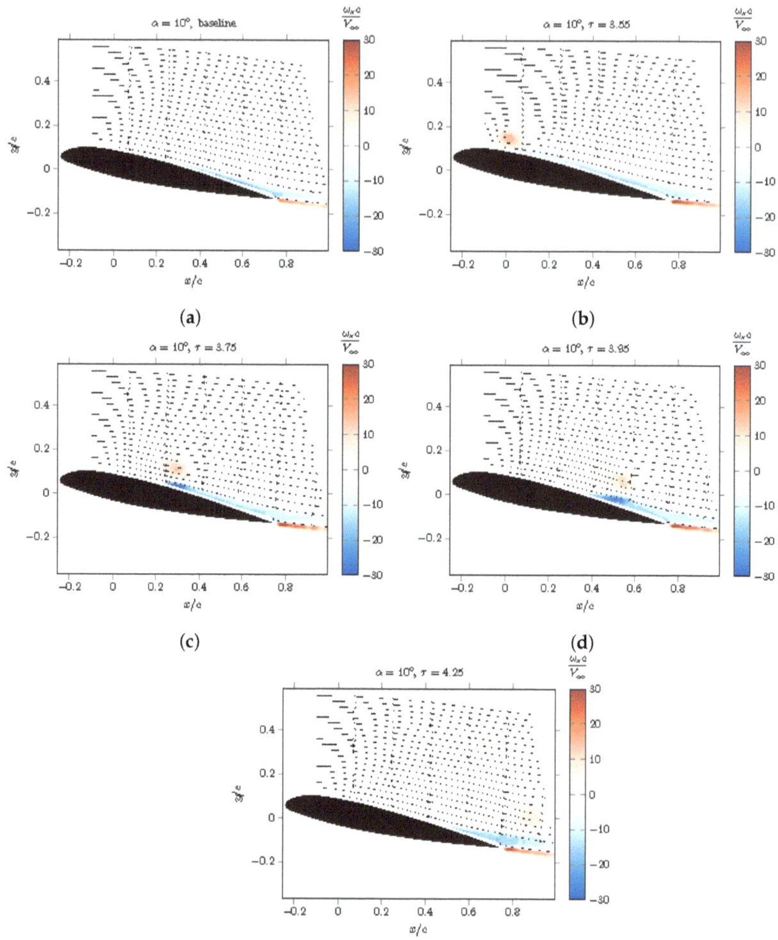

Figure 7. Vorticity contours from PIV measurements of the flow field during the vortex interaction with the static airfoil at 10° and comparison with baseline: (**a**) baseline, (**b**) $\tau = 3.55$, (**c**) $\tau = 3.75$, (**d**) $\tau = 3.95$, (**e**) $\tau = 4.25$.

More insight on the effects of the interaction can be gained from the time histories of the aerodynamic coefficient computed from the pressure measurements. To better show this, the variation ΔC_l in the lift coefficient is introduced as

$$\Delta C_l = \frac{C_l - \overline{C}_l}{\overline{C}_l}, \tag{3}$$

where $\overline{C}_l$ is the lift coefficient in the baseline condition; the same reasoning leads to the definition of the variation in the drag coefficient ΔC_d. Of course, the above definitions are meaningful only where $\overline{C}_l \gg 0$ (and similarly for the $\overline{C}_d$), which is always the case in the proximity of the maximum vortex interaction, as shown below.

Concerning the lift coefficient C_l during the blade–vortex interaction, as seen in Figure 8 and from Table 1, a substantial impulsive increase is evident for all the incidences. This effect is consistent with the behaviour of BVI as described in the literature, and it is usually explained in terms of the induced upwash of the approaching vortex, followed by a corresponding downwash effect, as already mentioned above. Comparison with the PIV measurements shows that the peak in the lift coefficient is found in correspondence with the vortex reaching the leading edge portion of the airfoil, which is also in accord with the findings of similar works [10,20]. The time τ_{peak} of the peak occurrence is also very similar for all three incidences, with the slight differences accounting for the variations in position of the airfoil and in induced flow. The lift increase is inversely proportional to the blade incidence, with the highest variation being +38% for $\alpha = 10°$. This effect can be expected as the vortex strength is the same for all three cases, while the magnitude of the induced velocity field of the airfoil increases with its incidence, so that the influence of the vortex, in terms of its induced velocity, is proportionally smaller.

Immediately after the peak, the lift falls briskly to values below the baseline: it can be noticed that this difference with respect to the baseline is greater for $\alpha = 15°$, which reaches $\Delta C_l = -14.2\%$, while it is similar, although smaller in magnitude, for the two other cases. This trend, which contrasts with the interpretation given above for the lift peak, can be explained by the occurrence of flow separation as observed from the PIV measurements, which differentiates the behaviour at the highest incidence.

The return to the baseline $\overline{C}_l$ values, for all cases, is noticeably slower than the sharp peak, taking several time units before reaching a steady state; it can also be noticed how in the case of $\alpha = 15°$, the lift coefficient remains very slightly higher than its baseline value. This persistence of the disturbance following the interaction could be attributed to the relatively slow evolution and eventual disappearance of the secondary vortical structures in the boundary layer of the airfoil. To compute an estimate of this settling time, the following approach was chosen: firstly, the difference ΔC_l between the lift measured during the interaction and its baseline value is computed for all times; then, starting from the time instant corresponding to the peak induced by the interaction, the settling time $\Delta\tau$ is determined as the time interval after which the maximum variation in ΔC_l keeps under 2% of the maximum value of the baseline C_l:

$$\Delta\tau = \tau_{2\%} - \tau_{\text{peak}}, \tag{4}$$

where $\tau_{2\%}$ is the time value such that

$$\frac{\max \Delta C_l - \min \Delta C_l}{\max \overline{C}_l} < 2\% \tag{5}$$

for all $\tau > \tau_{2\%}$, up to a suitably large time. This strategy allows for a comparison of the settling times between the static airfoil interactions and the following oscillating airfoil interactions, while also accounting for any discrepancies between the steady state and the baseline values.

The computed settling times for the static airfoil interaction cases are reported in the last column of Table 1. The largest value of $\Delta\tau$ is found in correspondence of $\alpha = 10°$, which might reflect the greater perturbing effect induced by the interaction at this lower incidence. The difference between the settling times in the other two cases can be accounted for by considering the larger flow separation occurring for $\alpha = 15°$. Despite the different behaviours among the three cases, as shown by the PIV measurements, the values of $\Delta\tau$ are relatively similar and the time histories show a comparable trend.

Concerning the drag coefficient C_d, a greater variation in the magnitude of the behaviour can be seen among the three cases. In particular, in the case $\alpha = 10°$ the interaction has the first effect of reducing the drag value, with a downward peak which exactly corresponds in time to the lift peak, representing the same impulsive variation typical of BVI. Drag falls significantly, to almost $\Delta C_d = -60\%$. This behaviour can be explained as a suction effect of the low-pressure field associated to the vortex approaching the leading edge of the airfoil. After this peak, the value of C_d rises sharply with a quick succession of two peaks at around $\Delta C_d = +16.5\%$, before returning to the baseline value. This trend is qualitatively similar in the other two cases: at first the drag is reduced, more briskly, but to a lesser extent with respect to the previous case, which again indicates the greater relative influence of the interaction at low incidences. Then, the drag rises abruptly in a much more significant way: in particular, for $\alpha = 15°$, the variation is as high as $\Delta C_l = +34\%$. This increase in drag, and its dependence on the incidence, can be associated with the occurrence of separated flow, as also indicated by the PIV measurements, as well as with the suction effect mentioned above. It is to be noticed that the appearance of two definite peaks can be attributed to the relative coarse chord-wise spacing between the pressure taps, which causes a loss of spatial resolution when dealing with a very localised phenomenon such as the vortex passage. Nonetheless, it is interesting to point out that the second peak occurs at around $\tau = 4.25$, that is, after the vortex has passed the trailing edge: the load variation can therefore be attributed to the downstream convection of the vorticity bubble or, in general, to secondary structures generated by the vortex interaction.

In conclusion, by examining the pressure data, the effect of the parallel BVI is confirmed as impulsive variations in the aerodynamic loads, with drag being particularly affected. A comparison of the aerodynamic loads time history during the interaction can be made with the results by [15] for the case of the higher relative encounter between vortex and airfoil: the results, for both lift and drag, are in good agreement, at least qualitatively, while a quantitative comparison is difficult to make, given the differences in the test conditions already mentioned above. From their analysis, moreover, secondary spikes in the loads are found to be associated with the vortical structures generated in the boundary layer because of the BVI, which supports the observations made earlier.

Table 1. Vortex interaction with static airfoil at incidence α: minimum and maximum lift ΔC_l and drag ΔC_d variation, time of lift peak occurrence τ_{peak} and settling time $\Delta\tau$.

α [deg]	ΔC_l [%]		ΔC_d [%]		τ_{peak}	$\Delta\tau$
10	−10.3	+38.0	−59.2	+16.5	3.18	9.33
14	−10.6	+29.0	−26.9	+22.0	3.17	7.19
15	−14.2	+27.1	−25.3	+34.3	3.20	8.23

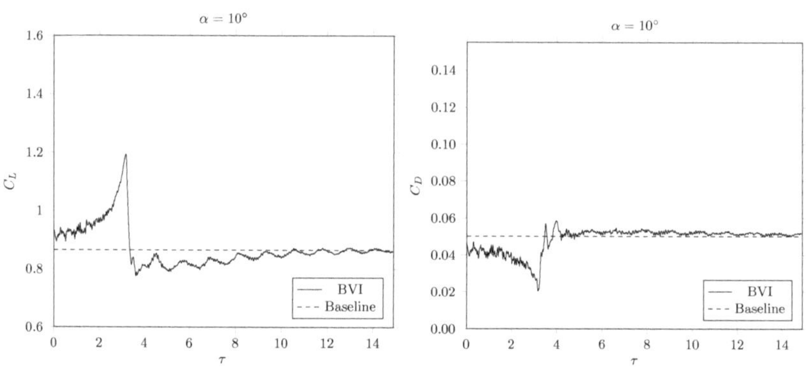

(a) (b)

Figure 8. *Cont.*

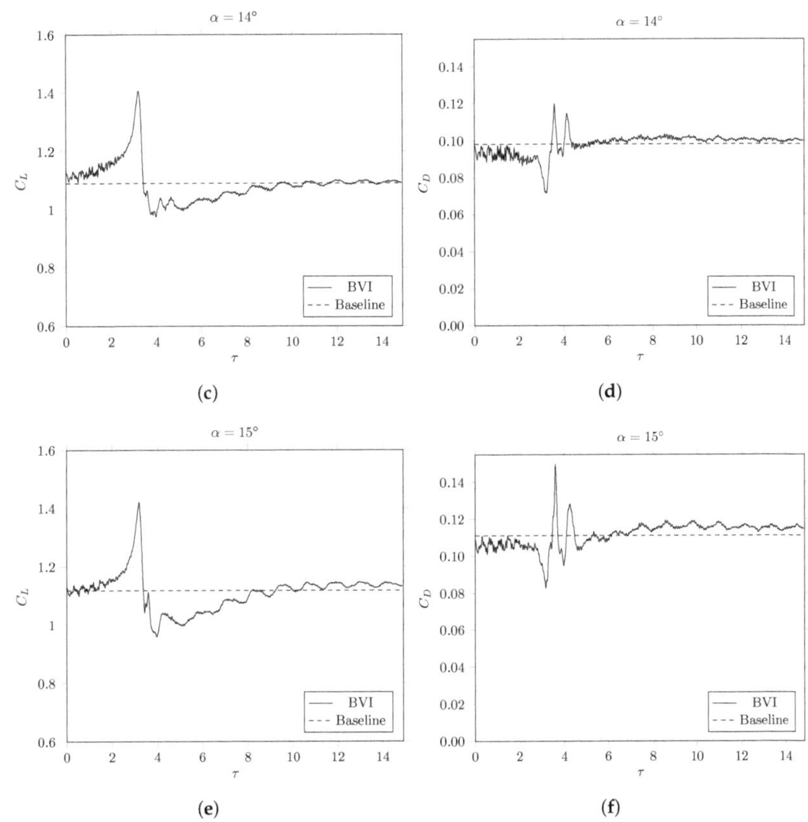

Figure 8. C_l (left) and C_d (right) time histories during the vortex interaction with the static airfoil at different incidences: (**a**,**b**) $\alpha = 10°$, (**c**,**d**) $\alpha = 14°$, (**e**,**f**) $\alpha = 15°$.

3.4. Vortex Interaction—Oscillating Airfoil

From a qualitative point of view, the flow behaviour during the interaction with the oscillating airfoil shows a similarity with the interaction in the case of the static airfoil discussed above. Figures 9–11 present the results from the PIV measurements of the HII cases, showing the thickening of the boundary layer, resulting from the passage of the vortex, with respect to the baseline conditions. A region of separated flow near the trailing edge can be identified, especially for $\alpha_0 = 7°$, but its extent does not appear to be significantly different when compared to the static airfoil cases at the corresponding incidences α; in particular, there is no indication of large separations or other phenomena related to dynamic stall.

Similar remarks can be made for the LII cases, of which Figure 12 shows the one with $\alpha_0 = 7°$: the perturbation induced by the vortex appears to be less significant than the corresponding HII case at the same α_0, which is consistent with the fact that the interaction is occurring at a lower incidence, although, at the same time, by comparing with the static case at $\alpha = 10°$ a slightly larger trailing edge separation can be seen.

The results of the unsteady pressure measurements during the interaction with the oscillating airfoil are reported in Figures 13 and 14 in terms of lift coefficient values as a function of blade model incidence. It can be noticed how the data show a small hysteresis cycle for all baseline cases tested (dashed lines in figure); moreover, the fact that a gap is visible in the BVI graph is to be attributed to the different position of the vortex generator during the blade model oscillation cycle, as discussed above.

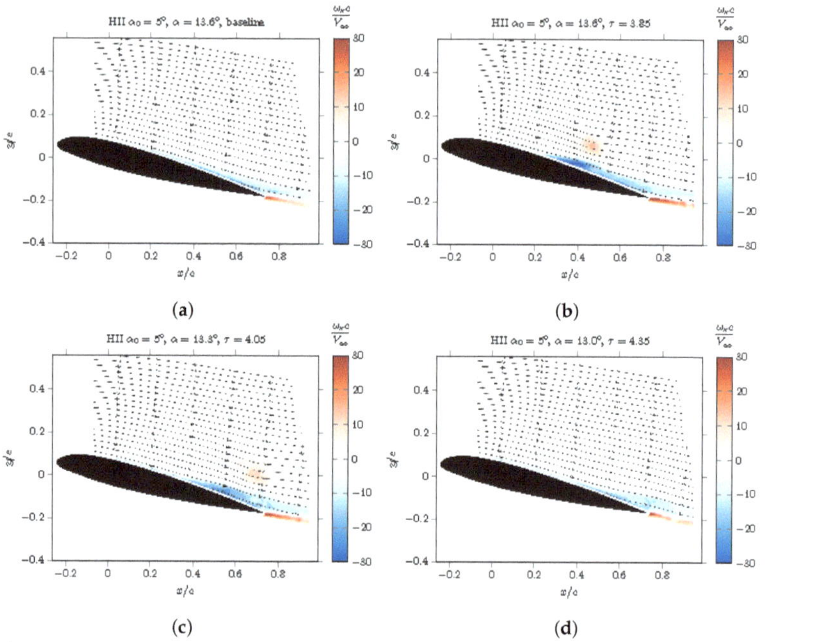

Figure 9. PIV measurements of the flow field during the HII vortex interaction with the oscillating airfoil at $\alpha_0 = 5°$ and comparison with baseline. The actual airfoil incidence α is reported for each time τ: (**a**) baseline, (**b**) $\tau = 3.85$, (**c**) $\tau = 4.05$, (**d**) $\tau = 4.35$.

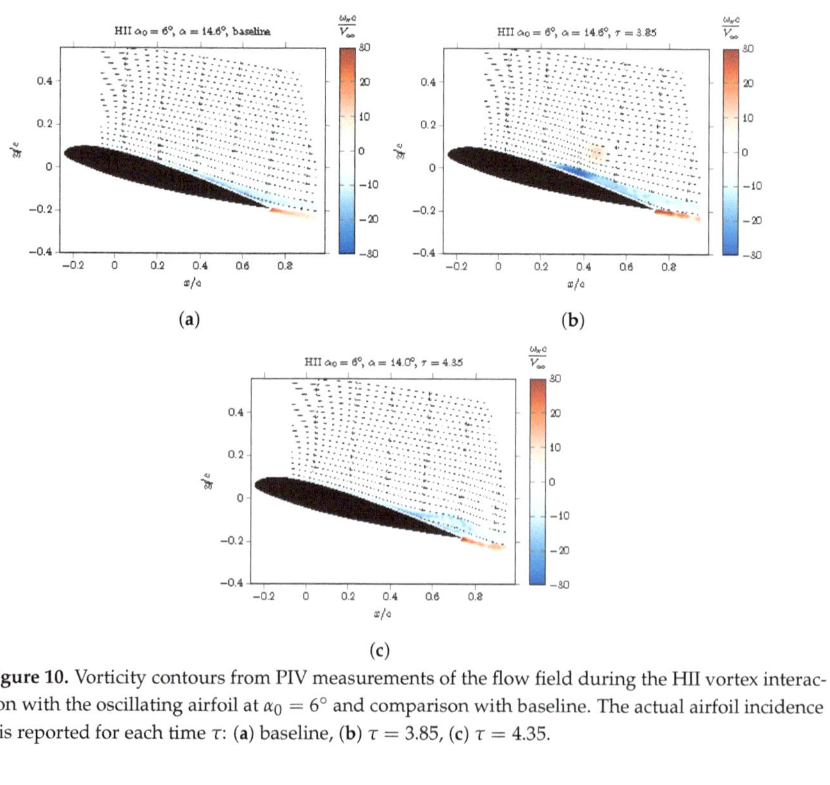

Figure 10. Vorticity contours from PIV measurements of the flow field during the HII vortex interaction with the oscillating airfoil at $\alpha_0 = 6°$ and comparison with baseline. The actual airfoil incidence α is reported for each time τ: (**a**) baseline, (**b**) $\tau = 3.85$, (**c**) $\tau = 4.35$.

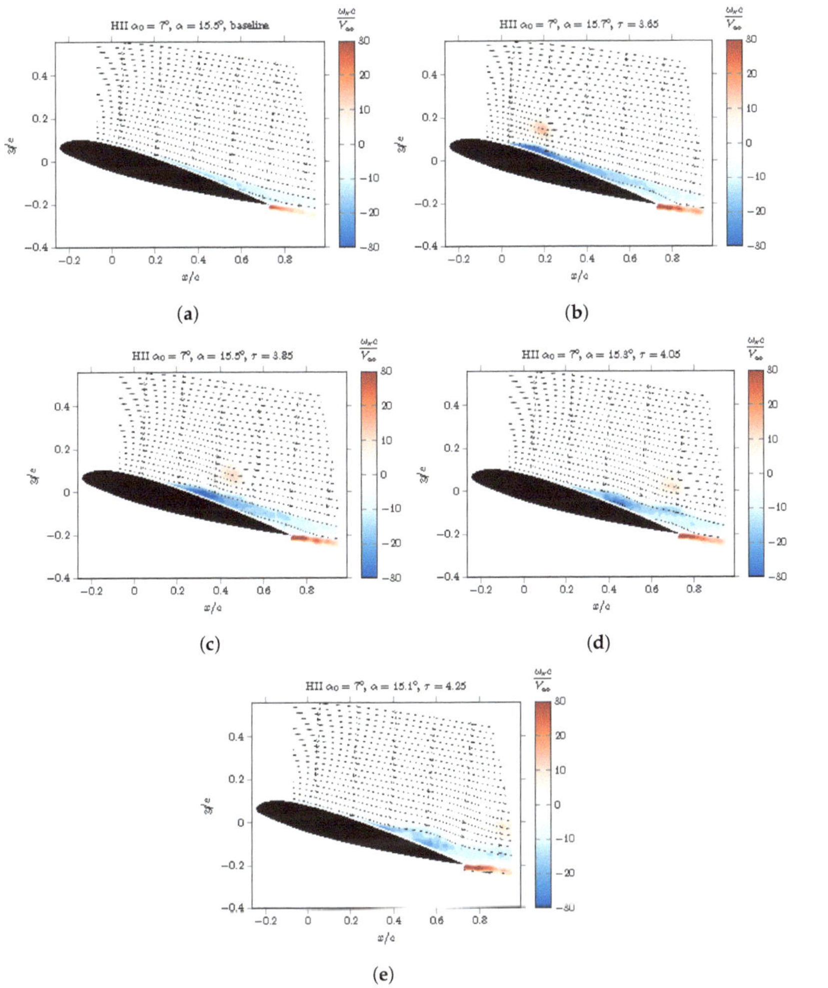

Figure 11. Vorticity contours from PIV measurements of the flow field during the HII vortex interaction with the oscillating airfoil at $\alpha_0 = 7°$ and comparison with baseline. The actual airfoil incidence α is reported for each time τ: (**a**) baseline, (**b**) $\tau = 3.65$, (**c**) $\tau = 3.85$, (**d**) $\tau = 4.05$, (**e**) $\tau = 4.25$.

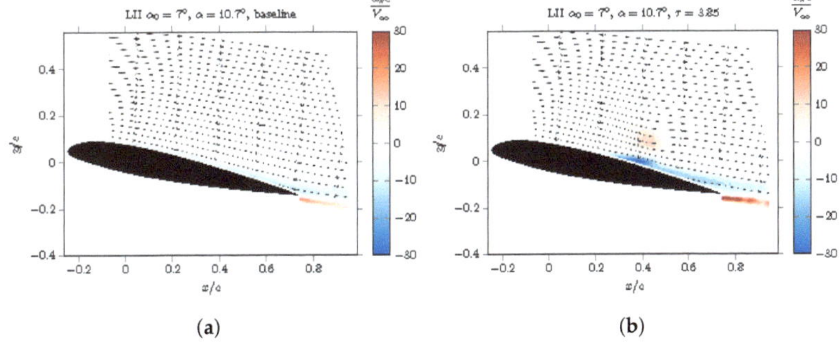

Figure 12. *Cont.*

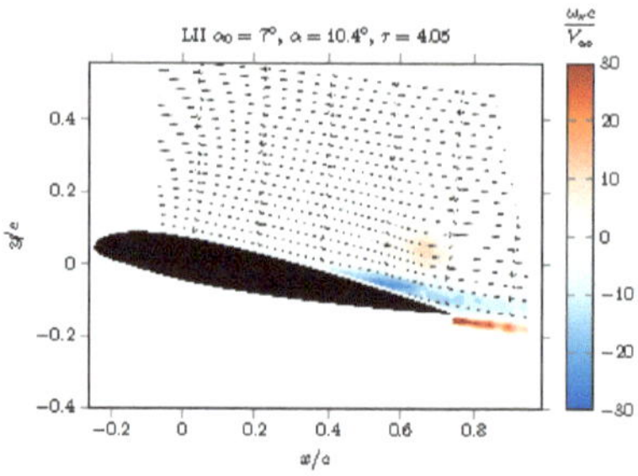

(c)

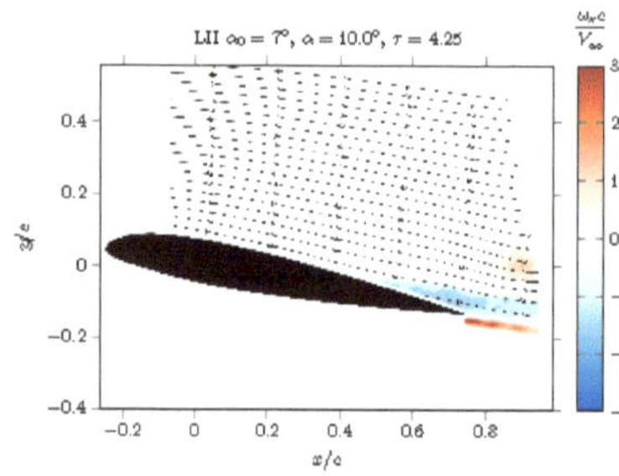

(d)

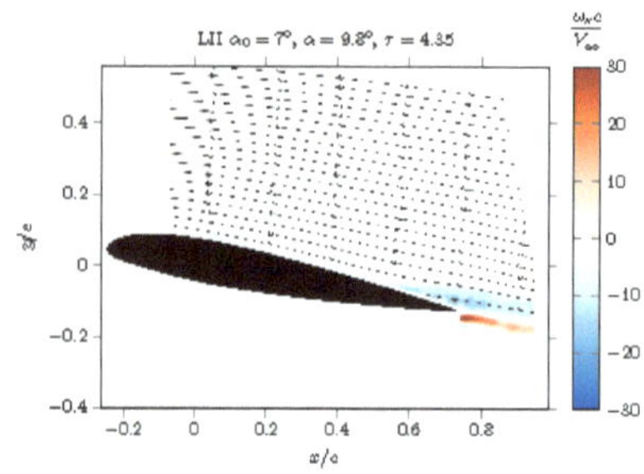

(e)

Figure 12. Vorticity contours from PIV measurements of the flow field during the LII vortex interaction with the oscillating airfoil at $\alpha_0 = 7°$ and comparison with baseline. The actual airfoil incidence α is reported for each time τ: (**a**) baseline, (**b**) $\tau = 3.85$, (**c**) $\tau = 4.05$, (**d**) $\tau = 4.25$, (**e**) $\tau = 4.35$.

By looking at the measurements for the HII, a sharp peak in C_l can be seen, as expected, followed by an interval where the lift is lower than the baseline values, effectively widening the hysteresis cycle in correspondence of the downstroke motion. This effect is similar to, although not as severe as, dynamic stall behaviour, and it is more evident the higher the incidence. Table 2 reports the relative variations in C_l and C_d along with information on the time of occurrence of the lift peak and the settling time. By comparing with the data in Table 1 the same trends are generally found: τ_{peak} is very similar for all three HII cases, and the maximum C_l increase in correspondence of the peak is inversely proportional to the airfoil incidence; the minimum ΔC_l, however, shows a much greater variation at the highest incidences with respect to the static interaction cases, dropping to almost $\Delta C_l = -26\%$ for $\alpha_0 = 7°$, which is also the case exhibiting a larger hysteresis cycle. Concerning the drag values, the decrease in C_d is similar to the static $\alpha = 14°$ and $15°$ cases, while there is a much more significant increase in drag, reaching $+75.5\%$ for $\alpha_0 = 7°$; this behaviour can be related to the insurgence of flow separation, as the highest values are recorded in correspondence of the vortex passing over the region close to the trailing edge of the airfoil. In terms of the settling time $\Delta\tau$, very little difference is found among the three HII cases; comparing to Table 1, moreover, it can be seen that these cases present a $\Delta\tau$ lower than in the static interactions.

Table 2 also reports the data from the measurements in the LII cases, while the corresponding trends in C_l are shown in Figure 14. Concerning the airfoil lift, the variations for $\alpha_0 = 5°$ and $6°$ are similar, and slightly larger in magnitude than the previously described cases, which is expected since the interaction is occurring at lower incidences. The case $\alpha_0 = 7°$ shows a much more detrimental effect, which, again, can be tied to the trailing edge flow separation as shown by the PIV measurements. By looking at the C_d variations, it can be seen that a very large drop is recorded for all LII cases, as low as $\Delta C_d = -82.5\%$; while such values are comparable to the ΔC_d for the $\alpha = 10°$ static airfoil case, the following peak is considerably higher than the corresponding case. The settling times are significantly larger than in the HII cases, also affecting part of the upstroke motion of the blade model; this difference could hint to a possible restoring effect by the down-stroke motion dynamics itself. For both kinds of interaction, however, the trends confirm that the effects of BVI can be considered to have vanished in less than a period of oscillation, justifying the choice of the motion history for the vortex generator.

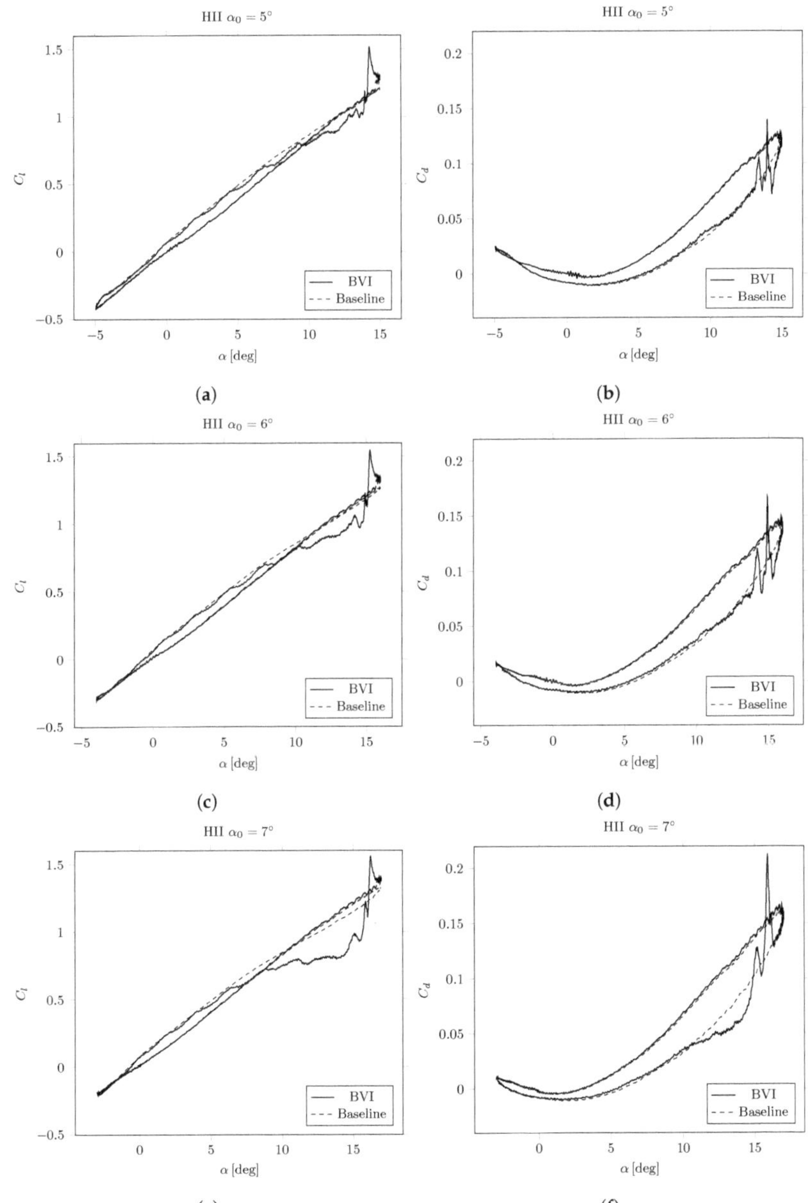

Figure 13. Lift coefficient C_l (left) and drag coefficient C_d (right) as function of blade model incidence α, comparison between HII and baseline for different base incidence α_0: (**a**,**b**) $\alpha_0 = 5°$, (**c**,**d**) $\alpha_0 = 6°$, (**e**,**f**) $\alpha_0 = 7°$.

By extrapolating the trends described above in the case of the interactions with the static airfoil, it can be suggested that the BVI behaviour is not fundamentally different from that seen in the case of the oscillating blade model when compared at a similar incidence at the moment of interaction. The most noticeable difference, a general increase in drag, can be associated with the insurgence of flow separation at lower incidences than would be the case if the airfoil were static, although the pitching motion seems to affect only weakly the overall effects.

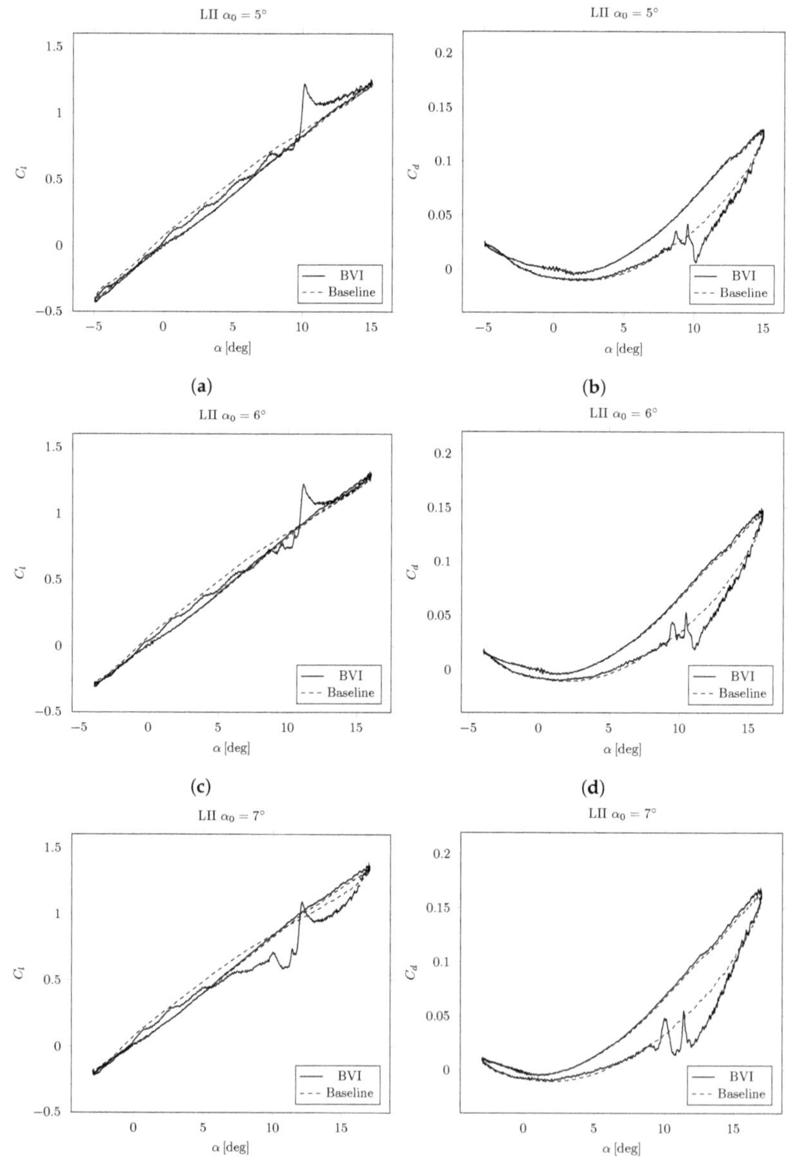

Figure 14. Lift coefficient C_l (left) and drag coefficient C_d (right) as function of blade model incidence α, comparison between LII and baseline for different base incidence α_0: (**a**,**b**) $\alpha_0 = 5°$, (**c**,**d**) $\alpha_0 = 6°$, (**e**,**f**) $\alpha_0 = 7°$.

Table 2. Vortex interaction with oscillating airfoil with base incidence α_0: minimum and maximum lift ΔC_l and drag ΔC_d variation, time of lift peak occurrence τ_{peak} and settling time $\Delta\tau$.

Case	α_0 [*deg*]	ΔC_l [%]		ΔC_d [%]		τ_{peak}	$\Delta\tau$
	5	−11.5	+30.3	−29.0	+48.8	3.08	7.45
HII	6	−15.5	+28.9	−20.9	+53.6	3.08	7.03
	7	−25.9	+26.0	−28.6	+75.5	3.09	7.07
	5	−22.5	+39.5	−82.5	+38.9	3.08	10.50
LII	6	−14.8	+31.9	−59.0	+37.1	3.08	12.51
	7	−34.2	+13.1	−66.3	+47.2	3.09	13.58

4. Conclusions

The aim of this study was to afford an experimental insight into the mechanism of parallel blade–vortex interaction and its effects, in particular in terms of the blade's overall aerodynamic performance, both in the case of interaction with a static airfoil and in the case of interaction with a sinusoidally pitched airfoil. Unsteady pressure measurements on the blade model were performed, as well as PIV measurements of the flow field in the suction side region of the airfoil. The design of the experimental test rig, with an upstream pitching airfoil model acting as vortex generator, allowed one to obtain a well-developed vortex with the desirable qualities of a reduced size with respect to the blade model and ease and repeatability of the process of generation, which was a key factor for the intensive test campaign.

The results of the measurements concerning the static interaction confirmed the expectations from the literature on the subject, with the blade being subjected to impulsive variation in the aerodynamic loads, which can be related to the upwash and downwash effect of the flow induced by the vortex downstream and upstream of its position, respectively. This unsteady effect is very clearly visible, especially in the comparisons between the baseline values and time history of the lift coefficient, with peaks up to +38%, and of the drag coefficient, which reduces by −59.2% in the most affected case. These variations show a dependence on the blade airfoil incidence, with lower ΔC_l and higher ΔC_d values measured in correspondence with the highest incidences, which can be explained by BVI-induced flow separation and also by considering that the perturbing effect of the vortex would be proportionally lower the higher the incidence. The corresponding PIV data showed a perturbation of the boundary layer over the suction side of the blade airfoil, which thickens as vorticity accumulates in a bubble in correspondence of the passage of the vortex and moves along the suction side following it at a slower speed. For the cases $\alpha = 14^\circ$ and 15°, the vortex passage was seen to trigger trailing-edge flow separation over the aft portion of the airfoil, up to about 0.5 c. Similar flow behaviour is in agreement with the description found in the literature on vortex–airfoil interaction, and vortex–surface interaction in general. Moreover, a significant delay in the return to baseline, undisturbed conditions was recorded, up to $\Delta\tau = 9.33$, which corresponds to several chord lengths after the vortex has passed the blade trailing edge; this could indicate that the unsteady perturbation of the boundary layer dissipates slowly.

The analysis of the interactional behaviour for the case of the oscillating blade model, which represents the main novelty of the present work, showed similarities with the static interaction case. In particular, the variation of the aerodynamic loads with respect to the airfoil incidence presents hysteresis cycles, with a reduction of the lift which affects most of the downstroke phase. This effect is particularly evident for the highest incidence of pitching motion and, for the same incidence, it is greater when the interaction happens as the airfoil downstroke motion has just begun. An increase in the settling time was recorded for the LII cases, with respect to the interaction in the static case. The flow field analysis of the PIV data showed that the BVI behaviour resembles qualitatively the one for the static case, but a region of separated flow can also be seen at low airfoil incidences. In the tested conditions, there was no evidence of dynamic stall behaviour induced by the vortex interaction.

In conclusion, the wind tunnel test campaign confirmed that parallel BVI is an important source of transient aerodynamic loads, which can lead to strong vibrations and impact on the generated noise. From the tested conditions, this kind of interaction does not appear to be able to produce catastrophic flow separation and stall, although, in the case of the oscillating airfoil, a significant hysteresis behaviour with a reduction in the lift coefficient was measured. Ultimately, this behaviour could be explained by the inherent transiency of the parallel interaction, as the potentially detrimental effect of the vortex induced velocity is more limited in time during the interaction, with respect to other conditions such as perpendicular BVI. Nonetheless, the observed effects of the interaction on the airfoil boundary layer and the formation of secondary vortical structures highlight the importance of viscous effects in parallel BVI, which should be taken into account when studying and modelling this kind of interaction.

It is to be noticed that, in the present study, neither the effect of surface roughness nor the influence of vortex strength as a parameter were considered. With respect to the former, no significant variation is expected in the mechanism of the vortex interaction itself, provided that any difference in the stalling behaviour of the airfoil, depending on surface roughness, is taken into account. As to the latter, the strength of the generated vortex was less than ideal for the comparison with BVI in a helicopter rotor environment. It cannot be excluded, therefore, that a stronger vortex could trigger the stall of the blade at low incidences, or induce more severely stalled conditions. The limitation on the strength of the produced vortex is imposed by the current experimental setup, as it depends on the execution time of the impulsive pitching motion of the vortex generator: to overcome it, therefore, the test rig should be updated, e.g., with a more powerful motor, a different vortex generator model, or an altogether different actuation method. Alternatively, a computational approach could be followed, for example, employing high-fidelity LES, which would allow to arbitrarily modify the vortex characteristics and would afford a detailed description of the boundary layer behaviour during the interaction.

Author Contributions: Conceptualisation, A.C., A.Z. and G.G.; methodology, A.C., A.Z. and G.G.; software, A.C.; validation, A.C.; formal analysis, A.C.; investigation, A.C.; resources, A.Z. and G.G.; data curation, A.C.; writing—original draft preparation, A.C.; writing—review and editing, A.Z. and G.G.; visualisation, A.C.; supervision, A.Z. and G.G.; project administration, G.G.; funding acquisition, A.Z. and G.G. All authors have read and agreed to the published version of the manuscript.

Funding: This research received no external funding.

Institutional Review Board Statement: Not applicable.

Data Availability Statement: Dataset available on request from the authors.

Acknowledgments: The authors wish to acknowledge the steadfast support kindly provided by the technical team of the Aerodynamic Laboratory of the Department of Aerospace Science and Technology.

Conflicts of Interest: The authors declare no conflicts of interest.

Abbreviations

The following abbreviations are used in this manuscript:

BVI	Blade–Vortex Interaction
PIV	Particle Image Velocimetry
MAV	Micro Aerial Vehicle
eVTOL	electric Vertical Take-Off and Landing
c	blade model chord
c_g	vortex generator chord
V_∞	wind-tunnel free-stream speed
ρ	air density
ω_z	z-component of the vorticity vector
L	airfoil lift per unit span

D	airfoil drag per unit span
α	blade model incidence
α_g	vortex generator incidence
C_l	airfoil lift coefficient
C_d	airfoil drag coefficient
f	pitch oscillation frequency
k	pitch oscillation reduced frequency
α_0	mean pitch for oscillating airfoil
HII	high-incidence interaction
LII	low-incidence interaction
τ	non-dimensional time
τ_{peak}	time of lift peak
$\Delta\tau$	settling time
$\bar{\cdot}$	baseline value
$\Delta\cdot$	relative variation with respect to baseline value
$\tau_{2\%}$	threshold time value

References

1. Leishman, G.J. *Principles of Helicopter Aerodynamics with CD Extra*; Cambridge University Press: Cambridge, UK, 2006.
2. Widnall, S. Helicopter Noise due to Blade-Vortex Interaction. *J. Acoust. Soc. Am.* **1971**, *50*, 354–365. [CrossRef]
3. Booth, E.R., Jr. Experimental observations of two-dimensional blade-vortex interaction. *AIAA J.* **1990**, *28*, 1353–1359. [CrossRef]
4. Caradonna, F.; Lautenschlager, J.; Silva, M. An experimental study of rotor-vortex interactions. In Proceedings of the 26th Aerospace Sciences Meeting, Reno, NV, USA, 11–14 January 1988; p. 45.
5. Rockwell, D. Vortex-body interactions. *Annu. Rev. Fluid Mech.* **1998**, *30*, 199–229. [CrossRef]
6. Gibertini, G.; Mencarelli, A.; Zanotti, A. Oscillating aerofoil and perpendicular vortex interaction. *Proc. Inst. Mech. Eng. Part G J. Aerosp. Eng.* **2014**, *228*, 846–858. [CrossRef]
7. Droandi, G.; Gibertini, G.; Zanotti, A. Perpendicular blade–vortex-interaction over an oscillating airfoil in light dynamic stall. *J. Fluids Struct.* **2016**, *65*, 472–494. [CrossRef]
8. Chaderjian, N.M. Navier-Stokes simulation of UH-60A rotor/wake interaction using adaptive mesh refinement. In Proceedings of the AHS International Annual Forum & Technology Display, Fort Worth, TX, USA, 9–11 May 2017; number ARC-E-DAA-TN40776.
9. Richez, F. Analysis of dynamic stall mechanisms in helicopter rotor environment. *J. Am. Helicopter Soc.* **2018**, *63*, 1–11. [CrossRef]
10. Wilder, M.; Telionis, D. Parallel blade–vortex interaction. *J. Fluids Struct.* **1998**, *12*, 801–838. [CrossRef]
11. Rival, D.; Manejev, R.; Tropea, C. Measurement of parallel blade–vortex interaction at low Reynolds numbers. *Exp. Fluids* **2010**, *49*, 89–99. [CrossRef]
12. Peng, D.; Gregory, J.W. Vortex dynamics during blade-vortex interactions. *Phys. Fluids* **2015**, *27*. [CrossRef]
13. Weingaertner, A.; Tewes, P.; Little, J.C. Parallel vortex body interaction enabled by active flow control. *Exp. Fluids* **2020**, *61*, 137. [CrossRef]
14. Carlson, B.; Ullah, A.H.; Estevadeordal, J. Experimental Investigation of Vortex-Tube Streamwise-Vorticity Characteristics and Interaction Effects with a Finite-Aspect-Ratio Wing. *Fluids* **2020**, *5*, 122. [CrossRef]
15. Barnes, C.J.; Visbal, M.R. Counterclockwise vortical-gust/airfoil interactions at a transitional Reynolds number. *AIAA J.* **2018**, *56*, 2540–2552. [CrossRef]
16. Martínez-Muriel, C.; Flores, O. Analysis of vortical gust impact on airfoils at low Reynolds number. *J. Fluids Struct.* **2020**, *99*, 103138. [CrossRef]
17. Alaminos-Quesada, J.; Fernandez-Feria, R. Effect of the angle of attack on the transient lift during the interaction of a vortex with a flat plate. Potential theory and experimental results. *J. Fluids Struct.* **2017**, *74*, 130–141. [CrossRef]
18. Qian, Y.; Wang, Z.; Gursul, I. Interaction of quasi-two-dimensional vortical gusts with airfoils, unswept and swept wings. *Exp. Fluids* **2022**, *63*, 124. [CrossRef]
19. Seath, D.; Kim, J.M.; Wilson, D. Investigation of the parallel blade-vortex interaction at low speed. *J. Aircr.* **1989**, *26*, 328–333. [CrossRef]
20. Straus, J.; Renzoni, P.; Mayle, R. Airfoil pressure measurements during a blade vortex interaction and a comparison with theory. *AIAA J.* **1990**, *28*, 222–228. [CrossRef]
21. Świrydczuk, J. A visualization study of the interaction of a free vortex with the wake behind an airfoil. *Exp. Fluids* **1990**, *9*, 181–190. [CrossRef]
22. Zanotti, A. Retreating Blade Dynamic Stall. Ph.D. Thesis, Politecnico di Milano, Milano, MI, Italy, 2012.
23. Colli, A.; Gibertini, G.; Krishna, P.U.; Marchesi, V.; Zanotti, A. Parallel blade-vortex interaction on pitching airfoil. In Proceedings of the 47th European Rotorcraft Forum (ERF 2021), Glasgow, UK, 7–9 September 2021; pp. 1–13.
24. PIVTEC, G. PIVview 2C Version 3.9, User Manual, 2021. Available online: https://www.pivtec.com/ (accessed on 1 March 2024).
25. Raffel, M.; Willert, C.; Wereley, S.; Kompenhans, J. *Particle Image Velocimetry-A Practical Guide*; Springer: New York, NY, USA, 2007.
26. Barlow, J.B.; Rae, W.H.; Pope, A. *Low-Speed Wind Tunnel Testing*; John Wiley & Sons: New York, NY, USA, 1999.

27. Goerttler, A.; Braukmann, J.N.; Schwermer, T.; Gardner, A.D.; Raffel, M. Tip-vortex investigation on a rotating and pitching rotor blade. *J. Aircr.* **2018**, *55*, 1792–1804. [CrossRef]
28. Bandyopadhyay, P.; Balasubramanian, R. Vortex Reynolds number in turbulent boundary layers. *Theor. Comput. Fluid Dyn.* **1995**, *7*, 101–117. [CrossRef]
29. Ramasamy, M.; Leishman, J.G. A reynolds number-based blade tip vortex model. *J. Am. Helicopter Soc.* **2007**, *52*, 214–223. [CrossRef]
30. Vatistas, G.H.; Kozel, V.; Mih, W. A simpler model for concentrated vortices. *Exp. Fluids* **1991**, *11*, 73–76. [CrossRef]
31. Jeong, J.; Hussain, F. On the identification of a vortex. *J. Fluid Mech.* **1995**, *285*, 69–94. [CrossRef]
32. Wenzinger, C. *Wind-Tunnel Investigation of Ordinary and Split Flaps on Airfoils of Different Profile*; Report (United States. National Advisory Committee for Aeronautics); NACA: Washington, DC, USA, 1935.
33. Loftin Jr, L.K.; Poteat, M.I. *Aerodynamic Characteristics of Several NACA Airfoil Sections at Seven Reynolds Numbers from 0.7 × 10 (exp 6) to 9.0 × 10 (exp 6)*; Technical Report; NACA: Washington, DC, USA, 1948.
34. Althaus, D.; Wortmann, F.X. *Stuttgarter Profilkatalog: Messergebnisse aus dem Laminarwindkanal des Instituts für Aerodynamik und Gasdynamik der Universität Stuttgart*; Vieweg: Decatur, IL, USA, 1981; Volume 1.
35. Wenzinger, C. *Pressure Distribution Over an N.A.C.A 23012 Airfoil with an N.A.C.A. 23012 External Airfoil Flap*; Report (United States. National Advisory Committee for Aeronautics); NACA: Washington, DC, USA, 1938.
36. Harvey, J.; Perry, F.J. Flowfield produced by trailing vortices in the vicinity of the ground. *AIAA J.* **1971**, *9*, 1659–1660. [CrossRef]
37. Walker, J. The boundary layer due to rectilinear vortex. *Proc. R. Soc. Lond. A. Math. Phys. Sci.* **1978**, *359*, 167–188.
38. Doligalski, T.; Smith, C.; Walker, J. Vortex interactions with walls. *Annu. Rev. Fluid Mech.* **1994**, *26*, 573–616. [CrossRef]

fluids MDPI

Article

Analyzing the Influence of Dean Number on an Accelerated Toroidal: Insights from Particle Imaging Velocimetry Gyroscope (PIVG)

Ramy Elaswad [1,*], Naser El-Sheimy [2] and Abdulmajeed Mohamad [1]

[1] Department of Mechanical and Manufacturing Engineering, Schulich School of Engineering, University of Calgary, 2500 University Dr. NW, Calgary, AB T2N 1N4, Canada
[2] Department of Geomatics Engineering, Schulich School of Engineering, University of Calgary, 2500 University Dr. NW, Calgary, AB T2N 1N4, Canada
* Correspondence: ramy.elaswad1@ucalgary.ca

Abstract: Computational Fluid Dynamics (CFD) simulations were utilized in this study to comprehensively explore the fluid dynamics within an accelerated toroidal vessel, specifically those central to Particle Imaging Velocimetry Gyroscope (PIVG) technology. To ensure the robustness of our simulations, we systematically conducted grid convergence studies and quantified uncertainties, affirming the stability, accuracy, and reliability of our computational grid and results. Comprehensive validation against experimental data further confirmed our simulations' fidelity, emphasizing the model's fidelity. As the PIVG is set up to address the primary flow through the toroidal pipe, we focused on the interaction between the primary and secondary flows to provide insights into the relevant dynamics of the fluid. In our investigation covering Dean numbers (D_e) from 10 to 70, we analyzed diverse aspects, including primary flow, secondary flow patterns, pressure distribution, and the interrelation between primary and secondary flows within toroidal structures, offering a comprehensive view across this range. Our research indicated stability and fully developed fluid dynamics within the toroidal pipe under accelerated angular velocity, particularly for low D_e. Furthermore, we identified an optimal Dean number of 11, which corresponded to ideal dimensions for the toroidal geometry with a curvature radius of 25 mm and a cross-sectional diameter of 5 mm. This study enhances our understanding of toroidal fluid dynamics and highlights the pivotal role of CFD in optimizing toroidal vessel design for advanced navigation technologies.

Keywords: computational fluid dynamics (CFD); inertial navigation sensor (INS); gyroscope; particle imaging velocimetry gyroscope (PIVG); fluid-based gyroscope; toroidal; Dean number (D_e)

Citation: Elaswad, R.; El-Sheimy, N.; Mohamad, A. Analyzing the Influence of Dean Number on an Accelerated Toroidal: Insights from Particle Imaging Velocimetry Gyroscope (PIVG). *Fluids* **2024**, *9*, 103. https://doi.org/10.3390/fluids9050103

Academic Editors: Kambiz Vafai, Mingming Ge, Xin-Lei Zhang and Guangjian Zhang

Received: 18 March 2024
Revised: 14 April 2024
Accepted: 22 April 2024
Published: 25 April 2024

1. Introduction

Inertial measurement units (IMUs) are integral components of navigation systems; they provide essential information about a moving platform's position, velocity, and attitude. IMUs rely on the accurate measurement of linear acceleration and angular rate. However, the traditional gyroscopes used in commercial IMUs suffer from bias instability, which affects their performance and introduces errors in attitude estimation. To overcome this limitation, novel gyroscope technologies are being explored to improve the accuracy and reliability of IMUs [1–3].

One such emerging technology is the Particle Imaging Velocimetry Gyroscope (PIVG). The PIVG is a fluid-based gyroscope that offers several advantages over conventional gyroscopes. It is nearly drift-free and exhibits minimal bias instability, even for navigation-grade IMUs. Additionally, the PIVG boasts a high signal-to-noise ratio (SNR), making it an attractive alternative for various navigation applications. Moreover, its low cost makes it a cost-effective solution in comparison to traditional gyroscopes. The PIVGs are constructed with three channels that are arranged in perpendicular axes, each with a circular, toroidal shape. A significant process employed in this configuration is Particle Imaging Velocimetry (PIV). This involves utilizing digital cameras to capture images of particles while they move through the water within toroidal channels. After capture, the images undergo processing; determining the displacement of each particle provides a measurement indicating fluid velocity. This information allows us to calculate the angular acceleration of the toroidal structure and provides valuable insights into the rotational dynamics of the PIVG. This three-axis design allows the gyroscope to comprehensively measure angular motion in multiple dimensions. However, for some applications, the design can be simplified to a single-axis system by using only one circular channel, making the construction more streamlined and less complex [1].

The study of fluid flow behavior in curved geometries is significant in various engineering applications, such as pipeline systems, ducting in internal combustion engines, heat exchangers, biological system flows, and microfluidic devices [4]. This behavior, which is complex in nature, exhibits secondary flows and vortices. Our literature review focuses on two phenomena: pressure-driven flow and rotating or spinning curved pipes. By leveraging insights from these studies, we can enhance our understanding of the intricate fluid behavior within the toroidal chamber of the PIVG. These phenomena, despite their different driving mechanisms, share common dynamics and characteristics, underscoring the importance of their study in the context of fluid flow in curved structures.

In pressure-driven research, Dean proposed an analytical solution for Newtonian flow in a curved conduit with a circular cross-section in 1928 [5,6]. He considered the secondary flow as a minor perturbation upon the primary flow, like Poiseuille flows in a straight pipe. Dean's study established that the axial velocity within a curved pipe and the stream function of the secondary radial flow depend on the curvature parameter (k), which is summarized in the equation:

$$\mathrm{k} = 2\mathrm{R_e}^2 \frac{a}{\mathrm{R}} \tag{1}$$

where "a" is the channel's radius, "R" is the torus's radius, and "$\mathrm{R_e}$" is the Reynolds number. Subsequently, White introduced the term "Dean's number ($\mathrm{D_e}$)":

$$\mathrm{D_e} = \mathrm{R_e}\sqrt{\frac{a}{\mathrm{R}}} \tag{2}$$

The Dean number ($\mathrm{D_e}$) and the curvature number (k) provide valuable insights into the flow behavior in curved pipes. The Dean number measures the relative influence of centrifugal forces compared to viscous forces, indicating the fluid flow's behavior. Simultaneously, the curvature number quantifies the importance of the pipe's curvature in relation to the fluid's viscosity. A crucial finding emerged in the insightful studies conducted by Ligrani [7] under laminar flow conditions. It was revealed that a critical limit exists for the Dean number, specifically in the $\mathrm{D_e} \leq 40$~60 range, under which the influence of secondary flows on the primary flow remains minimal. This vital understanding emphasizes the critical balance necessary to maintain the stability and integrity of the fluid dynamics within the given parameters.

On the other hand, in 1963 Greenspan and Howard [8], who were researching spinning curved pipes, employed linear techniques to investigate spin-up phenomena. They recognized three unique phases: the formation of Ekman boundary layers, secondary flow within the body of the fluid, and viscous decay of residual motion. Additionally, they applied linear theory to estimate the characteristic spin-up time for two cylindrical geometries. In the first scenario, the cylinder was bounded by a flat disk, while in the second, it was enclosed by a conical section. Notably, the disk and the conical sections rotated along the cylinder walls. Through experimentation, Greenspan and Howard validated their theoretical predictions in each case. Greenspan (1968) [9] provided specifics of the methods involved, and a significant portion of the early work in this field is described.

Moreover, Cowley et al. [10] studied the fluid inside a curved pipe rotated around its curvature axis. In the limit of a high $\mathrm{D_e}$, they demonstrated that boundary-layer collision and unsteady separation can occur impulsively and can begin to flow through a stationary curved conduit. A more thorough explanation of this phenomenon is given by Lam [11], who also includes estimates of the flow caused by the impulsive rotation of a curved pipe around its curvature axis.

The work that aligns most closely with our research, albeit with different objectives, is the study conducted by Madden and Mullin in 1994 [12]. They explored how fluid flow behaved within a toroidal sealed pipe under the impact of inertial angular rotation. Their study aimed to examine fluid velocity affected by inertia and the effects of ocean flow on the earth's surface. They performed experiments on a toroidal pipe filled with water. However, Laser Doppler Velocimetry (LDV) was used to track the fluid movement inside the toroidal pipe. Therefore, tracer particles were added to the stream. The tracer particles were observed to calculate the fluid flow velocity field in a Eulerian frame. Additionally, the study made sure that the toroidal pipe was not subjected to any external forces aside from the inertial force that was created when the pipe was spun about its vertical axis from rest up to a constant angular velocity with a linear acceleration ramp, ensuring the pipe reached its top speed in 0.8 s. The experimental setup measured the absolute velocity field in relation to a stationary LDV setup. Moreover, the study conducted a numerical simulation for various fluid flow channel dimensions to produce a definite conclusion regarding the fluid behavior under the effect of an angular rotation in the form of a spin-up from rest up to a specific angular velocity. However, the study's computational and experimental results showed that the fluid flow inside the toroidal pipe would have two main components to its velocity. A primary flow constitutes the majority of the fluid flow and moves in the opposite direction of the pipe flow. A second fluid flow also exists, and it moves in the radial direction of the toroidal pipe cross-section.

In this study, our aim is to conduct a thorough investigation of the fluid dynamics within accelerated toroidal geometry and its implications for PIVG systems. Two key criteria were evaluated with Computational Fluid Dynamics (CFD), namely domain validity and model fidelity. The range of Dean numbers (D_e) for reliable predictions of fluid velocities within the profile, known as domain validity, and model fidelity, which evaluates the mathematical model's accuracy in estimating these numbers needs to be considered when evaluating PIVG performance. Therefore, precise measurements of primary flow dynamics by integrated cameras are crucial for accurate angular acceleration measurement, which could be impeded by secondary flow phenomena. Hence, we examined primary and secondary flows across different toroidal channel dimensions and D_e ranges, and we analyzed pressure distributions during this study. According to empirical evidence, the Reynolds number is a key parameter, and laminar flow should not exceed 200, according to a literature review. Similarly, the Dean number plays a part in characterizing secondary flow effects; thus, a suggested range of 40–60 has been identified to reduce these effects while maintaining mostly one-dimensional profiles. Altogether, these criteria furnish a comprehensive framework for optimizing fluid dynamics within the PIVG, thereby ensuring its accurate operation and performance [1,2,13].

The structure of this paper is organized as follows. In Section 2, we present an overview of the governing equations that describe the fluid dynamics inside the toroidal geometry of the PIVG system. Section 3 presents our numerical setup for simulating fluid dynamics in the toroidal geometry, detailing the computational domain, boundary conditions, and solution methods for the governing equations. In Section 4, we ensure accuracy through grid convergence and validate our simulations against experiments, establishing credibility. We then present vital insights from our numerical analysis, following a systematic approach. Finally, in Section 5, we conclude the paper by summarizing the contributions of this study and highlighting the implications for the design and optimization of the toroidal geometry in fluid-based gyroscopes.

2. Governing Equations

In the study of fluid flow in toroidal pipes, a toroidal coordinate system (r, θ, ϕ), as shown in Figure 1, is commonly used. Here, 'r' is the coordinate along the radial direction, 'θ' represents the inclination of the cross-section in relation to a fixed axial plane, and ' ϕ ' is the coordinate along OZ (which corresponds to the axis of the toroidal). Additionally, the system is characterized by other key parameters: the cross-section radius of the pipe, denoted by '*a*', and the curvature radius of the toroidal, denoted by 'R'. Assuming the flow occurs in the direction of increasing ϕ due to angular acceleration around the z-axis, the velocity components can be defined as follows: 'u' in the r-direction, 'v' in the θ-direction, and 'w' in the ϕ-direction.

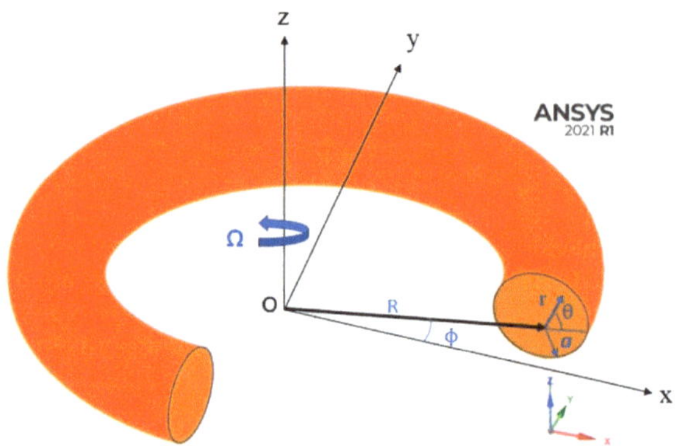

Figure 1. Toroidal coordinate system.

The equations of momentum and continuity [5,14] are:

$$\frac{\partial u}{\partial t}+u\frac{\partial u}{\partial r}+\frac{v}{r}\frac{\partial v}{\partial \theta}-\frac{v^2}{r}-\frac{w^2\sin\theta}{R+r\sin\theta}=-\frac{1}{\rho}\frac{\partial P}{\partial r}-\nu\left[\frac{1}{r}\frac{\partial}{\partial \theta}+\frac{\cos\theta}{R+r\sin\theta}\right]\left[\frac{\partial v}{\partial r}+\frac{v}{r}+\frac{1}{r}\frac{\partial u}{\partial \theta}\right] \tag{3}$$

$$\frac{\partial v}{\partial t}+u\frac{\partial v}{\partial r}+\frac{v}{r}\frac{\partial v}{\partial \theta}+\frac{uv}{r}-\frac{w^2\cos\theta}{R+r\sin\theta}=-\frac{1}{\rho}\frac{1}{r}\frac{\partial P}{\partial \theta}+\nu\left[\frac{\partial}{\partial r}+\frac{\sin\theta}{R+r\sin\theta}\right]\left[\frac{\partial v}{\partial r}+\frac{v}{r}-\frac{1}{r}\frac{\partial u}{\partial \theta}\right] \tag{4}$$

$$\frac{\partial w}{\partial t}+\frac{v}{r}\frac{\partial w}{\partial \theta}+\frac{uw\sin\theta}{R+r\sin\theta}+\frac{vw\cos\theta}{R+r\sin\theta}=-\frac{1}{R+r\sin\theta}\frac{1}{\rho}\frac{\partial P}{\partial \phi}+\nu\left\{\left[\frac{\partial}{\partial r}+\frac{1}{r}\right]\left[\frac{\partial w}{\partial r}+\frac{w\sin\theta}{R+r\sin\theta}\right]+\frac{1}{r}\frac{\partial}{\partial \theta}\left[\frac{1}{r}\frac{\partial w}{\partial \theta}+\frac{w\cos\theta}{R+r\sin\theta}\right]\right\}+f_\phi \tag{5}$$

$$\frac{\partial u}{\partial r}+\frac{u}{r}+\frac{u\sin\theta}{R+r\sin\theta}+\frac{1}{r}\frac{\partial v}{\partial \theta}+\frac{v\cos\theta}{R+r\sin\theta}=0, \tag{6}$$

where ν denotes the fluid's kinematic viscosity, ρ is the fluid's density, and f_ϕ is the body force, as shown in the equation. Equation (7) is valid when the radius of the toroidal pipe is much smaller than the radius of the toroidal ($a << R$) [15,16].

$$f_\phi = \rho R \alpha_z(t), \tag{7}$$

where α_z is the angular acceleration around the toroidal axis (OZ).

Equations (1)–(6), along with appropriate boundary conditions, describe an unsteady flow problem within a toroidal configuration. By assuming that the flow within the toroidal structure is fully developed, where the velocity components (u, v, and w) become independent of the angular coordinate ϕ, and the pressure gradient in the axial direction (ϕ) remains constant, we strategically simplify the complexities of our study [15,16].

3. Numerical Simulation Setup

3.1. Computational Domain and Mesh Generation

The computational domain was designed to accurately represent the geometry of the accelerated toroidal system. The toroidal pipe was carefully modeled to capture the dynamic flow, as shown in Figure 2. To discretize the domain, a structured mesh generation technique was employed. A structured Hex dominant mesh, capable of handling complex geometries, was created using the built-in software ICEM CFD 14.5 to discretize the computational domain. The mesh resolution was carefully chosen to ensure an accurate representation of the flow features and gradients within the system. Special attention was given to maintaining the mesh quality, ensuring proper element aspect ratios and skewness to prevent numerical instabilities and inaccuracies. The overall mesh size and density were determined through mesh independence studies to ensure that the results were not significantly affected by the mesh resolution.

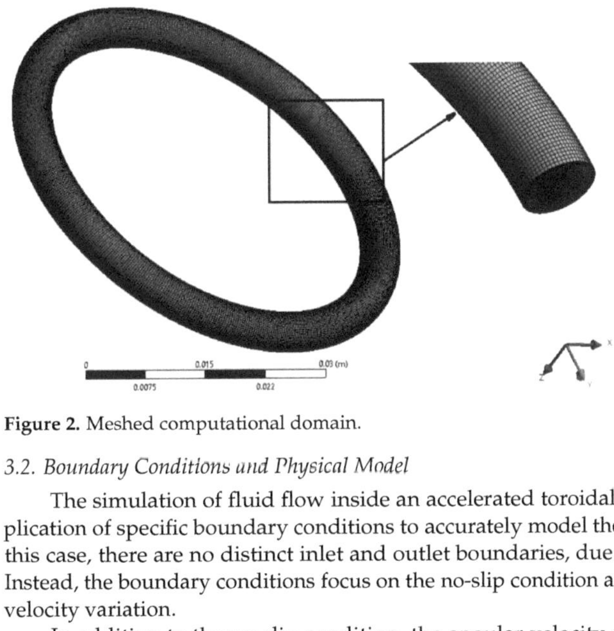

Figure 2. Meshed computational domain.

3.2. Boundary Conditions and Physical Model

The simulation of fluid flow inside an accelerated toroidal domain involves the application of specific boundary conditions to accurately model the behavior of the fluid. In this case, there are no distinct inlet and outlet boundaries, due to the toroidal geometry. Instead, the boundary conditions focus on the no-slip condition and the prescribed angular velocity variation.

In addition to the no-slip condition, the angular velocity of the toroidal domain is prescribed as a ramp function, as shown in Figure 3. The toroidal domain is at rest from 0 to t = 0.5 s, during which the angular velocity is 0 rad/s. Subsequently, the angular velocity gradually increases linearly until it reaches a constant value of 0.2793 rad/s (16 °/s) at t = 4.5 s. This ramp function represents the time variation of the angular velocity and is applied as a boundary condition at the toroidal surfaces. By incorporating the prescribed angular velocity, the simulation captures the effect of the acceleration on the fluid flow inside the toroidal domain.

In our simulation setup, we considered a transient flow regime with laminar flow characteristics, which mimicked the behavior of water as the fluid medium (density $\rho = 998.2$ kg/m^3), with a dynamic viscosity of $\mu = 0.001003$ kg/m.s. The walls were assumed to be stationary. To capture the rotational motion of the toroidal domain, we employed a rotating reference frame with angular accelerated motion. This setup allowed us to simulate the dynamic behavior of the toroidal vessel accurately. In terms of simulation initialization, a hybrid method was used, and the time step size was set to 0.1 s. The simulation was allowed a maximum of 100 iterations per time step, ensuring convergence within each step. Residuals for continuity and velocity were limited to 1×10^{-6} to guarantee accurate and stable results. The input angular velocity and corresponding angular acceleration profiles, which are crucial for the toroidal rotation, were carefully converted into an ASCII format in order to integrate them into the ANSYSFluent 2021R1 software. This approach to boundary conditions and the physical model accurately represented real-world fluid dynamics within the toroidal domain.

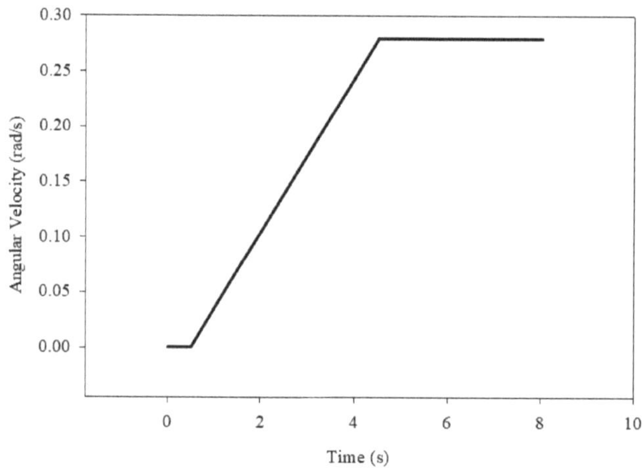

Figure 3. The input angular velocity for the toroidal.

4. Results and Discussion

4.1. Grid Independence Study

A grid convergence study was conducted to ensure that the numerical solution was independent of the mesh size and to estimate the numerical uncertainty associated with spatial discretization. The study utilized the Richardson extrapolation method, which is a widely accepted approach for assessing grid convergence [17]. In this work, three different meshes were generated by systematically refining the computational domain. The refinement ratio between the successive meshes was maintained at a value higher than 1.3, ensuring significant differences in grid resolution. The details of the three meshes are presented in Table 1, which showcases the varying levels of refinement achieved. This grid convergence study allows a reliable assessment of the numerical solution and provides valuable insights into the accuracy and reliability of the computed results. Additionally, the table provides visual representations of the three meshes employed in the grid convergence study, along with the corresponding total number of elements for each mesh. The slices of the meshes offer insights into the varying levels of refinement achieved.

To assess the uncertainty in the computed values of the tangential velocity, the relative tangential velocity (RTV) at a specific horizontal line on the cross-section is computed for each mesh, as depicted in Figure 4; to quantify the uncertainty resulting from spatial discretization, the grid convergence index (GCI) is calculated for the medium and fine meshes. The GCI is a measure used to estimate the uncertainty in the solution of a computational fluid dynamics (CFD) simulation due to spatial discretization. It is calculated by performing the simulation on two or more successively finer grids. As the grid is refined (the grid cells become smaller and the number of cells in the flow domain increases), the spatial discretization error should asymptotically approach zero [18]. The GCI values are summarized in Table 2. The table reveals that the medium mesh exhibits a higher uncertainty (GCI_{32}%) compared to the fine mesh (GCI_{21}%). The average uncertainty attributed to spatial discretization is 1.47% for GCI_{32}% and 0.27% for GCI_{21}%. These results indicate that as the number of elements increases, the solution becomes more independent of the grid refinement, indicating the consistency and reliability of the numerical scheme employed.

Table 1. Mesh refinement strategy.

	Coarse	**Medium**	**Fine**
Element Size	1.9309×10^{-4}	1.4586×10^{-4}	1.0319×10^{-4}
No. of Element (m)	427,680	992,160	2,803,600
Mesh Slice	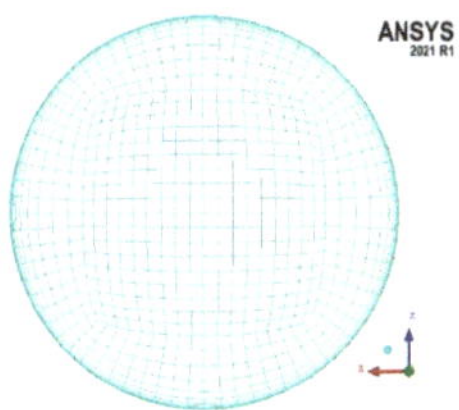	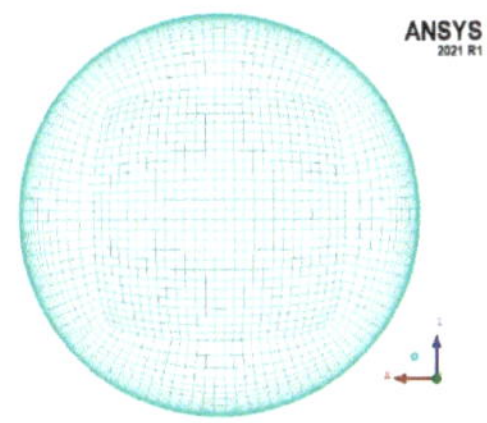	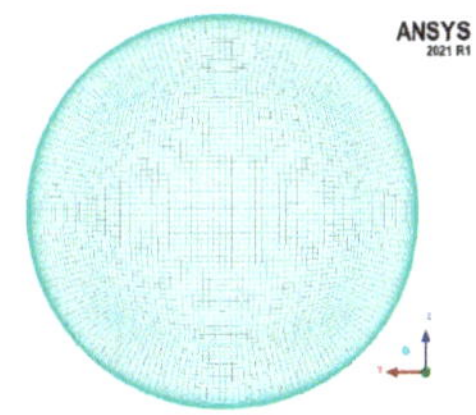

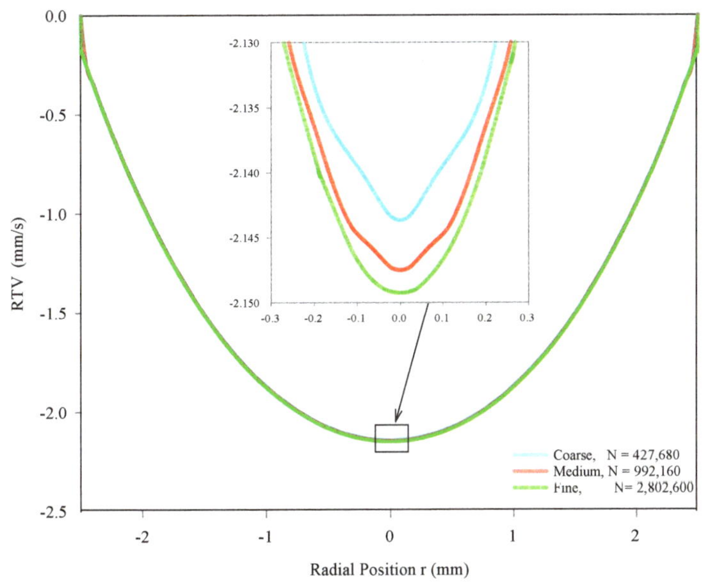

Figure 4. The relative tangential velocities function of different numbers of elements, at 2 s.

Table 2. Grid convergence index.

Radius (r)	V_3, Coarse	V_2, Medium	V_1, Fine	GCI_{21}%	GCI_{32}%
2.455	0.246183	0.250896	0.2519	0.731915	4.488632
2.005	0.93731	0.943939	0.9453	0.264389	1.678087
1.509	1.4937	1.49881	1.5009	0.255711	0.814675
1.0135	1.86093	1.86721	1.8681	0.087487	0.803668
0.518	2.07132	2.07658	2.0776	0.090156	0.605268
0.0225	2.14335	2.14728	2.1496	0.198192	0.437335

4.2. Validation

To validate our numerical simulation of the fluid-filled toroidal used in the PIVG sensor, we compare our results with the experimental work of Madden (1991) [12], who investigated the spin-up of a torus from rest. The experimental setup involved a torus made of Perspex with a radius of curvature of 125 ± 0.1 mm and a cross-sectional radius of 16 ± 0.1 mm, giving a radius ratio of $\delta = 0.128$. The torus was filled with distilled water. A powerful servo-controlled motor was used to rotate the torus, and its rotational speed was measured using an optical shaft encoder. Madden used a nonlinear ramp to increase the rotational speed from rest to a pre-selected rotation rate over 0.8 s. In this research, to compare the numerical results with the experimental data of Madden, a toroidal vessel was modeled with similar dimensions and operating conditions, including fluids with corresponding viscosities. The numerical model was solved using ANSYSFluent 2021R1 software, and the primary velocity time series at a specific point in the center of the pipe was extracted for validation purposes. These data were compared with previous findings by Madden to assess the accuracy and reliability of the numerical simulation.

The Reynolds number (R_e), which is a dimensionless quantity, plays a crucial role in characterizing fluid flow. It is defined as the ratio of inertial forces to viscous forces and provides insights into the flow regime. In the context of this study, the R_e is particularly relevant as it is influenced by the angular velocity of the toroidal vessel. At a specific angular velocity, the R_e can be calculated using the characteristic length scale and fluid properties, as shown in Equation (8) [19].

$$R_e = \frac{\Omega a R}{\nu}, \tag{8}$$

The numerical and experimental investigations were conducted at the same R_e of 300, indicating similar flow behavior in terms of inertia and viscosity. By maintaining a constant R_e, the focus was on comparing different modeling approaches. The agreement between the numerical and experimental outcomes at this R_e highlights the consistency and reliability of the numerical simulation. This agreement not only validates our model but also attests to its good domain fidelity performance, further enhancing the reliability of our study. The primary velocity, recorded at the pipe's center, was normalized to facilitate precise comparisons. This normalization method provides consistent scaling for meaningful analysis, as depicted in Figure 5. The dimensionless velocity profiles from the numerical simulation and experimental measurements exhibited close agreement, validating the accuracy of the numerical model in capturing the fluid flow characteristics in the accelerated toroidal.

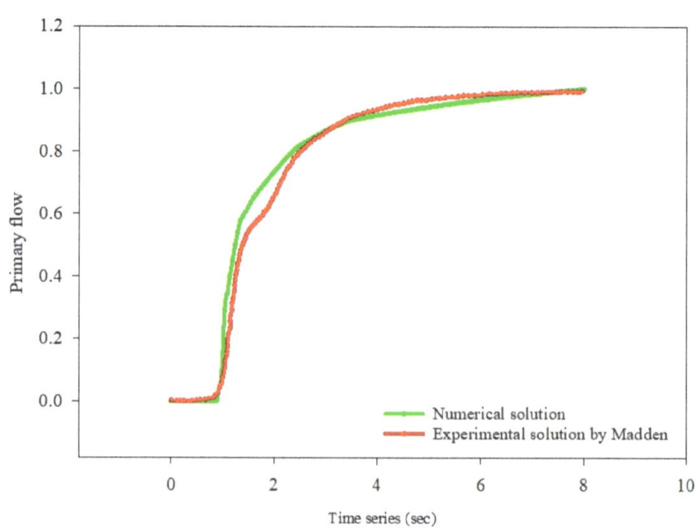

Figure 5. Primary velocity vs. time series.

4.3. Numerical Results

4.3.1. Dean Number (D_e) Study

The Dean number, denoted as D_e, serves as a fundamental parameter in fluid dynamics, particularly in the context of flow through curved conduits like toroidal pipes. In our examination, we systematically varied the D_e across a range from 10 to 70. This variation allowed us to analyze a complex interaction of factors that significantly influence fluid behavior within toroidal configurations. At the heart of our investigation lies the profound impact of the D_e on several critical aspects of fluid dynamics. First and foremost, it determines the RTV profile of the fluid within the toroid. This tangential velocity, a key parameter in the case of our PIVG sensor, plays a crucial role in determining the angular velocity [1,2]. Moreover, the D_e shapes the toroid's radial and axial flow profiles. Understanding how fluid moves in these directions is essential for optimizing toroidal designs in our PIVG application and for other contexts where precise fluid mixing, or efficient heat transfer is a priority. One of the most exciting outcomes of our study involved the development of secondary flows. By comprehensively exploring the Dean number's role in shaping secondary flows, we gain insights into how to utilize or reduce its effects in practical settings. In the context of PIVG, measuring angular velocity hinges on obtaining accurate relative tangential velocity data inside the toroidal while minimizing the impact of secondary flows.

Before delving into the analysis of Figures 6–8, let us first define some of the key parameters that these graphs represent: D is the curvature diameter of the toroid, d is the cross-section diameter of the toroid, and δ is the ratio between the cross-section and curvature diameters. Figures 6–8 visually represent the RTV, radial, and axial contour within the toroidal pipe at different time steps of 2, 4, and 8 s. These contour maps illustrate the distribution of tangential velocity, radial, and radial velocity across the cross-section of the toroid. Red lines indicate areas with lower velocities, while blue lines signify regions with higher velocity due to the negative sign associated with velocity calculation. These maps showcase how the D_e, a key parameter in our study, influences the rotational flow patterns within the toroid.

At a lower Dean number (D_e = 11), the maximum relative tangential velocity (RTV) is observed at the center of the cross-section, indicating a uniform flow dominated by viscous forces at 2, 4, and 8 s (as shown in Figures 6a, 7a and 8a). However, as the Dean number increases to 70, the RTV exhibits distinct behaviors over time intervals. At 2 s, and Dean number 70, the RTV peaks, forming an elliptical shape (as shown in Figure 6a). Subsequently, at 4 s, the maximum velocity shifts towards the inner curvature of the toroid, indicating a transition in flow dynamics. Finally, at 8 s, the maximum velocity shifts towards the outer diameter of the toroidal curvature. This shift is attributed to the heightened influence of centrifugal forces at higher Dean numbers, which intensify the secondary flow and decelerate the RTV.

The observed changes in RTV behavior at different time intervals and D_e underscore the complexities of fluid dynamics within toroidal structures. Specifically, the shift in maximum velocity positions presents challenges for PIVG measurements, particularly at higher Dean numbers like 70. As the PIVG relies on capturing primary flow dynamics, alterations in RTV due to secondary flow effects can impact measurement accuracy. At Dean 70, where the RTV position shifts, it becomes challenging to precisely capture these changes using the camera, highlighting the importance of understanding and controlling secondary flow phenomena for accurate PIVG operation.

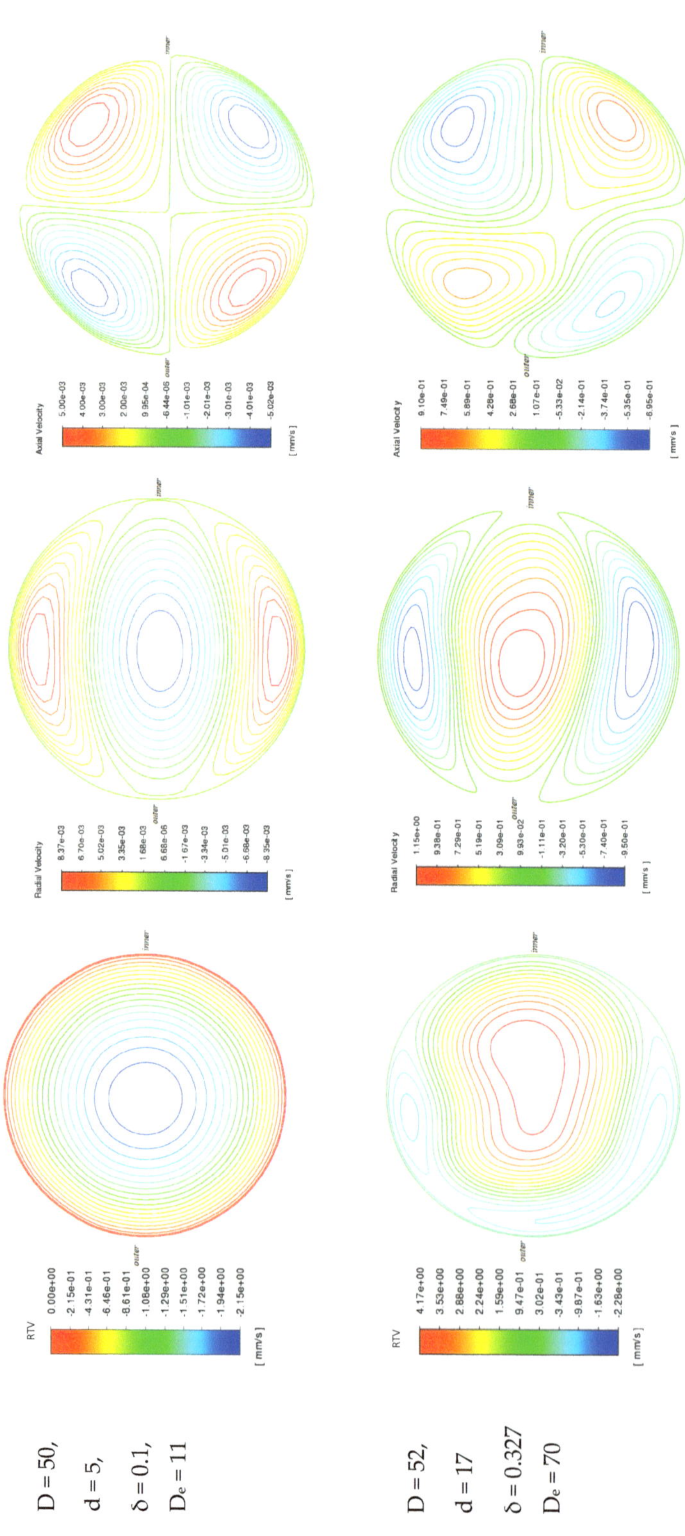

Figure 6. **(a)** RTV contour, **(b)** radial contour, **(c)** axial contour at 2 s.

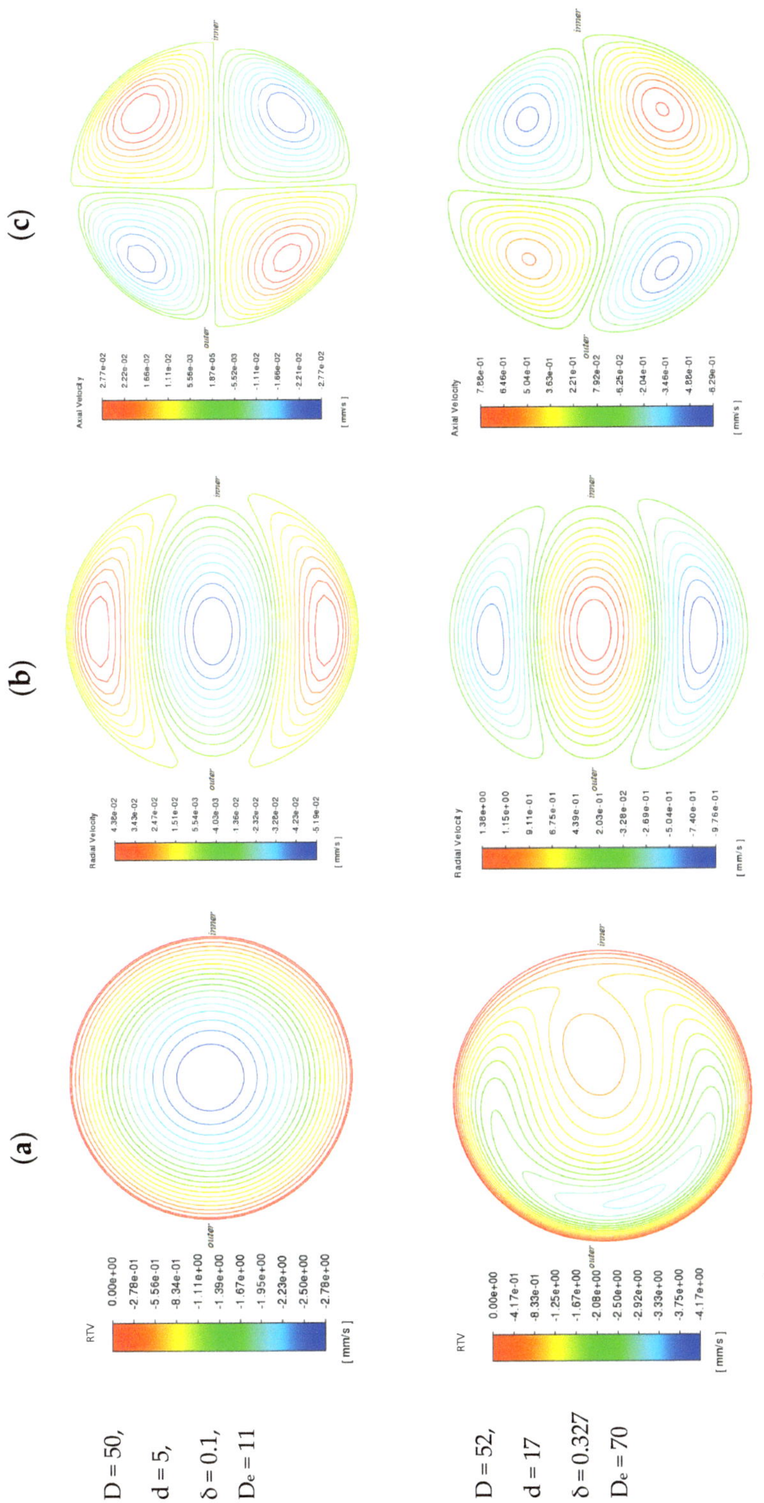

Figure 7. (**a**) RTV contour, (**b**) radial contour, (**c**) axial contour at 4 s.

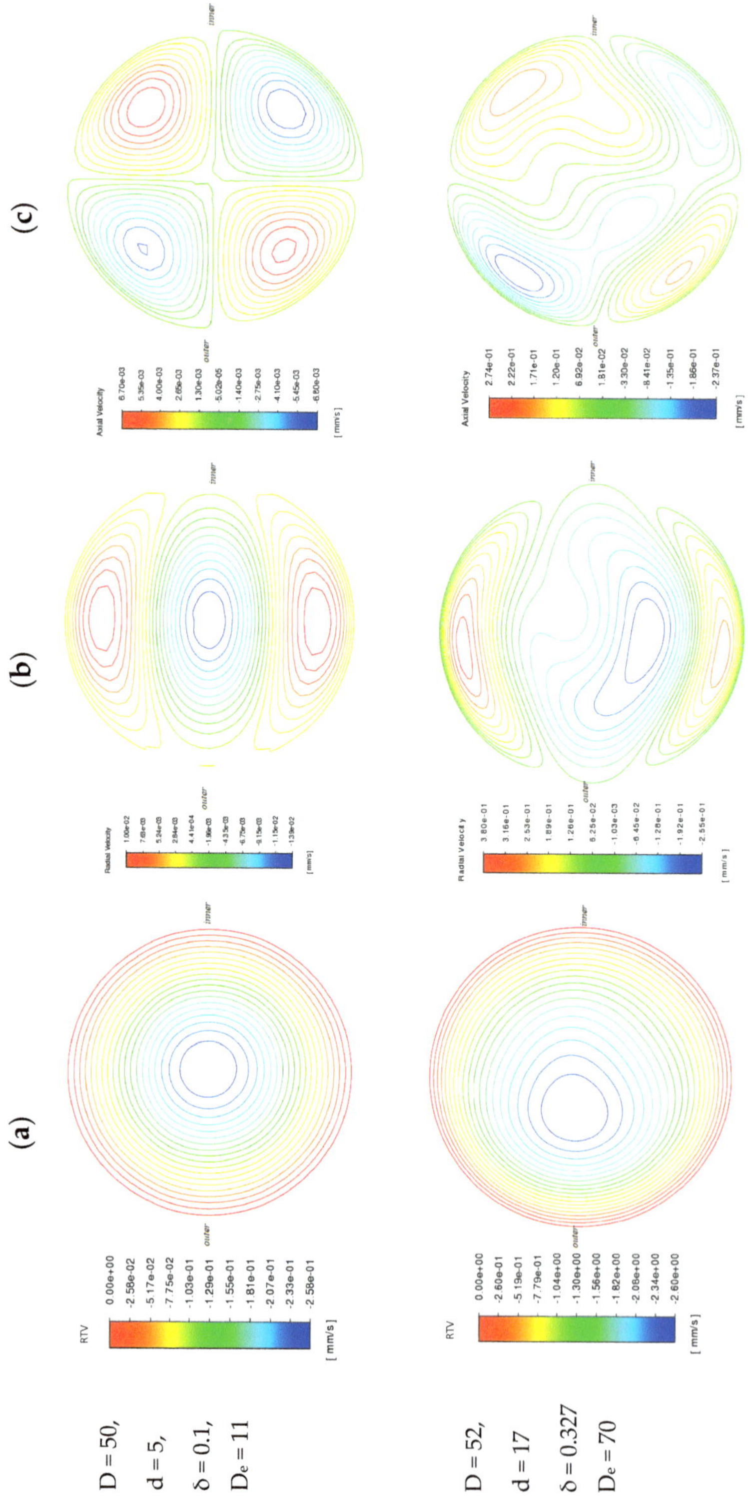

Figure 8. (**a**) RTV contour, (**b**) radial contour, (**c**) axial contour at 8 s.

The radial contours are depicted in Figures 6b, 7b and 8b for 2, 4, and 8 s, respectively, while the axial contours are illustrated in Figures 6c, 7c and 8c at 2, 4, and 8 s. The radial flow, representing the cross-sectional fluid movement, exhibits an elliptic dimple pattern with a central vortex, which is indicative of this secondary flow. Concurrently, the axial flow, which is parallel to the rotation axis, exhibits four distinct vortices, two near the inner wall and two near the outer wall. Notably, the flow pattern at D_e 11 demonstrates a higher degree of uniformity compared to D_e 70; this is crucial factor for ensuring consistent and predictable fluid behavior, particularly in systems like the PIVG. The results for other Dean numbers are provided in Appendix A for further reference.

Figure 9 illustrates the RTV at different D_e over specific time intervals. The graphs are normalized, simplifying the comparison of trends across different Dean numbers. At 2 s, the fluid appears notably unstable, forming a parabolic shape only at $D_e = 11$, suggesting a specific stability threshold at this early stage. Moving to 4 s, $D_e = 10$, 20, and 30, exhibiting stability and forming parabola shapes which are indicative of potential full development, while the other Dean numbers do not follow this pattern. By 6 s, all the Dean numbers, except for 70, demonstrate stability, suggesting a transition to a more consistent flow pattern. Finally, at 8 s, all the Dean numbers stabilize and form parabolic shapes, indicating a fully developed and stable flow regime—characterized by consistent and predictable behavior over time, without sudden fluctuations.

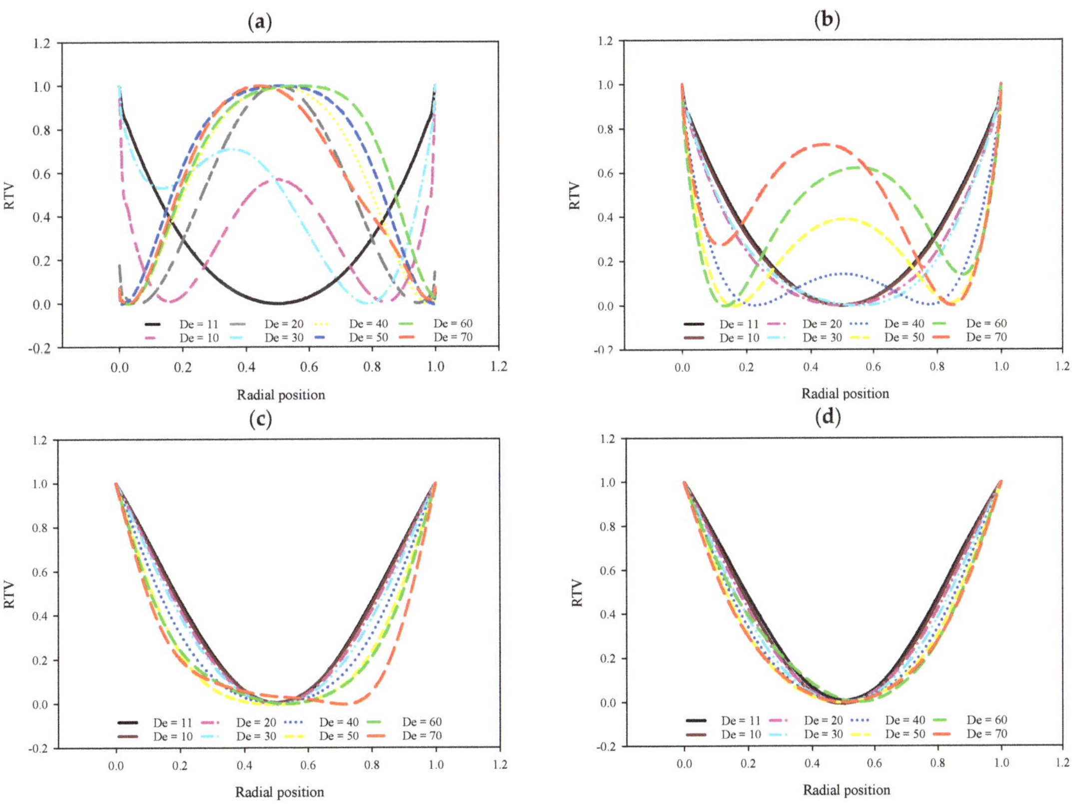

Figure 9. RTV at different D_e (**a**) 2 s, (**b**) 4 s, (**c**) 6 s, (**d**) 8 s.

In addition, it is noteworthy that at all time intervals, the D_e of 11 consistently exhibits a uniform and predictable profile, aligning closely with the input velocity shown in Figure 4. This stability ensures that the PIVG measurement can accurately capture the angular velocity at each time step. In contrast, higher Dean numbers exhibit fluctuations, resulting in less reliable measurements and potentially compromising the accuracy of the PIVG system. This highlights the significance of selecting an optimal Dean number, such as 11, to ensure consistent and reliable performance in capturing fluid dynamics within the toroidal pipe.

The normalized axial and radial velocity profiles are shown in Figure 10, which highlights the complex secondary flow patterns within the toroidal pipe at various Dean numbers. As mentioned, normalization enhances comparability, allowing a detailed analysis of these patterns. The axial velocity profile showcases consistent flow patterns at lower Dean numbers, resembling sinusoidal trends. This sinusoidal behavior signifies stable fluid movement along the toroid's length. However, as the Dean numbers increase, especially around 70, the sinusoidal pattern becomes more frequent, indicating a higher number of oscillations within a given time frame. This increase in frequency implies intensified oscillatory behavior in the fluid's axial and radial movements. Similarly, the radial velocity profiles, exhibit sinusoidal trends at lower Dean numbers, indicating predictable radial fluid movements. Yet, at higher Dean numbers, particularly around 70, the sinusoidal pattern becomes more rapid, signifying a higher frequency of radial oscillations. This observation emphasizes the complex fluid dynamics that accompany higher Dean numbers, leading to rapid and complex movements. Understanding these complications is crucial when researching secondary flows in many engineering contexts.

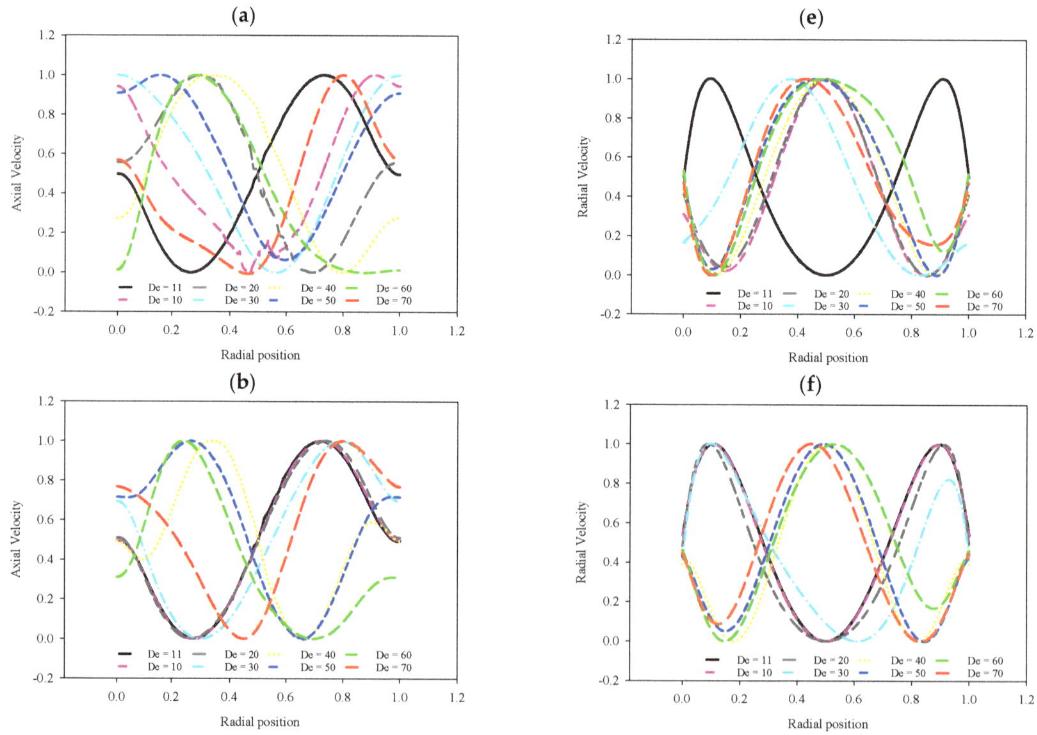

Figure 10. *Cont.*

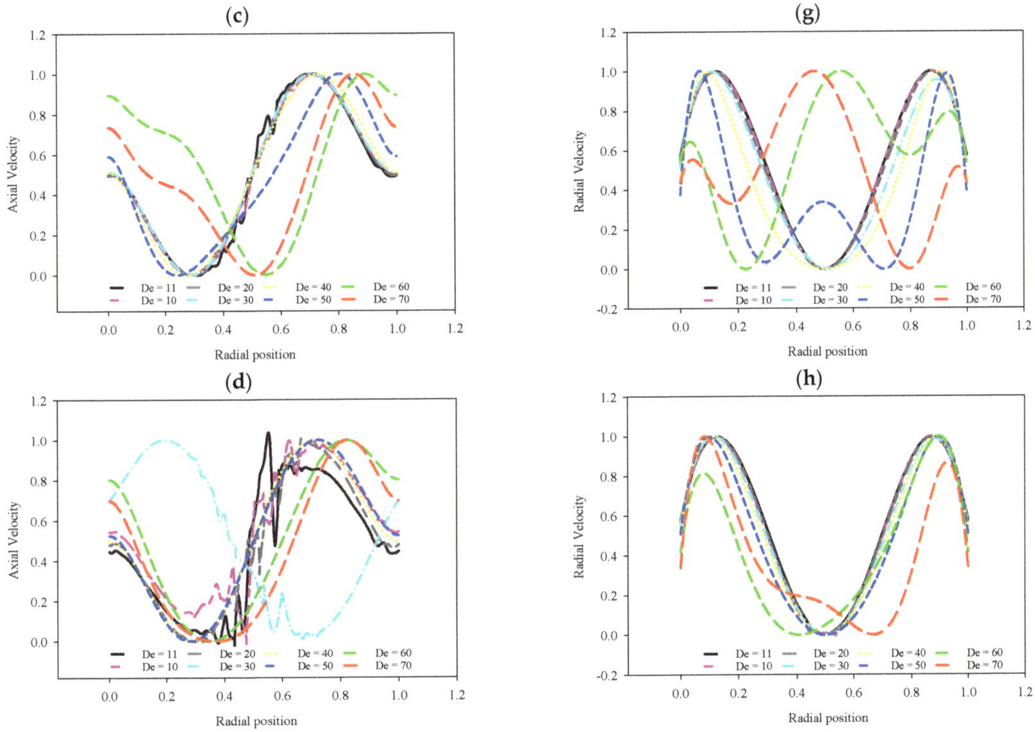

Figure 10. Axial velocity (**a**) 2 s, (**b**) 4 s, (**c**) 6 s, (**d**) 8 s and radial velocity (**e**) 2 s, (**f**) 4 s, (**g**) 6 s, (**h**) 8 s, at different D_e.

4.3.2. Optimal Fluid Dynamics at $D_e = 11$ for PIVG Operation

In this part, we consider the vital decision regarding selecting a D_e of 11 for the PIVG system. Our purpose is to illuminate the dependability and uniformity of fluid flow phenomena under these specific circumstances and to provide useful thoughts about the design and operation.

- The RTV and Secondary Flow

In this section of results, attention is directed towards the optimization of the PIVG system, with a specific emphasis on the most pertinent Dean number, $D_e = 11$. This choice is pivotal for ensuring the ideal conditions for PIVG functionality and marks a crucial step towards enhancing the reliability and precision of the PIVG sensor. Additionally, it is important to note that the results presented in this section are not normalized, and they allow a clear distinction between the values of the primary and secondary flow.

In this 3D (Figure 11) graph the exponential increase in velocity until 4.5 s, followed by an exponential decrease until the end, confirms the correlation between the velocity distribution and the input angular acceleration rate. These findings further support the consistency between the numerical simulation and the expected behavior of the fluid flow inside the toroidal system. The stability is vital for the PIVG sensor's accurate operation; regardless of the PIV's position, it will yield consistent results. Furthermore, this finding underscores the suitability of the toroidal dimensions and fluid behavior for integration into the PIVG system.

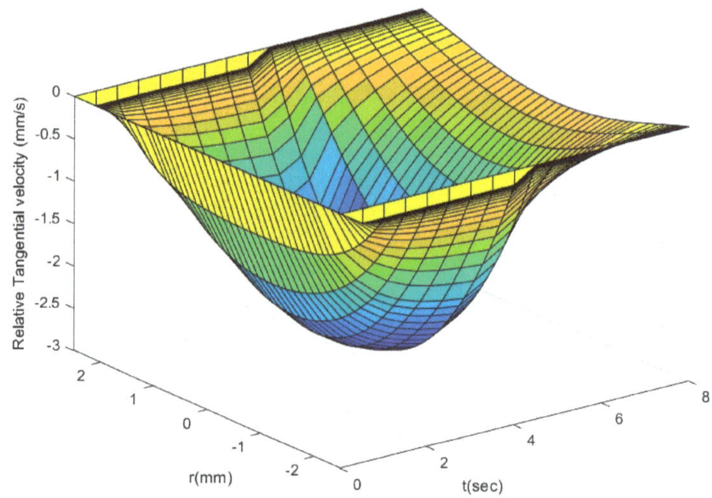

Figure 11. The relative tangential flow velocity profile changes over time.

The secondary flow at the $D_e = 11$ is shown here, and the velocity vector field on the radial cross-section of the toroidal pipe is displayed, offering insights into flow dynamics at different time steps, as shown in Figure 12. Velocity vectors intensify until 4 s, reflecting the toroidal system's increasing angular velocity. A transformation occurs from 6 to 8 s; while the counter vortices are always present at all times, their intensity notably increases between 6 and 8 s. This evolution from outwards to inwards aligns with Dean's curved pipe study [4,6,20], underlining the enduring relevance of his insights. Additionally, in Figure 13, the temporal evolution of the secondary flow (SF) at time intervals measured at radial positions r = 0 (center), r = 1, and r = 2 within the toroidal pipe is illustrated. Clearly, for all three radial positions, the secondary flow exhibits a consistent pattern of gradual augmentation from 2 to 4 s, culminating in a prominent peak at the 4 s mark. This peak is observed both at the center (r = 0) and progressively towards the outer radial positions (r = 1 and r = 2), accentuating the synchronized nature of this phenomenon across the toroidal pipe's cross-section. Beyond this peak, the secondary flow gradually decreases at 6 and 8 s, indicating a consistent behavior across various radial positions.

The graph (Figure 14) presents a comparative visualization of two crucial flow aspects within a toroidal system: secondary flow and RTV. Remarkably, it is discernible that the secondary flow exhibits notably weaker intensity in contrast to the comparatively more dominant RTV. The noticeable difference in strength implies that the secondary flow can be disregarded, especially when using particle image velocimetry (PIV) to measure RTV, which is crucial for the function of the PIVG. The reduced impact of the secondary flow allows a simplified measurement process (i.e., u = v = 0). This simplification enables a direct and precise determination of fluid velocity specifically in the direction of relative tangential motion, rather than measuring in three dimensions using PIV techniques.

- Pressure Distribution

The examination of pressure along the curvature of the toroidal geometry in Figure 15 offers valuable insights into the fluid dynamics within the PIVG system. By measuring pressure at various angles along the curvature, we gain an understanding of how centrifugal forces are distributed within the toroidal vessel. The consistent pattern observed in the pressure gradient graph, with the gradient remaining close to zero across different angles, indicates a balanced distribution of centrifugal forces. This balance is attributed to the uniform curvature of the toroidal cross-section, which ensures consistent fluid behavior regardless of the angle of measurement. Importantly, the near-zero pressure gradient

($\partial P/\partial \phi = 0$) signifies a stable fluid flow pattern, which is crucial for accurate and reliable performance of the PIVG. Understanding the pressure distribution along the toroidal curvature helps optimize the design of the PIVG system, ensuring uniform fluid behavior and minimizing disturbances that could affect angular velocity measurements.

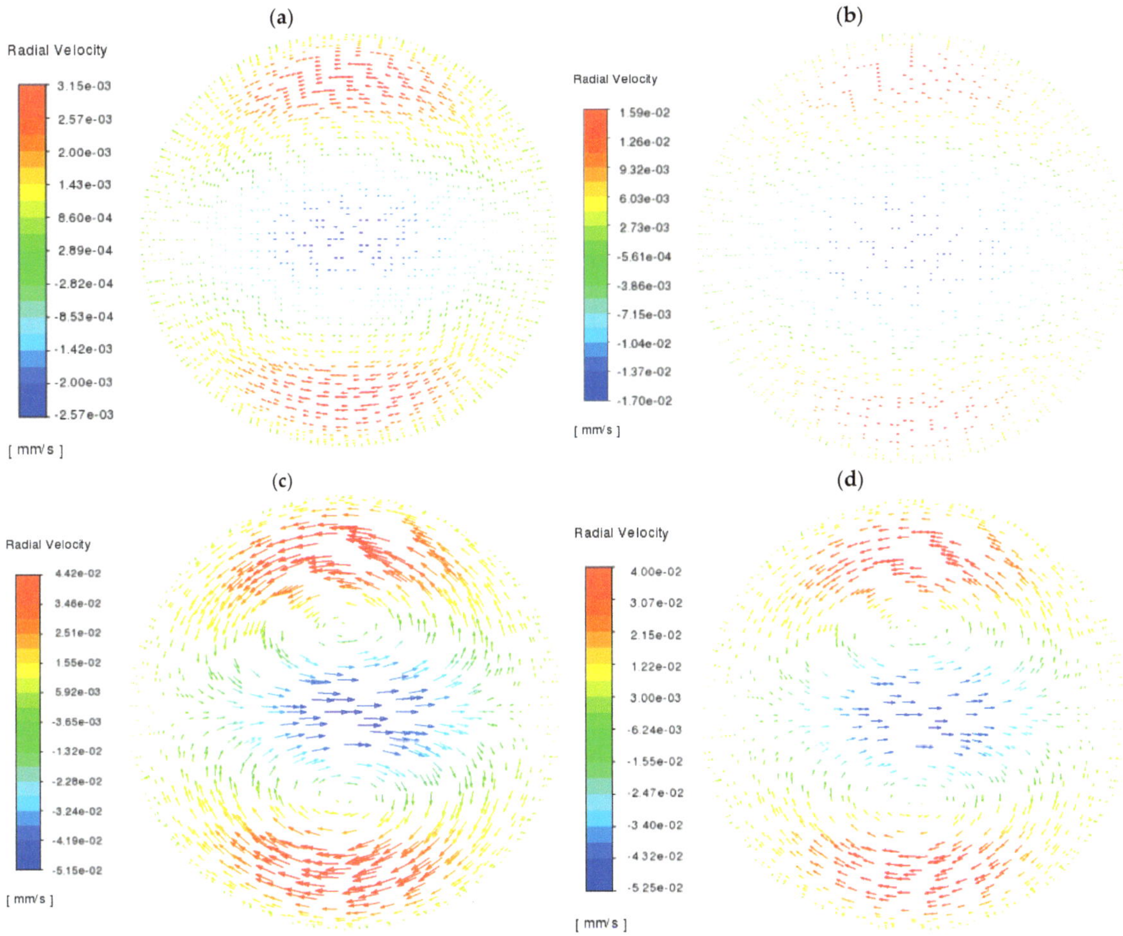

Figure 12. The relative velocity vector colored by radial velocity, outer (**left** side), inner (**right** side) (**a**) 2, (**b**) 4, (**c**) 6, and (**d**) 8 s.

From the CFD results, it can be inferred that the mathematical model, represented by the Navier–Stokes equation, for the PIVG sensor provides a wide domain of validity with high fidelity. However, ensuring that the fluid behavior aligns with the prescribed limits, particularly in terms of laminar flow with a low Dean number, is crucial for optimal PIVG performance. After analyzing a range of Dean numbers from 10 to 70, our study identified a D_e of 11 as optimal for achieving favorable fluid dynamics within the toroidal vessel. This finding has significant implications for the design of the toroidal geometry, as we pinpointed ideal dimensions characterized by a curvature radius of 25 mm and a cross-sectional diameter of 5 mm. These dimensions were found to promote stable and well-developed fluid flow patterns, which are particularly suitable for PIVG operation. With these dimensions identified, we can tailor the packaging of the gyroscope to ensure optimal performance and functionality, offering valuable guidance for decision making in the design and implementation of PIVG systems, thereby enhancing their effectiveness in inertial navigation applications.

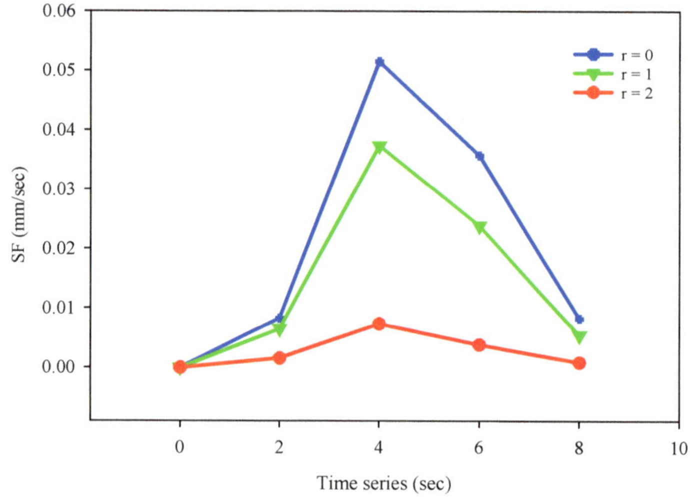

Figure 13. RTV time series at different cross-section radii.

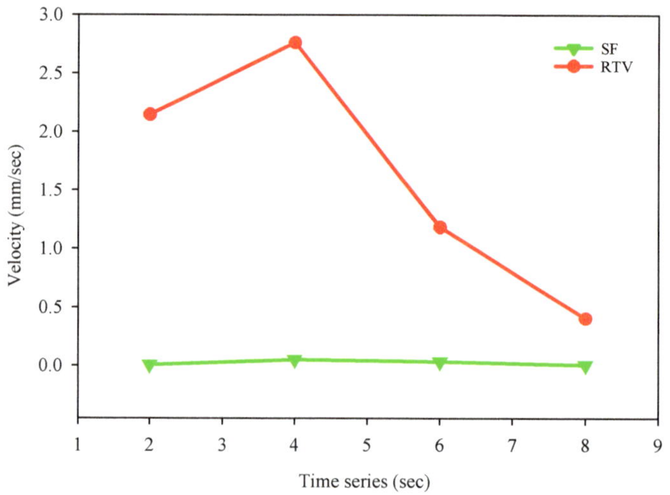

Figure 14. SF and RTV at r = 0 (center).

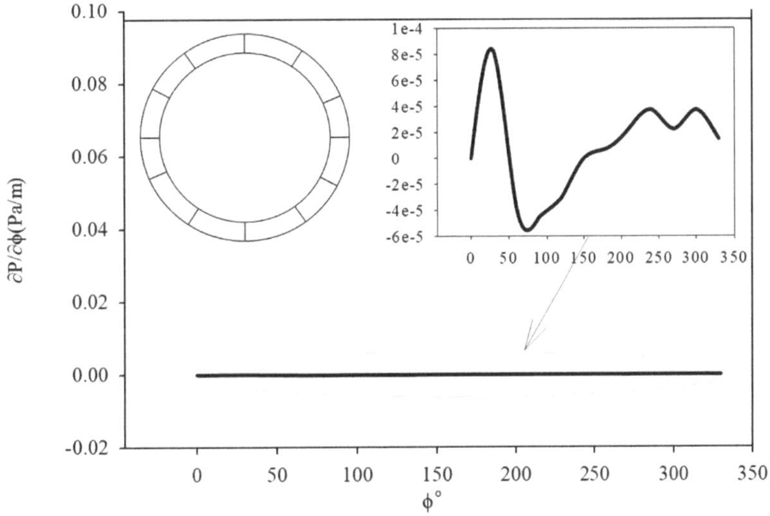

Figure 15. Pressure gradient at 2 s.

5. Conclusions

In conclusion, our study, which is specifically tailored to the accelerated toroidal pipes utilized in PIVGs, stands as a cornerstone in the ongoing enhancement of navigation sensor technologies. The study examined several aspects, including a convergence study, validation, evaluation of the pressure gradients, and a detailed examination of relative tangential and secondary flows. The convergence study revealed notable differences, with the medium mesh exhibiting a higher uncertainty (GCI_{32}%) than the fine mesh (GCI_{21}%). The average uncertainty attributed to spatial discretization is 1.47% for GCI_{32}% and 0.27% for GCI_{21}%. In the validation phase, the study showed remarkable agreement with the experimental data, validating the accuracy of our simulations. This agreement establishes the domain fidelity crucial for PIVG performance. Our simulations have provided a viewpoint on an accelerated toroidal geometry, highlighting how the fluid behaves under this acceleration and different conditions, with Dean numbers ranging from 10 to 70, and we utilized a ramp profile as the input angular velocity for the toroidal pipe. However, the most stable configuration, which was particularly beneficial for PIVG applications, was identified at $D_e = 11$ due to its stable dynamic fluid behavior, aligning with the toroid's specific dimensions. By selecting $D_e = 11$, we not only identify suitable dimensions for toroidal construction but also mitigate secondary flow effects, thus enhancing the precision and reliability of PIVGs in inertial navigation applications. Moreover, by selecting $D_e = 11$ and minimizing secondary flow, our study suggests the potential for a reduction in the mathematical model complexity and a focus on PIVG measurements based solely on primary flow dynamics; this offers a promising avenue for future research and practical implementation.

Author Contributions: Conceptualization, R.E.; Methodology, R.E.; Validation, R.E.; Writing—original draft, R.E.; Writing—review & editing, A.M.; Supervision, N.E.-S. and A.M. All authors have read and agreed to the published version of the manuscript.

Funding: This research received no external funding.

Data Availability Statement: The original contributions presented in the study are included in the article, further inquiries can be directed to the corresponding author.

Acknowledgments: The authors acknowledge the National Science Engineering Research Council of Canada (NSERC) for their support.

Conflicts of Interest: The authors declare no conflicts of interest.

Appendix A. The Results at Different De

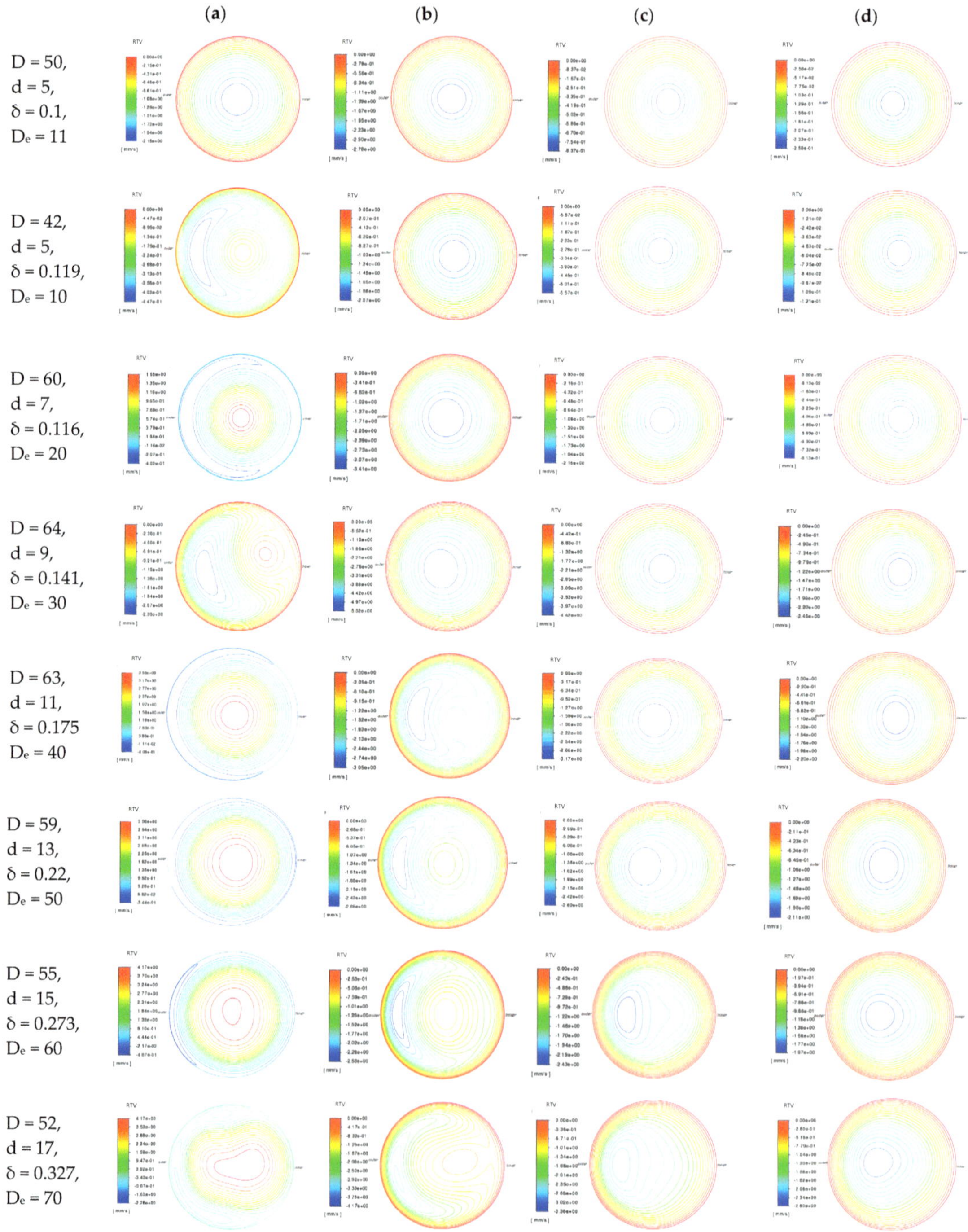

Figure A1. RTV contour at (**a**) 2 s, (**b**) 4 s, (**c**) 6 s, and (**d**) 8 s and different values of D_e.

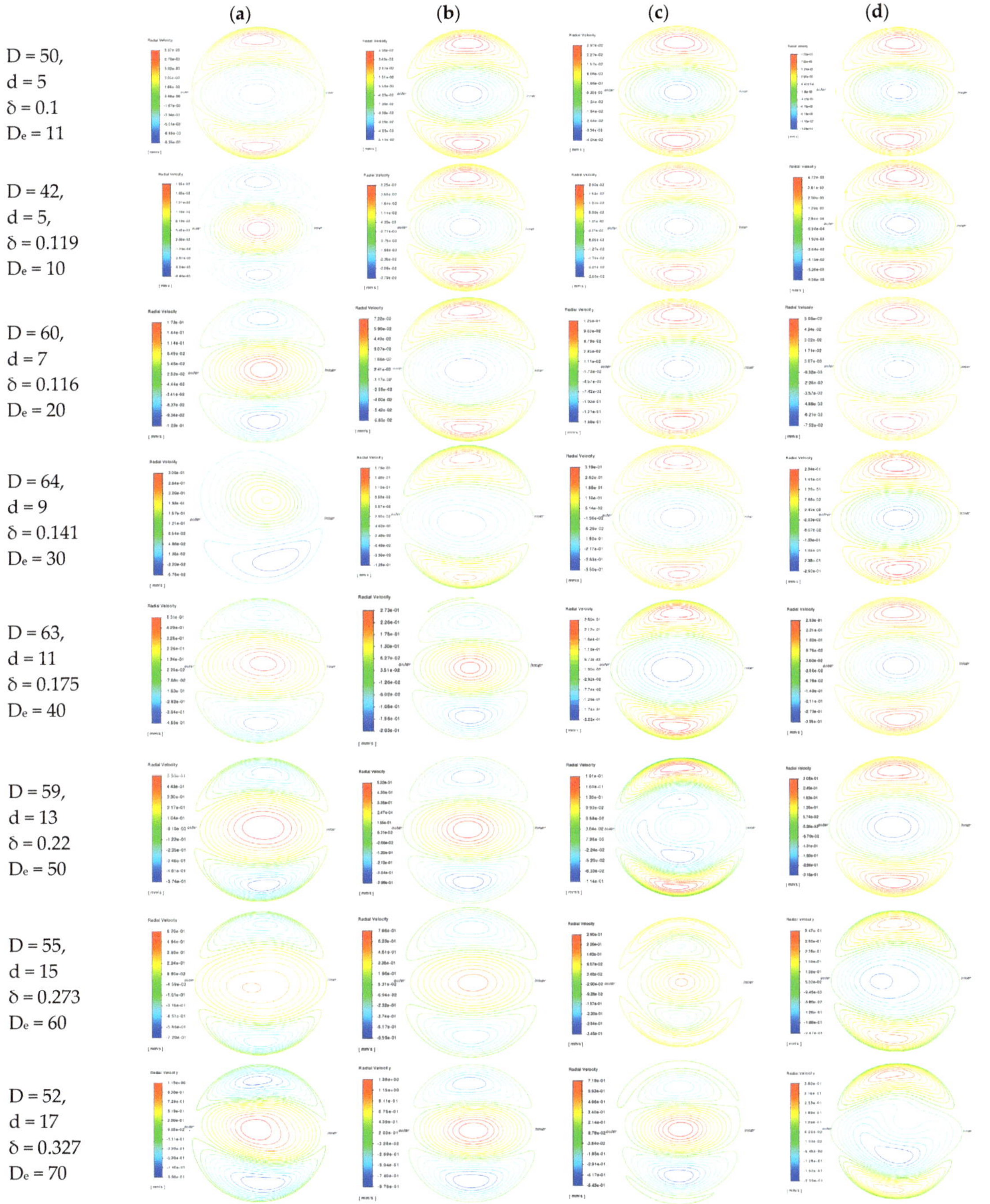

Figure A2. Radial velocity contour at (**a**) 2 s, (**b**) 4 s, (**c**) 6 s, and (**d**) 8 s and different values of D_e.

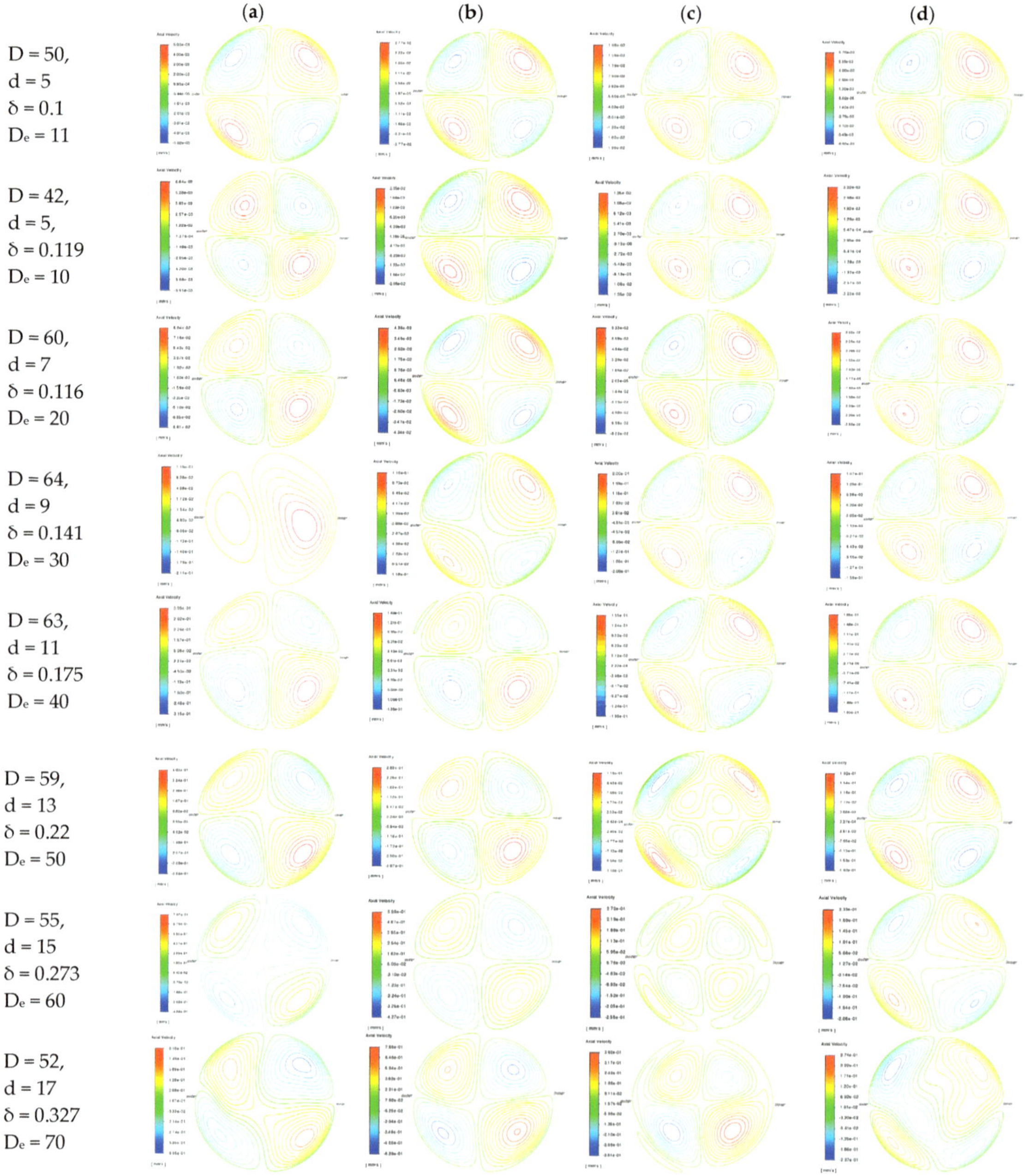

Figure A3. Axial velocity contour at (**a**) 2 s, (**b**) 4 s, (**c**) 6 s, and (**d**) 8 s and different values of D_e.

Appendix B. The Pressure Distribution at Different De

Figure A4 presents the pressure distribution profiles at various time points and different D_e, offering a comprehensive insight into the pressure behaviors within the toroidal vessel. Incredibly, the pressure trends mirror the patterns observed in the RTV profiles, indicating a significant correlation between fluid pressure and rotational dynamics. Notably, all the profiles have been normalized, ensuring a consistent basis for comparison. These findings have pivotal implications for applications such as the PIVG sensor, where understanding pressure dynamics is integral.

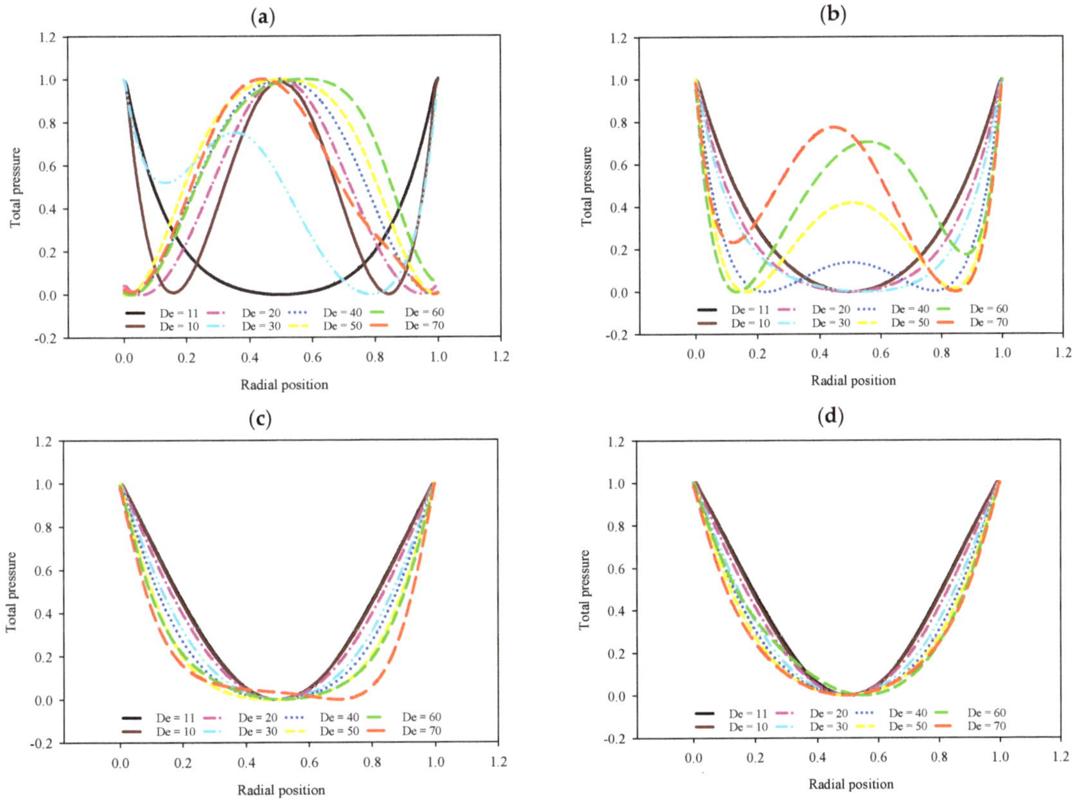

Figure A4. Pressure distribution at (**a**) 2 s, (**b**) 4 s, (**c**) 6 s, and (**d**) 8 s and at different D_e.

Appendix C. The Detailed RTV at Different Curvature Angle and Different Radius

The stability of the fluid dynamics at $D_e = 11$ is confirmed by a different approach. Figure A5a, representing the relative tangential velocity at the time interval of 2 s at varying curvature angles, offers an insight into the fluid dynamics within our toroidal system. A consistent trend emerges across the diverse angles examined, revealing a uniformity in the RTV within the toroidal pipe. As well as Figure A5b shows the RTV against time series at different cross-section radii.

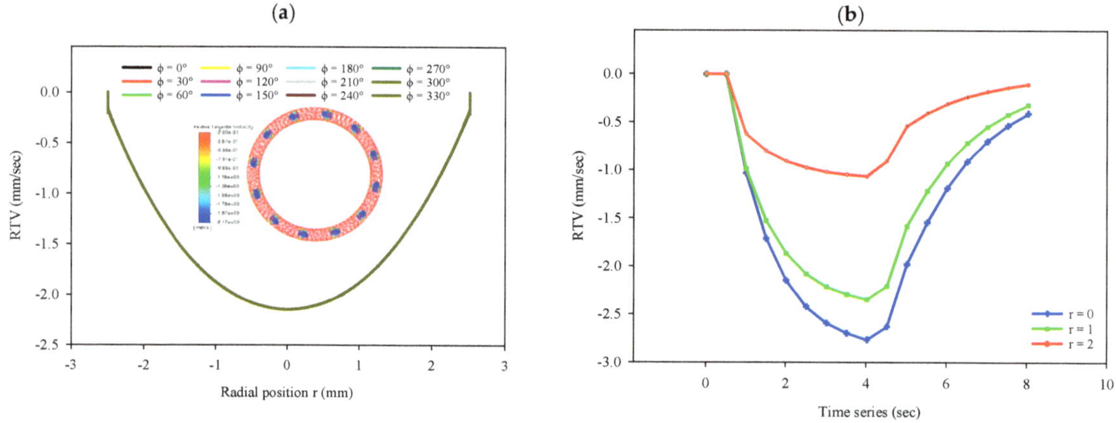

Figure A5. (**a**) RTV at the different circumferential positions at 2 s, (**b**) RTV vs. time series at different cross-section radii.

References

1. Youssef, A.A.; El-Sheimy, N. Particle imaging velocimetry gyroscope. *Sensors* **2019**, *19*, 4734. [CrossRef] [PubMed]
2. Elaswad, R.; El-Sheimy, N.; Mohamad, A.A. Review—Basic and Advanced Inertial Navigation Fluid-Based Technology. *ECS Sens. Plus.* **2023**, *2*, 023201. [CrossRef]
3. El-Sheimy, N.; Youssef, A. Inertial sensors technologies for navigation applications: State of the art and future trends. *Satell. Navig.* **2020**, *1*, 2. [CrossRef]
4. Norouzi, M.; Sedaghat, M.; Shahmardan, M. An analytical solution for viscoelastic dean flow in curved pipes with elliptical cross section. *J. Non-Newtonian Fluid Mech.* **2014**, *204*, 62–71. [CrossRef]
5. Dean, W.R.; Hurst, J.M. Note on the Motion of Fluid in a Curved Pipe. *Mathematika* **1959**, *6*, 77–85. [CrossRef]
6. Dean, W.R., LXXII. The stream-line motion of fluid in a curved pipe (Second paper). *Lond. Edinb. Dublin Philos. Mag. J. Sci.* **1928**, *5*, 673–695. [CrossRef]
7. Ligrani, P.M. *A Study of Dean Vortex Development and Structure in a Curved Rectangular Channel with an Aspect Ratio of 40 at Dean numbers up to 430*; AD-A286 159 NASA Army Research Laboratory Contractor Report 4607; Army Research Laboratory: Adelphi, MD, USA, 1994.
8. Greenspan, H.P.; Howard, L.N. On a time-dependent motion of a rotating fluid. *J. Fluid Mech.* **1963**, *17*, 385–404. [CrossRef]
9. Weissenberg, K. The Theory of Rotating Fluids. *Phys. Bull.* **1968**, *19*, 426–427. [CrossRef]
10. Van Dommelen, L.L.; Cowley, S.J. On the Lagrangian description of unsteady boundary-layer separation. Part 1. General theory. *J. Fluid Mech.* **1990**, *210*, 593–626. [CrossRef]
11. Lam, S.T. On High-Reynolds-Number Laminar Flow through a Curved Pipe, and Past a Rotating Cylinder. 1988. Available online: https://hdl.handle.net/10044/1/47142 (accessed on 17 March 2024).
12. Madden, F.N.; Mullin, T. The Spin-Up from Rest of a Fluid-Filled Torus. *J. Fluid Mech.* **1994**, *265*, 217–244. [CrossRef]
13. Elaswad, R.; El-Sheimy, N.; Mohamad, A. Trajectory dynamics of particles in accelerated toroidal pipe: A computational study using CFD-DPM simulations. *J. Comput. Sci.* **2024**, *78*, 102285. [CrossRef]
14. Zhou, Y.; Shah, S. Fluid Flow in Coiled Tubing: A Literature Review and Experimental Investigation. *J. Can. Pet. Technol.* **2004**, *43*, 52–61. [CrossRef]
15. van Tiem, J.; Groenesteijn, J.; Sanders, R.; Krijnen, G. 3D printed bio-inspired angular acceleration sensor. In Proceedings of the 2015 IEEE Sensors, Busan, Republic of Korea, 1–4 November 2015; pp. 1–4.
16. Groenesteijn, J.; Droogendijk, H.; de Boer, M.J.; Sanders, R.G.P.; Wiegerink, R.J.; Krijnen, G.J.M. An angular acceleration sensor inspired by the vestibular system with a fully circular fluid-channel and thermal read-out. In Proceedings of the 2014 IEEE 27th International Conference on Micro Electro Mechanical Systems (MEMS), San Francisco, CA, USA, 26–30 January 2014; pp. 696–699.
17. Celik, I.B.; Ghia, U.; Roache, P.J.; Freitas, C.J.; Coleman, H.; Raad, P.E. Procedure for Estimation and Reporting of Uncertainty Due to Discretization in CFD Applications. *J. Fluids Eng.* **2008**, *130*, 078001. [CrossRef]
18. Roache, P.J. *Fundamentals of Verification and Validation*; Hermosa Publishers: Socorro, NM, USA, 2009.

19. del Pino, C.; Hewitt, R.E.; Clarke, R.J.; Mullin, T.; Denier, J.P. Unsteady fronts in the spin-down of a fluid-filled torus. *Phys. Fluids* **2008**, *20*, 124104. [CrossRef]
20. Dean, W., XVI. Note on the motion of fluid in a curved pipe. *Lond. Edinb. Dublin Philos. Mag. J. Sci.* **1927**, *4*, 208–223. [CrossRef]

fluids

MDPI

Article

Application of a Combinatorial Vortex Detection Algorithm on 2 Component 2 Dimensional Particle Image Velocimetry Data to Characterize the Wake of an Oscillating Wing

Mathew Bussière, Guilherme M. Bessa, Charles R. Koch and David S. Nobes *

Department of Mechanical Engineering, University of Alberta, Edmonton, AB T6G1H9, Canada; bessa@ualberta.ca (G.M.B.); bob.koch@ualberta.ca (C.R.K.)
* Correspondence: david.nobes@ualberta.ca; Tel.: +1-780-492-7031

Abstract: To investigate the vortical wake pattern generated by water flow past an oscillating symmetric airfoil, using experimental velocity fields from particle image velocimetry (PIV), a novel combinatorial vortex detection (CVD) algorithm is developed. The primary goal is to identify and characterize vortices within the wake. Experimental flows introduce complexities not present in numerical simulations, posing challenges for vortex detection. The proposed CVD approach offers a more robust alternative, excelling in both vortex detection and quantification of essential parameters, unlike widely-used methods such as Q-criterion, λ_2-criterion, and Δ-criterion, which rely on subjective and arbitrary thresholds resulting in uncertainty. The CVD algorithm effectively characterizes the airfoil wake, identifying and analyzing vortices aligning with the Burgers model. This research enhances understanding of wake phenomena and showcases the algorithm's potential as a valuable tool for vortex detection and characterization, particularly for experimental fluid dynamics. It provides a comprehensive, robust, and non-arbitrary approach, overcoming limitations of traditional methods and opening new avenues for studying complex flows.

Keywords: vortex detection; vortex characteristics; vortex decay; oscillating airfoil; 2C2D PIV

Citation: Bussière, M.; Bessa, G.M.; Koch, C.R.; Nobes, D.S. Application of a Combinatorial Vortex Detection Algorithm on 2 Component 2 Dimensional Particle Image Velocimetry Data to Characterize the Wake of an Oscillating Wing. *Fluids* **2024**, *9*, 53. https://doi.org/10.3390/fluids9030053

Academic Editor: Stéphane Le Dizès

Received: 12 December 2023
Revised: 6 February 2024
Accepted: 13 February 2024
Published: 22 February 2024

1. Introduction

Vortices constitute fundamental flow features with widespread relevance in various fluid dynamic applications. Their formation in the wake of airfoils and streamlined bodies holds a paramount significance in both research and practical applications. In turbulent flow regimes, temporally evolving and spatially coherent structures, commonly identified as vortices [1], exert substantial influence. For instance, studying the airflow around a pitching airfoil not only aids in comprehending undesirable phenomena such as wing flutter [2] but also finds applications in understanding biologically inspired aquatic propulsion [3,4]. Researchers have extensively compared the simulation of pitching airfoils using full Navier–Stokes and vortex models [5–7]. In this context, precise detection and characterization of coherent wake structures emerge as central research objectives [2,8,9]. Notably, at a given Reynolds number, the generation of various wake schemes hinges on the positioning and organization of vortices in the wake [3].

The classification of wakes characterized by the presence of vortices can be effectively depicted in a phase diagram that illustrates significant wake transitions as functions of two independent nondimensional variables [3,10,11]. These parameters are the Strouhal number ($\mathrm{St} = fC/U$), which is based on the flapping frequency (f), the chord length (C), and the free stream velocity (U), and the dimensionless oscillation amplitude derived from the chord length. To construct accurate phase diagrams, it is crucial to identify vortices in the wake with high precision and reliability. Furthermore, beyond the fundamental task of vortex detection, it is important to compute several key parameters that are essential for describing the flow, especially in the context of aerodynamics. These parameters encompass crucial metrics such as the coordinates of vortex cores, their drift velocities, peak vorticity values, overall circulation, the radius defining vortex boundaries, and the highest circumferential velocity [12].

The existing methods for detecting vortices, such as the widely used Q-criterion [13], λ_2-criterion [1], Δ-criterion [14–16], and $\Gamma1$-criterion [17], have inherent limitations [18–20]. A primary challenge arises from their reliance on threshold values to discern vortices. Selecting an appropriate threshold is often subjective, and there may not be a universally optimal value across flow types. This is particularly pronounced in turbulent flow studies, marked by significant, multiscale, intermittent variations in both vorticity and strain rates. More significantly, the threshold sensitivity of these methods can lead to unreliable and inaccurate vortex identification [21,22]. Excessively low thresholds cause frequent false positives, incorrectly labeling non-vortex regions as vortices. Conversely, high thresholds result in many missed detections, failing to identify real vortices in the flow. This tradeoff hinders finding a robust threshold that generalizes across the wide variability in real-world flow dynamics [23,24].

One core limitation identified in the $\Gamma1$-criterion, developed by Laurent Graftieaux et al. (2001) [17], is that the $\Gamma1$ function used to locate vortex centers lacks Galilean invariance, rendering it susceptible to the constant advection of the flow field. This inherent lack of robustness becomes particularly evident in scenarios where the overall flow undergoes consistent shifts. Additionally, Graftieaux's theoretical relations linking the $\Gamma1$ and $\Gamma2$ functions to vortex properties assume axisymmetry in the swirling flow structure. While this assumption simplifies the theoretical framework, real unsteady flows often deviate from the idealized axisymmetric case, introducing complexities that challenge the method's applicability to diverse flow conditions. Moreover, directly applying the $\Gamma1$-criterion to flows with multiple, interacting vortices poses challenges. Ambiguity can arise in attributing vortex indices, especially in scenarios where vortices dynamically interact, merge, or evolve over time. The method's reliance on theoretical relations may lead to difficulties in precisely characterizing complex vortex structures within the flow. Consequently, the method's effectiveness diminishes in situations where accurate identification and differentiation of interacting vortices are crucial [17].

To provide a robust approach for vortex detection, researchers often use combinations of criteria or data-driven techniques such as machine learning to adaptively determine optimal thresholds for a given flow [25–27]. However, subjectivity and extensive parameter tuning may still be required. Furthermore, while the Q-criterion, λ_2-criterion, and Δ-criterion are effective in binary classification, categorizing regions as vortical or non-vortical, they do not inherently provide direct quantification of vortex strength or intensity [28]. Evaluating vortex strength is often crucial for in-depth analysis of vortex dynamics, comparative studies between vortices, and various related objectives [29,30]. Researchers typically need to perform supplementary analyses to calculate circulation, maximum vorticity, core radius, or other relevant parameters to obtain strength information [29,30].

Recently, advanced vortex identification techniques such as the Ω-method [24] and 'Rortex' method [31,32] have shown promise in addressing some of the limitations of traditional criteria. The Ω-method defines vortices as regions where vorticity overtakes deformation, providing a clear physical interpretation. It does not require arbitrary threshold selection and can simultaneously capture strong and weak vortex structures [24]. The 'Rortex' method decomposes vorticity into rotational and non-rotational parts, excluding shear contamination. As a vector quantity, 'Rortex' identifies both vortex axis orientation and strength [31,32]. These methods offer greater objectivity and physical insight compared to conventional techniques. However, the Ω-method and 'Rortex' method have thus far been predominantly applied to direct numerical simulations (DNSs) data [24,28,31–34]. Their effectiveness in processing experimental measurements with inherent noise remains to be rigorously validated.

Substantial effort has been dedicated to the development and validation of advanced techniques using DNS data [1,35,36]. DNS provides high-resolution flow field information for testing novel vortex identification strategies, especially in three-dimensional turbulent flows such as boundary layers [37,38]. However, DNS is limited to relatively simplified flow conditions and small domains due to computational constraints. Therefore, while beneficial for initial development, additional rigorous validation of new methods with experimental data remains essential before broader utilization across complex flows. DNS aids technique refinement, but ultimate validation still requires experiments.

Detecting vortices in particle image velocimetry (PIV) data poses unique challenges due to inherent measurement noise and experimental uncertainties [39]. Typically, mitigating measurement noise involves prior filtering before applying a detection algorithm [12]. However, selecting appropriate thresholds becomes crucial to avoid overlooking weak or small vortices [39]. Additionally, the flow field near the vortex core exhibits substantial gradients, potentially causing difficulties in particle seeding and correlating PIV data [12].

In contrast to DNS data that can utilize local mesh refinement for better spatial detail in areas like vortex centers, PIV analysis methods must seek alternative strategies to effectively manage the pronounced gradients encountered around the vortex core [40].

Due to camera resolution limits, PIV analysis often requires integrating images from several cameras, necessitating image de-warping and stitching, which might lead to discontinuities in the flow-field representation, potentially leading to the erroneous identification of elevated local vorticity as a vortex [1].

Given the increasing availability of PIV data in experimental fluid mechanics [12], the need for developing dedicated and effective algorithms for vortex detection and characterization in PIV datasets has emerged as a pivotal area of research. The precise definition of a vortex remains elusive [1], making their detection in practical flows a complex task. The most common and intuitive conception of a vortex revolves around the perception of swirling fluid motion centered around a focal point [35]. However, translating this intuitive notion, rooted in human perception, into a rigorous numerical characterization detectable by algorithmic methods poses considerable challenges.

Lugt (1979) [41] proposed that "A vortex is the rotating motion of a multitude of material particles around a common center", aligning with the fundamental notion of swirling motion. While this concept is intuitively appealing, it lacks specific criteria for describing vortical structures and does not easily translate into a workable vortex identification algorithm. Building upon Lugt's concept, Robinson (1991) [36] offered a more geometrically precise definition: "A vortex exists when instantaneous streamlines, projected onto a plane perpendicular to the vortex core, exhibit a roughly circular or spiral pattern when observed from a reference frame moving with the center of the vortex core". While this definition offers clarity in geometric characterization, it necessitates a priori information about the vortex core's location, posing challenges for systematic vortex detection. In contrast, Jeong and Hussain (1995) [1] put forth the notion that "A vortex core must possess net vorticity, hence net circulation". This concept offers a direct method to pinpoint potential vortex regions without requiring prior knowledge of the vortex core's location or drift velocity. By computing scalar vorticity fields from velocity vectors, researchers can generate and label regions of interest. Combining Jeong and Hussain's definition [1] with Robinson's geometric criteria [36] forms a solid groundwork for creating an effective vortex detection algorithm.

Vortex identification techniques are often designed for specific data types and unique flow fields, aiming to identify well-defined vortex structures based on precise definitions [35]. These algorithms exhibit a high degree of sensitivity to various parameters, including vortex spacing, particularly when multiple vortices coexist within a given flow field, vortex size, angular velocity, and the presence of shear flow [12].

Sadarjoen and Post (2000) [42] proposed a classification of vortex detection algorithms found in the literature, distinguishing them into two primary categories. The first category encompasses traditional vortex detection algorithms, which rely on the evaluation of physical flow-field properties at specific points. Commonly assessed properties include velocity, vorticity, pressure, and the velocity gradient tensor. However, one notable limitation of these algorithms, as emphasized by Sadarjoen and Post (2000) [42], is their reduced sensitivity to large, slowly rotating vortices. For instance, the maximum vorticity method defines a vortex core based on local vorticity maxima [35]. Inherently, this method disregards regions where vorticity falls below a predefined threshold, potentially leading to the failure of detecting weak, slowly rotating vortices.

The second category comprises geometric methods, representing a more recent set of vortex detection algorithms. These methods prioritize the examination of flow patterns and trajectories, such as streamlines and pathlines, over the analysis of scalar attributes measured at distinct grid locations [35]. Although geometric methods demand more computational resources, they offer a significant advantage over traditional techniques by effectively differentiating between actual vortices and false positives, addressing a common limitation in earlier approaches [35,42].

The limitations of the most widely adopted vortex identification criteria, including their sensitivity to threshold values and their inability to quantify vortex strength directly, underscore the pressing need for complementary techniques, modifications to existing criteria, combinations of criteria, and the exploration of emerging methods to enable accurate and comprehensive vortex detection across a wide range of fluid mechanic applications.

This work introduces a vortex detection algorithm that combines selected features from three individual algorithms to interrogate PIV data. The primary aim of this algorithm is to efficiently process data, detecting and locating vortices while calculating characteristic vortex parameters. Given the potential magnitude of the datasets associated with the PIV results, the algorithm also incorporates automated procedures for identifying false positive vortices without the need for human intervention. The algorithm's performance is evaluated by analyzing vortex behavior in the extensively studied wake behind a symmetrically pitching airfoil [2,10].

2. Background on Existing Vortex Detection Algorithms

The development of an effective vortex detection algorithm relies on the fundamental principles and methodologies established in the existing literature. This background reviews three pivotal methods that serve as the basis for the proposed combinatorial vortex detection (CVD) framework tailored for the analysis of PIV data. These methods encompass the maximum vorticity (MV) method [43], the cross-sectional lines (CSL) method [12], and the winding angle (WA) method [44]. The following sections delve into the core principles of each method and discuss key aspects leveraged in developing the CVD algorithm.

2.1. Maximum Vorticity (MV) Method

The MV method, proposed by Strawn et al. (1999) [43], assumes that a vortex core exists at locations characterized by a local maximum in vorticity magnitude. This approach proves effective in detecting vortices in situations where their cores may overlap [35]. However, a notable limitation of the MV method is its propensity to not only identify vortices but also regions exhibiting shearing activity [1]. Consequently, devising a vortex detection algorithm solely based on vorticity fields becomes more challenging, particularly in non-free shear flow [1,35]. An alternative utility of vorticity fields lies in outlining regions of interest (ROIs) where potential vortices may be located, enabling the use of other algorithms to evaluate these ROIs individually.

To enhance the performance of the MV method within the framework of the CVD method, a Gaussian low-pass spatial filter is applied to the velocity data to mitigate measurement noise and experimental uncertainties [39]. This filtering process produces a smoother field, effectively eliminating small-scale fluctuations that might otherwise lead to localized spikes in the vorticity field. Scalar vorticity, ω, calculated numerically at each grid point, is defined within the two-dimensional velocity field, $\vec{v}$, as $\omega = \vec{\nabla} \times \vec{v}$.

In the pursuit of detecting weak vortices characterized by relatively low vorticity, it becomes necessary to apply a low vorticity threshold. However, this approach carries the inherent risk of grouping stronger, nearby vortices into a single ROI, effectively evaluating them as a singular vortex structure. To mitigate this, multiple threshold levels are used, and the indexed image is further analyzed using logical operations and image morphology (IM) techniques to distinguish between closely situated vortices [45].

The application of multilevel thresholds serves the dual purpose of allowing the identification of weak vortices while ensuring that even when strong vortices are close together or their cores overlap, they can still be distinctly identified and analyzed as separate entities. Figure 1 provides a visual representation of this concept, depicting two vorticity field test cases where a three-level thresholding approach is applied. The resulting indexed images, accompanied by the corresponding desired ROI maps, are presented for these sample vorticity fields. The left-hand side of Figure 1 illustrates the indexed images, where index levels are denoted as i = 1, 2, and 3. On the right, the desired ROI map is displayed, signifying the regions to be extracted from the corresponding indexed image.

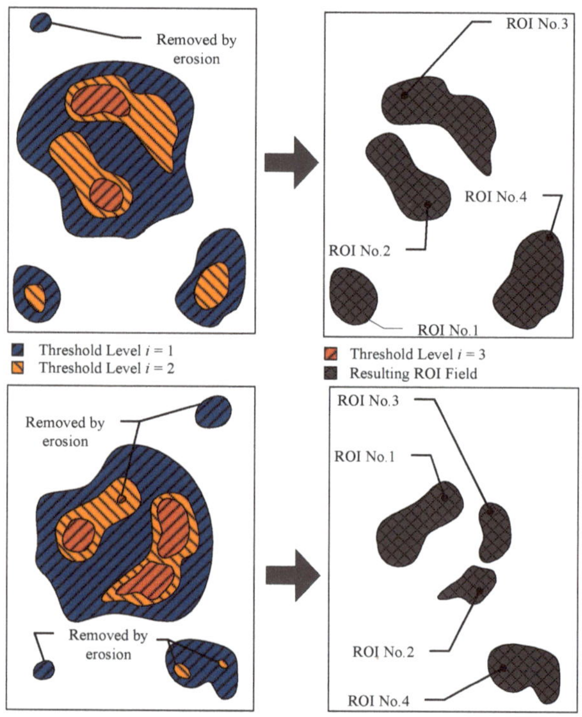

Figure 1. Schematic illustrating the application of a multilevel threshold technique for identifying vortices in two vorticity fields. (**left**): indexed images with levels i = 1, 2, and 3. (**right**): corresponding desired ROI maps for vortex extraction.

Without the utilization of multilevel thresholds, a single threshold level would be chosen, leading to one of two scenarios. When solely threshold level i = 1 is applied, as seen in both examples, the broader vorticity region, clearly exhibiting two distinct peaks of vorticity, is incorrectly grouped into one single ROI, consequently treated as a single potential vortex. However, this area features two distinct peaks and should rightfully be considered as two individual ROIs. Conversely, when exclusively threshold level i = 3 is employed, the weaker vortices failing to meet the vorticity ω_{peak} level i = 3 threshold are excluded from consideration as ROIs and remain unidentified as potential vortices.

To address the notable limitations associated with the use of a single-level threshold for identifying ROIs, the algorithm employs a three-level threshold to generate an ROI map. In this approach, the vorticity field is segmented, resulting in a classified image using threshold values specified by the threshold intensity vector (TIV):

$$\text{TIV} = \langle TI_1,\ TI_2,\ TI_3 \rangle \tag{1}$$

Subsequently, the algorithm executes morphological opening for each of the thresholds specified in Equation (1). Morphological opening combines two fundamental image morphology techniques, erosion and dilation, performed in a specific sequence. Initially, the image undergoes erosion using a diamond-shaped structuring element of size S_e. This operation effectively eliminates clusters of pixels smaller than S_e, while reducing the size of clusters larger than S_e.

Following erosion, a dilation step is performed with the same structuring element S_e to bring the pixel clusters, which remained post-erosion, back to their original size. This combination of erosion and dilation, performed in a well-defined sequence, plays a pivotal role in refining the ROI map and contributes to enhancing the algorithm's efficacy in vortex detection.

Connected pixel groups in the vorticity intensity indexed image are categorized by intensity levels (I_1, I_2, I_3). Groups associated with index I_1, denoting the lowest vorticity intensity level, undergo an initial morphological opening procedure, employing an IM diamond-shaped structuring element of size S_{e1}. Subsequently, these groups are labeled and individually investigated. Structures corresponding to index I_2 also undergo morphological opening, but this time utilizing an IM diamond-shaped structuring element of size S_{e2}. They are then sub-labeled within each of the previously labeled groups from index I_1.

The algorithm follows a conditional path: If none of the groups from the second threshold index I_2 are found within a group from the first level I_1, the algorithm concludes, and the group of pixels from index level I_1 is designated as an ROI. Conversely, if at least one I_2 group is found within a group from index I_1, the algorithm progresses further. Pixels corresponding to index I_3 undergo morphological opening using an IM diamond-shaped structuring element of size S_{e3} and are further categorized within each of the index I_2 groups.

At this stage, the algorithm enumerates the group counts at each threshold level. If an I_2 cluster encloses several I_3 groups, each individual index I_3 group becomes a distinct ROI. However, if only one index I_3 group is counted, the algorithm reverts to a lower level, considering the number of index I_2 clusters. Once again, if multiple index I_2 groups are detected, each individual index I_2 group is recognized as an ROI. However, if only one index I_2 label is counted, the group from index I_1 becomes the designated ROI. This process ensures that the ROI encompasses the largest possible area while partitioning into multiple ROI's when distinct, higher vorticity peaks are detected.

2.2. Cross-Sectional Lines (CSL) Method

The CSL method is designed to identify potential vortex cores within a designated ROI by evaluating the velocity component perpendicular to parallel straight cross-sectional lines intersecting the ROI at arbitrary angles [12]. In this context, the CSL method is integrated into the CVD framework for its proven ability to accurately pinpoint vortex cores within ROIs labeled using the MV algorithm.

It is essential to clarify that the CSL method does not serve as a vortex detection mechanism itself. Instead, its primary function revolves around locating potential vortex cores and defining boundary radii within the ROIs. The method provides two-dimensional coordinates for a core, whether or not an actual vortex exists within the current ROI.

Each labeled ROI undergoes a CSL algorithm akin to the one described by Vollmers (2001) [12]. A schematic outlining the fundamental CSL procedure is presented in Figure 2, featuring a sample vortex for illustrative purposes. The schematic employs concentric ellipses to represent vorticity contours, with the vortex core identified by an '×'.

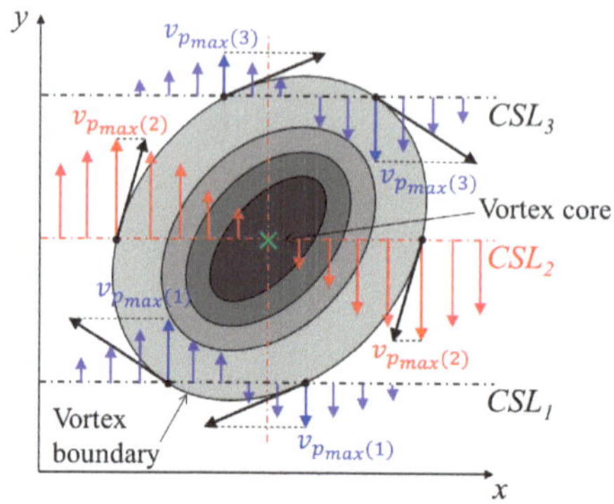

Figure 2. Schematic representation of the cross-sectional lines (CSL) procedure with four concentric ellipses illustrating vorticity contours and the vortex core indicated by an '×'.

In the context of a vector velocity field with dimensions $A \times B$, the y-component of velocity, represented as $v_y(i,j)$, is assessed for columns $i = 1, 2, 3, \ldots, A$ and rows $j = 1, 2, 3, \ldots, B$ along each row within the defined grid. During the discussion of the CSL method, this velocity component, $v_y(i,j)$, is referred to as the "perpendicular velocity," denoted as $v_p(i,j)$.

The determination of maximum and minimum perpendicular velocities for each row, designated as j, within a specified ROI is a pivotal aspect of the analysis. These extreme values are calculated as follows:

$$v_{p,\max\,(j)} = \max\left(v_p(A_s,j),\ v_p(A_s+1,j),\ldots,v_p(A_e,j)\right) \quad (2)$$

$$v_{p,\min\,(j)} = \min\left(v_p(A_s,j),\ v_p(A_s+1,j),\ldots,v_p(A_e,j)\right) \quad (3)$$

The indices A_s and A_e are used to denote the first and last velocity vectors within a ROI on a given row, respectively. Within the ROI, the row ψ that exhibits the difference $\left|v_{p,\max\,(j)} - v_{p,\min\,(j)}\right|$ is identified as the critical cross-section line and is represented as CSL_ψ. This 'critical line' corresponds to the y-coordinate of a potential vortex core, should one exist within the ROI under evaluation.

The perpendicular velocity along this critical row is denoted as $v_p(i,\psi)$. The x-coordinate of the vortex core is determined by the location along CSL_ψ where the velocity satisfies the condition

$$v_p(\varsigma,\psi) = \frac{\left(v_{p,\max\,(\psi)} + v_{p,\min\,(\psi)}\right)}{2} \quad (4)$$

and the coordinates of the vortex core are, therefore, represented as (ς,ψ).

For the sake of simplicity, Figure 2 illustrates a hypothetical vortex with only three rows/lines of the v_y velocity field. In this example, the critical perpendicular line is denoted as $CSL_\psi = CSL_2$. Among the three lines, CSL_2 exhibits the most significant difference between its maximum and minimum perpendicular velocities, $v_{p,\max\,(2)}$ and $v_{p,\min\,(2)}$, respectively. Consequently, the y-coordinate for the hypothetical vortex core is determined as $\psi = 2$.

Using Equation (4) along CSL_2, the velocity $v_p(\varsigma,2)$ is evaluated. The algorithm searches for the point along CSL_2 where $v_p(\varsigma,2)$ matches, thus finding the value of ς. These two computed coordinates effectively pinpoint the vortex core of a potential vortex within the ROI. However, it is important to note that this process alone does not confirm the presence of an actual vortex within the ROI.

The CSL method also offers an estimation of vortex size by fitting a circle with a radius r_v centered at the vortex core, which is determined as

$$r_v = \frac{\left|\varsigma_{v,\max\ (\psi)} - \varsigma_{v,\min\ (\psi)}\right|}{2} \tag{5}$$

Here, $\varsigma_{v,\max\ (\psi)}$ represents the column position on row ψ where the velocity vector $v_{p,\max\ (\psi)}$ is located, and $\varsigma_{v,\min\ (\psi)}$ refers to the column position on row ψ where the velocity vector $v_{p,\min\ (\psi)}$ is found.

The CSL method proves to be particularly effective for identifying vortex cores and estimating vortex size and drift velocity in flow fields where available information regarding the vortical structures present is scarce. Additionally, Vollmers (2001) [12] proposed that the transverse and streamwise components of the vortex drift velocity, represented as $\vec{v}_{drift} = \left(v_{drift,x}, v_{drift,y}\right)$, correspond to the x and y components of velocity evaluated at the core coordinate (ς, ψ). This algorithm operates on a predefined ROI, which is determined using the MV method. The algorithm comprises two main nested loops: one iterating over individual ROIs and the second for individual cross-sectional lines within each ROI. The input data include the 2C2D velocity field and the labeled ROI map.

A drawback of the CSL method lies in its inherent inability to reject false positives. It may identify a vortex core in an ROI that does not actually contain a vortex. To address this challenge, a hybrid detection algorithm is proposed that combines physical quantity-based methods, such as MV and CSL, with geometric methods integrated into the CVD framework, as geometric methods are known for their effectiveness at identifying and rejecting false positives, a capability often overlooked by other methods.

2.3. Winding Angle (WA) Method

The WA method, initially proposed by Portela in 1999 [44], represents a geometric approach to vortex detection. This method involves the assessment of discretized streamlines to determine their affiliation with a vortex. According to the criteria suggested by Sadarjoen and Post in 2000 [46], a streamline must meet two conditions to be classified as part of a vortex.

Firstly, for a streamline denoted as S_k, the winding angle $\alpha_{w,k}$ must satisfy the equation $\left|\alpha_{w,k}\right| = n2\pi$, where n is a positive integer [42]. The signed angle α_{k_i} between vectors $\vec{V_1}$ and $\vec{V_2}$ on a given streamline is expressed as

$$\alpha_{k_i} = \frac{cos^{-1}\left(\vec{V_1}\cdot\vec{V_2}\right)}{\left|\vec{V_1}\right|\cdot\left|\vec{V_2}\right|}\left(\frac{\vec{V_n}\cdot\left(\vec{V_1}\times\vec{V_2}\right)}{\left|\vec{V_n}\cdot\left(\vec{V_1}\times\vec{V_2}\right)\right|}\right) \tag{6}$$

where

$$\vec{V_1} = \left\langle\left(P_{x,(i-1)} - P_{x,(i-2)}\right), \left(P_{y,(i-1)} - P_{y,(i-2)}\right), 0\right\rangle \tag{7}$$

and

$$\vec{V_2} = \left\langle\left(P_{x,i} - P_{x,(i-1)}\right), \left(P_{y,i} - P_{y,(i-1)}\right), 0\right\rangle \tag{8}$$

and

$$\vec{V_n} = \langle 0, 0, 1\rangle \tag{9}$$

Here, P_x and P_y are the respective (x,y) location on a given streamline and $\vec{V_n}$ is a unit vector normal to the flow-field of interest. Figure 3 demonstrates how the individual angles $\alpha_{w,i}$ are calculated on a section from streamline S_k. The winding angle $\alpha_{w,k}$ for a streamline

S_k is determined by summing the angle contributions for all points along the streamline, as expressed in the following equation:

$$\alpha_{w,k} = \sum_{i=2}^{N} \alpha_{k_i} \tag{10}$$

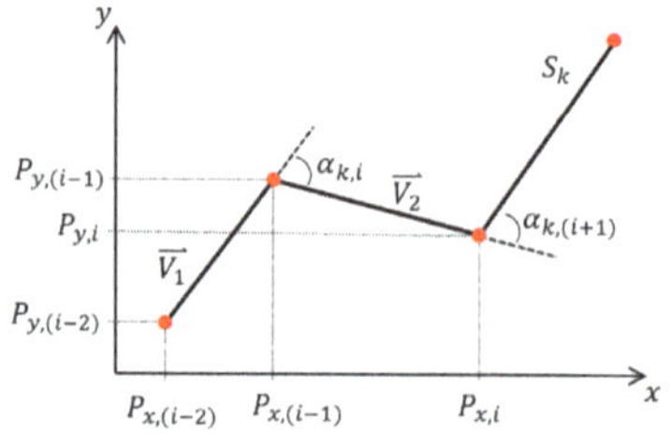

Figure 3. Computation of angle $\alpha_{w,i}$ on streamline S_k for the winding angle algorithm.

A crucial characteristic of a vortex-associated streamline is the formation of a closed semi-elliptical path [1,47,48]. However, relying solely on the winding angle criterion is insufficient to meet this requirement [46]. It is possible for a streamline to exhibit a substantial winding angle without tracing a closed path. Therefore, an additional condition is necessary.

The second requirement dictates that a streamline deemed part of a vortex must have its initial and terminal points in close proximity [42]. The distance between the starting point ($P_{x,1}$, $P_{y,1}$) and the endpoint ($P_{x,N}$, $P_{y,N}$) of a streamline S_k is calculated using

$$D_{\text{se}} = \sqrt{(P_{x,N} - P_{x,1})^2 + (P_{y,N} - P_{y,1})^2} \tag{11}$$

The initiation point of a streamline is predetermined, and each streamline consists of a fixed number of points with uniform spacing between them. Consequently, for a streamline tracing a closed path, its starting and ending points can be anywhere along the path, as the streamline's length is pre-defined. To address this challenge in establishing a D_{se} threshold, consider a scenario where streamlines follow a closed circular path. In such cases, the maximum separation between the starting and ending points of any streamline following a closed circular trajectory would equal the circle's diameter, which outlines the path. This presents a significant challenge in setting a D_{se} threshold, as streamlines associated with larger vortices would require a larger D_{se}. However, an excessively large D_{se} threshold may incorrectly associate non-vortical streamlines with relatively small vortices. To mitigate this issue, streamlines are split whenever the cumulative angle sum reaches the highest multiple of 2π. This technique ensures that the beginning and end points of a streamline on a closed loop remain as close as possible.

While the WA threshold is defined as $|\alpha_{w,k}| = n2\pi$, selecting a suitable value for the threshold D_{se} is necessary. The D_{se} threshold depends on the length scale of the vortical structures present, but the previously defined ROIs can aid in determining a sensible D_{se} threshold. Within each ROI, circles with a radius r_v are used to identify potential vortices. This parameter serves as a reasonable length scale for coherent structures in the flow. Therefore, an appropriate choice for the D_{se} threshold would be a value greater than $r_v/2$. This criterion ensures that only streamlines with minimal endpoint separation are considered as part of a vortex. Streamlines that extend significantly beyond the typical size of flow structures are excluded to maintain the accuracy of vortex identification.

The final step of the algorithm involves verifying whether individual streamlines adhere to the specified thresholds. Streamlines meeting all required criteria are marked and assigned unique identifiers. The sign of the winding angle $\alpha_{w,k}$ for each streamline determines the direction of rotation associated with the vortex corresponding to that streamline.

In situations where an ROI encompasses multiple vortices, it becomes essential to appropriately label all the closed streamlines associated with these vortices. This labeling process involves mapping each of the closed streamlines to a specific point and subsequently identifying clusters of closely situated points [42]. The mapping of the streamlines S_k belonging to a particular vortex is achieved by associating them with corresponding points MP_k, which are determined by

$$\mathrm{MP}_{x,k} = \frac{1}{N}\sum_{i=1}^{N} P_{x,i} \tag{12}$$

and

$$\mathrm{MP}_{y,k} = \frac{1}{N}\sum_{i=1}^{N} P_{y,i} \tag{13}$$

To ascertain the distance between two such mapped points, MP_k and $MP_{k'}$, the following computation is applied:

$$D_{MPk,k'} = \sqrt{\left(MP_{x,k} - \mathrm{MP}_{x,k'}\right)^2 + \left(\mathrm{MP}_{y,k} - \mathrm{MP}_{y,k'}\right)^2} \tag{14}$$

The labeling process begins with the first point, MP_1, serving as the foundation for the initial cluster group. When considering point MP_2, a crucial criterion is whether the distance between MP_1 and MP_2 falls within a specified tolerance. If it does, MP_2 is assigned to the existing group labeled as group 1. Conversely, if the distance exceeds the defined tolerance, MP_2 initiates the formation of a new cluster group.

Moving on to point MP_3, the algorithm evaluates two distances, namely $D_{MP_{3,1}}$, and $D_{MP_{3,2}}$. If neither of these distances meets the tolerance criteria, MP_3 takes the role of establishing yet another new cluster group. However, if either $D_{MP_{3,1}}$, or $D_{MP_{3,2}}$ falls within the tolerance range, MP_3 is assigned to the nearest existing group, ensuring optimal grouping based on proximity.

This meticulous labeling process is applied consistently to all other points, ensuring that each point is appropriately categorized within the designated cluster groups. Finally, after all the streamlines have been meticulously labeled, the algorithm proceeds to compute the arithmetic average for each cluster group.

The WA algorithm implemented in this study utilizes a three-tiered nested loop structure: the first loop investigates the ROIs, the second loop delves into the instantaneous streamlines within each ROI, and the third loop focuses on the individual grid points comprising each streamline. This intricate process operates on the input data, which consist of streamlines computed from local velocity fields with a reference velocity equal to the drift velocity of the corresponding ROI.

It is important to note that the WA method's accuracy is closely tied to the precise selection of the reference velocity for streamline computation. Deviations from the actual vortex core's drift velocity can lead to inaccuracies in estimating the vortex boundary. While the WA method may not excel in providing accurate estimates of vortex size and shape, it exhibits remarkable robustness in effectively eliminating false positives, as demonstrated by previous research [42]. Additionally, it serves as a valuable means of validating the accuracy of other vortex detection techniques.

Typically, verifying the presence of a vortex in a flow field involves visual inspection, but this approach becomes impractical when dealing with large datasets. As such, the WA method assumes the role of a swirling flow verification tool once potential vortex core locations and drift velocities have been computed through other detection methods.

2.4. Comparison of MV to other Detection Methods

Vorticity magnitude alone cannot distinguish between shear generated and free shear vorticity. However, in the current study, MV serves only as an initial means of identifying

ROIs. Detection methods such as Q-criterion, λ_2-criterion, or Δ-criterion are far less likely to identify a shear region, as a vortex, than the MV method and ultimately produce less false positives but are more complex [12,49]. The simplicity of the MV method makes it a good candidate for initially identifying ROIs in the flow and to effectively study the WA's ability to reject shear generated ROIs by deliberately identifying some of these shear regions as potential vortices. It is also important to consider that the vortices in the pitching airfoil flow-field have vastly different vorticity magnitudes. The Q, λ_2, and Δ methods have difficulty distinguishing between individual vortices with potentially overlapping cores and selecting the right threshold in order to distinguish individual vortices can be complex [49]. For example, in incompressible flows, the Q-criterion identifies vortices as regions where the vorticity magnitude prevails over the strain-rate magnitude [13,18]. It accomplishes this by finding connected regions in the flow field having a second invariant of the velocity gradient tensor that is less than a negative threshold value II_E [13] shown as

$$II < -II_E \tag{15}$$

The second invariant of the velocity gradient tensor (II) is expressed as

$$II = \frac{\partial u_i}{\partial x_j}\frac{\partial u_j}{\partial x_i} \tag{16}$$

As a result, vortices with overlapping cores will blend together and appear as a single vortex without the proper threshold value $-II_E$.

When the entire vortex identification procedure is carried out in a single method, and it is not guaranteed that the flow is free shear, then it is favorable to use a more elaborate method such as the Q-criterion, Δ-criterion, or λ_2-criterion.

In summary, MV is favored as an ROI identifier for several reasons:

- MV is not employed as a standalone detection method; instead, it is integrated with multilevel thresholds, CSL, and WA methods.
- The flow field comprises vortices with a wide range of magnitudes and potentially overlapping cores. Therefore, it is recommended that the use of multilevel thresholds and adjusting WA thresholds is more intuitive, given that the threshold unit is expressed as a vorticity magnitude.
- Unlike other methods such as Q-criterion, Δ-criterion, or λ_2-criterion, MV provides information about the direction of rotation. This additional feature aids in distinguishing closely spaced vortices by symmetrically extending the multilevel threshold.
- MV's simplicity is valuable in the development of a combinatorial method as it facilitates the isolation of each method's contribution.
- For the purpose of evaluating WA's effectiveness in identifying and eliminating false positives, it is useful to include some shear-generated ROIs (false positives).
- The computation of derivatives required by the Q-criterion, Δ-criterion, or λ_2-criterion methods is computationally demanding and sensitive to noise, which is often present in experimental data.

3. New Combinatorial Vortex Detection (CVD) Method

The CVD method has been developed through the integration of three distinct detection algorithms, MV, CSL, and WA, alongside straightforward image morphology techniques. The primary objective in crafting this method was to ensure its consistent and reliable capability to detect and characterize multiple vortices within velocity vector maps generated by PIV. The CVD method's functionalities encompass labeling each vortex within the flow field, pinpointing vortex cores, determining drift velocities, calculating circulation, assessing peak vorticity, and estimating boundary radii for individual vortical structures. Above all, the method serves the crucial purpose of reducing the dimensions of the original dataset while preserving essential vortex parameters accurately.

The CVD algorithm is succinctly outlined in the flow chart presented in Figure 4. Within the flowchart, the three individual detection algorithms are represented by green boxes, and the input and outputs are indicated in grey boxes, encompassing the 2C2D velocity vector field and the characterized vortex field, respectively. To begin, a 2D scalar vorticity field is derived from the global velocity map. This vorticity field undergoes indexing through the multilevel threshold algorithm, leading to the generation and labeling of sensible ROIs via the MV technique.

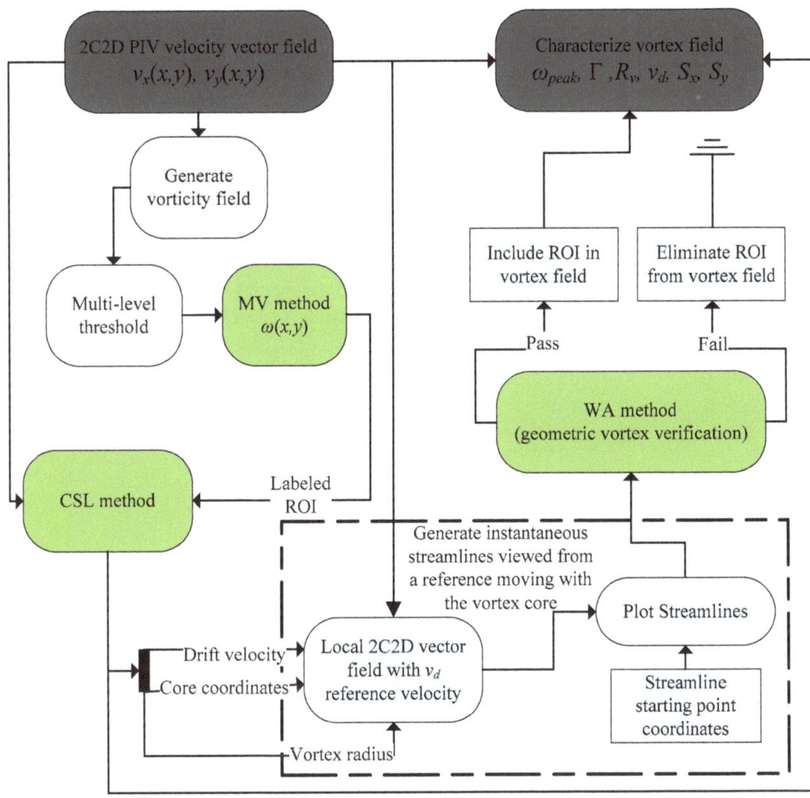

Figure 4. Flowchart of the combinatorial vortex detection (CVD) algorithm, utilizing three detection methods, MV, CSL and WA (green boxes), to process a 2C2D velocity vector field and generate characterized vortex data.

Subsequently, the CSL method is employed to evaluate each ROI individually, providing coordinates for the vortex core, boundary radius, and drift velocity. One of the CVD method's standout features is its reliance on Galilean invariant indicators for vortex core detection. The CSL method, which searches for critical points along perpendicular velocity profiles, ensures consistent and reliable identification of vortex cores, independent of uniform flow shifts. This robust approach significantly broadens the method's applicability in dynamic flow scenarios. However, it is essential to note that the CSL algorithm yields these vortex parameters regardless of the presence of an actual vortex, underscoring its inability to ascertain vortex existence within a given ROI.

The WA method serves the purpose of validating the presence of a vortex within an ROI. This is achieved by the WA method searching each individual ROI for closed streamlines and automatically confirming or denying the presence of one or multiple vortices. By identifying vortex-affiliated streamlines based on their inherent geometry, independently of any symmetry assumptions, this approach allows for the accurate detection of vortices in complex, asymmetric flows, even in scenarios involving significant vortex interactions. It is noteworthy that, for consistency with Robinson's definition [36], the WA method necessitates the computation of instantaneous streamlines from a reference frame moving with the vortex core [12]. Consequently, prior knowledge of the vortex, specifically its core coordinates and drift velocity, is required. This poses a unique challenge as, unlike CSL and MV, WA cannot be conducted within the global coordinate system. Instead, local velocity vector maps must be generated for each ROI, with the drift velocity, calculated from CSL, subtracted before computing the streamlines. Ultimately, the WA method either accepts an ROI and labels it as a vortex or rejects it.

The CVD algorithm has been implemented in MATLAB. The code is freely available on GitHub (link in Supplementary Material). This includes an example vector field generated from a commercial PIV software (DaVis 8.1, LaVision GmbH Göttingen Germany) and examples in the form of videos of the oscillating wing, determined vorticity field, and detected vortices.

4. Verification on a Simulated Flow Field

To determine the minimum required number of velocity vectors, obtained through PIV, for the CVD algorithm to accurately identify and describe a vortex, an investigation was conducted using a simulated flow field measuring 15 mm × 15 mm. This simulated field contained an axisymmetric Burgers vortex positioned at the center of the region. Various levels of discretization of the velocity field were employed, enabling an examination of vector density per vortex.

The Burgers vortex solution serves as a well-known model for illustrating key aspects of modern turbulence theory [50]. This vortex model offers an exact solution to the cylindrical Navier–Stokes equations, depicting the flow on a cylindrical vortex core inducing a circulation denoted as Γ_∞ at large distances [51]. For an axisymmetric Burgers vortex in a fluid with kinematic viscosity ν and radius r_v , the distribution of circumferential velocity v_θ and vorticity ω are [52]

$$v_\theta(r) = \frac{\Gamma_\infty}{2\pi r}\left(1 - e^{-\frac{\gamma r^2}{4\nu}}\right) \tag{17}$$

and

$$\omega(r) = \frac{\gamma\,\Gamma_\infty}{4\pi r}\left(e^{-\frac{\gamma r^2}{4\nu}}\right) \tag{18}$$

where the parameter γ represents the axial strain $(\partial w/\partial z)$ within a velocity field $\vec{u}$ described by an irrotational pure strain component denoted as $\vec{u}_s = (\alpha x,\ \beta y, \gamma z)$ and a rotational component confined to the x-y plane denoted as $\vec{u}_w = (u_x, u_y, 0)$. Specifically, for the case of an axisymmetric Burgers vortex, $\gamma > 0$ and $\beta = \alpha = -\gamma/2$.

The simulated flow field consists of (n,m) velocity vectors, denoted as $\vec{v}(i,j)$ with indices i in the x-direction and j in the y-direction, expressed by $\vec{v}(i,j) = \left(\vec{v}_x(i,j), \vec{v}_y(i,j)\right)$. In this simulation, an axisymmetric Burgers vortex is modeled, which travels in the negative y-direction at a freestream velocity of $U_\infty = -2.25$ mm/s. The vortex exhibits a peak vorticity of $\omega_{peak} = -3.96\ \mathrm{s}^{-1}$, a boundary radius of r_v =3.18 mm, and induces a circulation of Γ_∞ =100 mm^2/s at far distances.

The effects of changing the grid resolution of the flow field and introducing Gaussian white noise to the velocity vectors are investigated. This simulation aims to establish the minimum grid resolution and the acceptable level of velocity field noise that ensures the accurate application of the CVD algorithm to a universal flow field. Figure 5 illustrates the simulated vector field, with the background color map representing velocity magnitude. Additionally, Figure 6 displays the simulated velocity field after subtracting the freestream velocity, accompanied by a background color map representing vorticity.

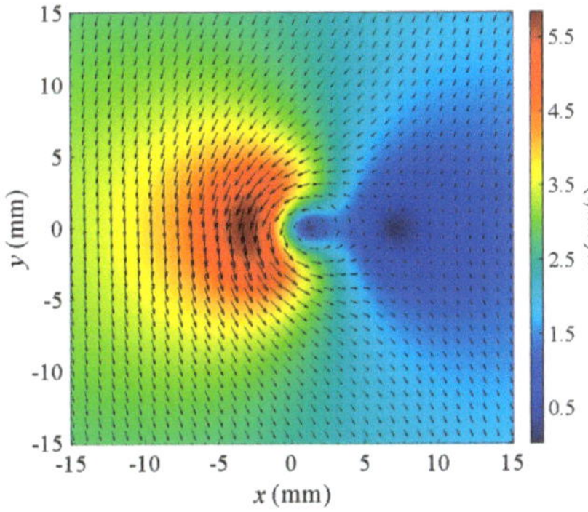

Figure 5. Simulated velocity vector field (200 × 200 resolution) with velocity magnitude represented by a color map, showing every 6th vector.

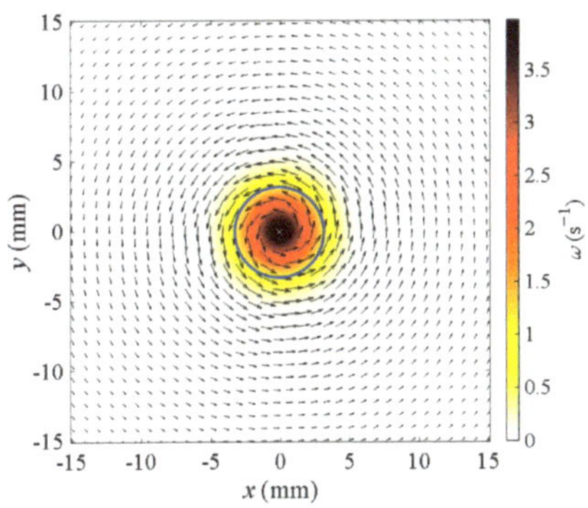

Figure 6. Simulated velocity vector field (200 × 200 resolution) with the freestream velocity (U_∞) subtracted, displaying every 6th vector, and featuring vorticity as a color map in the background.

The resolution is quantified by the number of velocity vectors (*n*,*m*) spanning the diameter of the vortex, which corresponds to $\varnothing_v = 2r_v$. This ratio, $nm/\varnothing_v$, defines the minimum necessary number of velocity vectors spanning a vortex's diameter for accurate identification and characterization. Figure 7 illustrates the vortex radius calculated by the CVD algorithm across different $nm/\varnothing_v$ values, alongside the known vortex radius employed in the Burgers model. Vortices with $nm/\varnothing_v < 5$ are categorized as Type II errors (as described later). Considering the expected radius of $r_v = 3.18$ mm, the computation of vortex radius for $r_v = 3.18 \text{ mm} \pm 5\%$ is achieved when $nm/\varnothing_v > 17$. Figure 7 also exhibits discretization artifacts originating from the simulation.

Figure 8 depicts the vortex circulation (Γ) calculated by the CVD algorithm across various $nm/\varnothing_v$ values. Circulation is not computed for cases with $nm/\varnothing_v < 5$, as these vortices are rejected by the WA method. The circulation derived using the CVD algorithm pertains specifically to the core region circulation, representing the circulation induced within the vortex boundaries within $-r_v \leq r \leq r_v$. This value is expected to be smaller than $\Gamma_\infty = 100 \text{ mm}^2/s$, which corresponds to the circulation induced by the vortex at far distances (as $r_v \to \infty$). Using $\Gamma_\infty = 100 \text{ mm}^2/s$ is impractical in this study due to interference from neighboring vortices, making it unobtainable experimentally. Measuring core region circulation is a more feasible approach to gauge vortex strength, as it can be calculated within a finite circle with a radius of r_v, within which most of the vorticity is concentrated [53]. Throughout this section, the term 'circulation Γ_{core}' refers to the core region circulation, and the expected core region circulation is determined as

$$\Gamma_{core} = \int_{\theta=0}^{\theta=2\pi} \int_{r=-r_v}^{r=r_v} \frac{\gamma\Gamma_\infty}{4\pi\nu}\left(e^{-\frac{\gamma r^2}{4\nu}}\right) r dr d\theta \quad (19)$$

By substituting $r_v = 3.18$ mm into Equation (19), the expected core circulation of the simulated vortex is determined to be $\Gamma_{core} = 71\ \text{mm}^2/s$. Computation of core circulation within $\Gamma_{core} = 71\ \text{mm}^2/s \pm 5\%$ range is achieved when $nm/\varnothing_v > 25$, as shown in Figure 8. For lower values of $nm/\varnothing_v$, the circulation is consistently underestimated, primarily due to discretization errors in the summation of vorticity pixels within the vortex boundaries. The grid resolutions used for calculating vorticity radius (r_v) and core circulation (Γ_{core}) are summarized in Table 1 for reference.

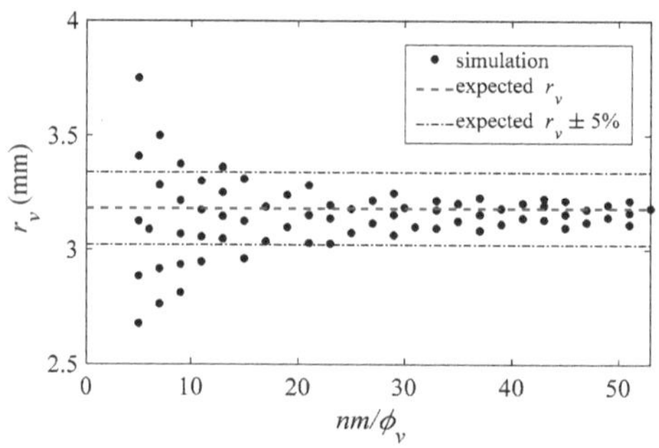

Figure 7. Vortex radius variations calculated by the CVD algorithm for different $nm/\varnothing_v$ values, compared with the expected radius ($r_v = 3.18$ mm $\pm 5\%$) in the Burgers model. Vortices with $nm/\varnothing_v < 5$ are disregarded. Discretization artifacts from the simulation are also evident.

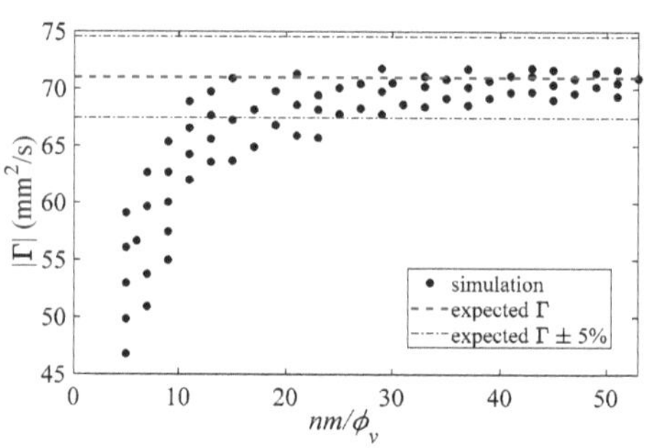

Figure 8. Circulation data are excluded for cases with $nm/\varnothing_v < 5$, as rejected by the WA method. The computed circulation pertains to the core region circulation, representing circulation induced within the vortex boundaries ($-r_v \leq r \leq r_v$).

Table 1. Grid resolution thresholds for simulated flow field.

Threshold	$nm/\varnothing_v$
Type II error	$nm/\varnothing_v < 5$
Computation of r_v within 5% of true value	$nm/\varnothing_v > 17$
Computation of Γ within 5% of true value	$nm/\varnothing_v > 25$

Having established the minimum grid resolution requirements using an ideal simulated flow field, the robustness of the CVD algorithm to noise is now examined. Specifically,

Gaussian white noise is deliberately introduced to the velocity vector components to determine the impact on vortex characterization. A simulated vector field with $n = m = 200$ was chosen, yielding a resolution of $nm/\varnothing_v = 43$, which satisfies the minimum requirements for accurate computation of circulation and radius, as determined previously. Gaussian white noise is generated by adding random real number vectors to the x and y components of each velocity vector, where both components follow a probability distribution with a zero mean and a predetermined variance. The noise vectors are statistically independent and have a standard deviation of ε_σ times that of the vector fields.

The impact of adding white noise on vortex characterization is assessed by calculating the standard deviation and mean values of both the core region circulation and the detected vortex radii across a range of noise levels. For this study, 10 noise levels ranging from $0 \leq \varepsilon_\sigma \leq 0.2$ with 100 samples per noise level were investigated. Beyond $\varepsilon_\sigma = 0.2$, the algorithm starts to fail in detecting the vortex. Figure 9 presents the mean circulation results, along with standard deviation error bars, and the expected circulation value for each of the 10 noise levels examined. Similarly, Figure 10 displays the mean vortex radius results for each of the noise levels.

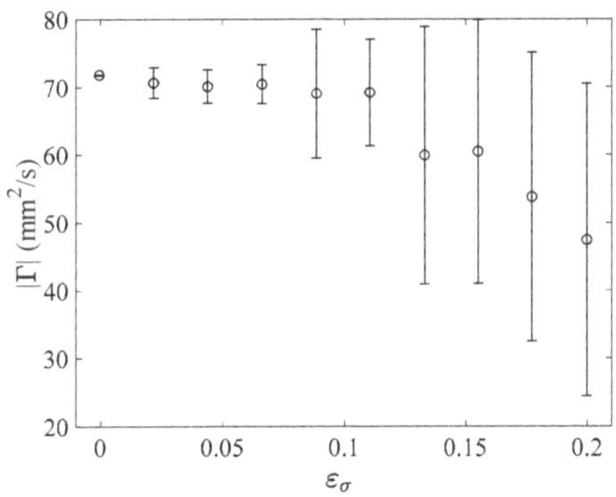

Figure 9. Mean circulation computed by the CVD for 100 simulations at various noise levels, along with standard deviation error bars.

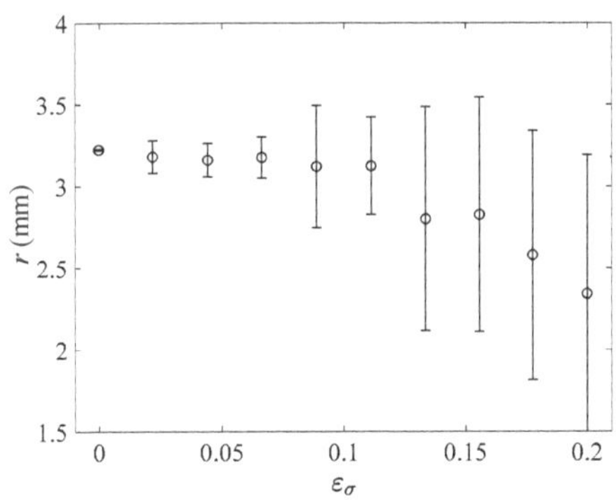

Figure 10. Mean radius computed by the CVD for 100 simulations at various noise levels, along with standard deviation error bars.

The velocity vectors near the center of the vortex exhibit larger magnitudes compared to those near the boundary. Consequently, the impact of adding white noise is less pronounced near the core. However, increasing ε_σ eventually results in the entire vector field appearing incoherent. Notably, this incoherence starts from the outer regions and progresses inward due to the increasing velocity magnitude in the negative radial direction.

Consequently, as ε_σ increases, the detection algorithm tends to underestimate the size of the vortex. It primarily identifies the center of the vortex, as this region is relatively less affected by the added noise (or is at least affected to a lesser extent, still resembling a vortex). Figure 11 illustrates the probability density function (PDF) of the boundary radius of a detected vortex based on 100 simulations conducted with ε_σ values of 0.05 and 0.1. For the higher noise level ($\varepsilon_\sigma = 0.1$), a distribution with a negative bias in its mean relative to the expected value of 3.18 mm is observed, indicating that the algorithm tends to underestimate the size of detected vortices at higher noise levels, as anticipated. This highlights the robustness of the CVD algorithm under controlled conditions. The CVD algorithm was able to detect and importantly return the characteristics of a vortex defined by Burger's model. The introduction of noise to the velocity field did affect that detection and determination of the characteristic circulation but was only detrimental at high noise levels.

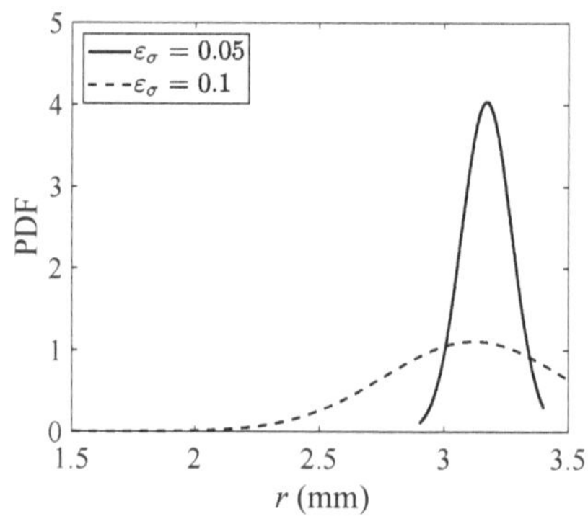

Figure 11. Probability density function of boundary radius for detected vortices based on 100 simulations at ε_σ values of 0.05 and 0.1, i.e., 5% and 10% noise, respectively.

5. Verification on an Experimental Flow Field

The velocity flow field immediately downstream from an oscillating NACA 0012 airfoil is captured using a PIV system with the experimental setup described below. The CVD approach is used to identify and characterize vortical structures.

5.1. Experimental Facility and PIV Setup

The experimental facility was a water flume measuring 0.7×0.4 m ($27.5'' \times 16''$), designed to exhibit low turbulence characteristics as detailed by Hilderman (2004) [54]. Within this setup, an airfoil is vertically suspended, allowing it to extend through the free surface into the water channel, positioned perpendicular to the upstream flow direction. The experimental setup and subsequent analysis were conducted in a two-dimensional domain, focusing on the planar cross-section of the wing's wake.

The experimental PIV setup, depicted in Figure 12, was equipped with four 2112×2072 pixel resolution, 14-bit, dual-frame CCD cameras (Imager Pro X 4M, LaVision). These cameras view the investigation plane through four independent local coordinate systems. Calibration of the cameras was carried out using a 300 mm $\times$ 800 mm calibration target featuring 1.3 mm diameter markers spaced at 3 mm intervals. Subsequently, the captured images were de-warped to account for variations in camera viewing angles and were stitched together with specified offsets to create a unified global field.

To mitigate surface refraction effects, an acrylic sheet was positioned at the free surface, with the cameras capturing images through it. In the illumination process, a double-pulse Nd:YAG laser (Solo III-15Hz, New Wave Research Fremont, California, United States) was employed to illuminate the flow. This flow was seeded with 18µm hollow glass spheres

(SPHERICEL®, Potters Industries, Malvern, PA, USA). The laser beam was focused into a thin sheet and directed upstream by a mirror and a submerged periscope.

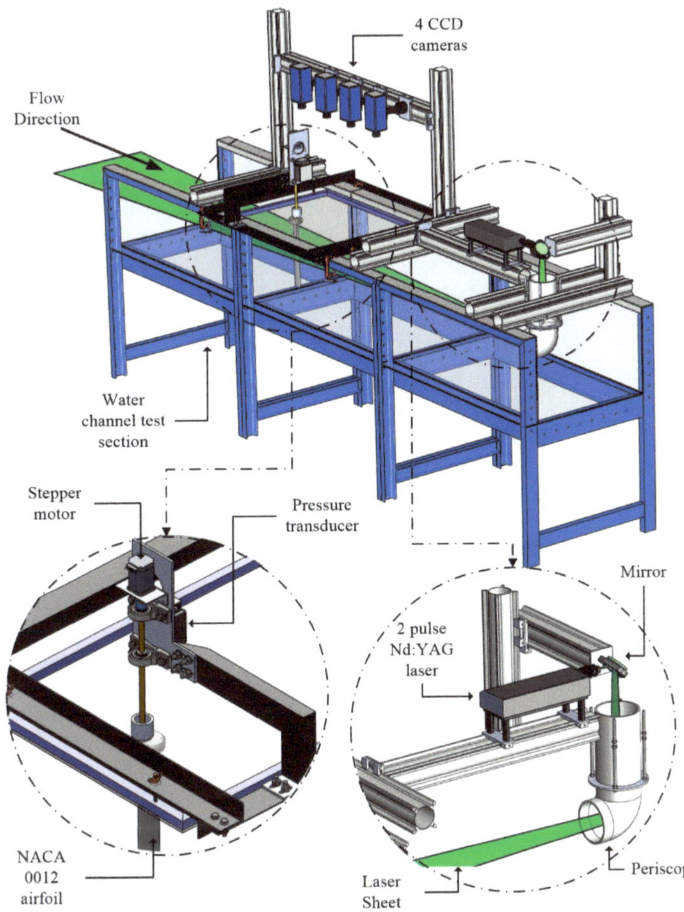

Figure 12. Experimental setup with a 0.7 × 0.4 m water channel designed for low turbulence. It includes a vertically suspended airfoil extending into the channel and a PIV setup with four CCD cameras capturing the measurement plane.

Before performing cross-correlation on image pairs, the raw images underwent preprocessing using commercial software (LaVision GmbH, DaVis 8.05). This preprocessing step was essential as it enhances particle intensity and shape, ultimately resulting in an improved correlation peak [55]. The strength of correlation can be affected by disparities in image intensity caused by factors such as non-uniform light sheets, shadows, reflections, or variations in particle size [55]. To enhance correlation strength, several image preprocessing methods were employed, including background intensity subtraction, sliding minimum subtraction, and min–max filtering for intensity normalization. In this study, the generation of the vector field involved three passes of a cross-correlation algorithm with window shifting. The first pass used a 64×64 pixel interrogation window, while the subsequent two passes employed a 32 ×32 pixel interrogation window with a 50% overlap.

An important consideration in this study is the error that arises from the disparity between the motion of seed particles (r) and the actual fluid motion. This error is primarily attributed to particle slip, wherein the seed particles lag behind the fluid motion by a finite quantity. To compute the slip velocity to first order, the approach outlined by Adrian and Westerweel (2011) [56] was utilized:

$$\left|v_p - v_f\right| = \left[\frac{(\overline{\rho} - 1)g\tau_0}{\overline{\rho}}\right] \quad (20)$$

In this equation, v_p represents the velocity of the seed particle, v_f is the fluid velocity, and g is the gravitational constant with a value of 9.81 m/s^2. The time constant, τ_0, is defined as

$$\tau_0 = \frac{\rho_p d_p^2}{18 v_f \rho_f} \quad (21)$$

where d_p is the diameter of the seed particles, set to 18 μm, and the densities of the seed particle and water are $\rho_p = 600$ kg/m^3 and $\rho_f = 998$ kg/m^3, respectively. Additionally, the density ratio, $\overline{\rho}$, is defined as

$$\overline{\rho} = \rho_p/\rho_f \quad (22)$$

The slip velocity error, which amounts to 0.47% relative to the freestream velocity ($U_\infty = 0.017$ mm/s), is initially approximated using Equation (20). To address variations in image magnification across the image domain, calibration with a target is employed. However, it is important to note that image magnification also varies along the thickness of the light sheet, introducing a magnification uncertainty typically around 0.3% in most PIV setups, as discussed by Adrian and Westerweel (2011) [56].

Furthermore, the measurement error associated with determining the precise location of the correlation peak for an 8-pixel particle displacement is approximately 1–2% of the full-scale velocity for similar planar PIV systems [56]. Lastly, in terms of event timing accuracy, the system achieves a resolution of 10 ns with a jitter of less than 1 ns.

In summary, the key measurement accuracy specifications of the present PIV system include a particle slip velocity error of 0.47% relative to the freestream velocity, 0.3% uncertainty in image magnification, 1–2% error in determining particle displacement correlation peaks, and 10 ns timing resolution with sub-ns jitter.

The aluminum airfoil used was a continuous extrusion featuring a cross-sectional shape conforming to the NACA 0012 airfoil profile, with a chord length (C) of 75 mm. The airfoil's cross-section and the parameters governing its motion are visually represented in Figure 13. The airfoil is suspended on a sturdy shaft, which is driven by a stepper motor (PK258-02DI, ORIENTAL MOTOR CO. LTD., Tokyo, Japan). This shaft passes through the airfoil's aerodynamic center, enabling precise pitch oscillations of any desired waveform.

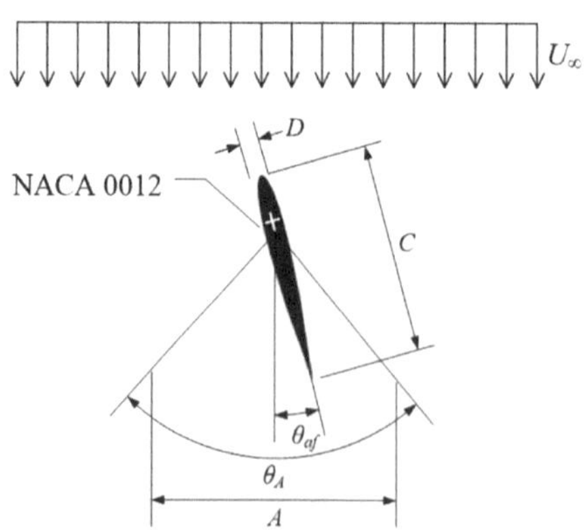

Figure 13. Representation of the aluminum airfoil, a continuous extrusion with an NACA 0012 cross-sectional shape and a chord length (C) of 75 mm. The figure also illustrates the motion parameters, including pitch oscillations, enabled by a sturdy shaft driven by a stepper motor passing through the airfoil's aerodynamic center.

To orchestrate these oscillations and maintain control, a commercial hardware system (DS1104, dSPACE Inc., Paderborn, Germany) was programmed. This system governs the motion of the stepper motor and generates output trigger signals when the airfoil's pitch reaches the desired angle. Consequently, this setup synchronizes the imaging system to capture data at predefined positions, facilitating the acquisition of both phase-averaged and time-averaged datasets.

To maintain a well-controlled flow and avoid the formation of disruptive leading-edge vortices in the wake, the airfoil, characterized by its chord length (C), is limited to small pitch oscillation amplitudes ($\theta_A \leq 8°$). The flow can be characterized by the Reynolds number (*Re*), defined in terms of the airfoil's chord thickness (*D*) as

$$Re = \frac{U_\infty D}{\nu} \quad (23)$$

In the experiments conducted in this study, a constant freestream velocity (U_∞) of 17 mm/s was maintained, using an airfoil thickness (*D*) of 8.6 mm and a kinematic viscosity of water (ν) equal to 1×10^{-6} m^2/s, resulting in a Reynolds number (*Re*) of 146.

The oscillation waveform of the airfoil is given by

$$\theta_{af}(t) = \frac{\theta_A}{2}\sin(2\pi f t) \quad (24)$$

By adjusting the oscillation amplitude (θ_A) and frequency (*f*), various wake conditions can be achieved.

5.2. Analysis of Oscillating Airfoil Wake

A schematic depicting the oscillating airfoil and the flow features observed in its wake is illustrated in Figure 14, highlighting the relationship between the airfoil and the generated vortices. The use of small airfoil oscillation amplitudes results in an orderly wake pattern, characterized by the shedding of precisely two counter-rotating vortices per oscillation cycle, as detailed by Bohl and Koochesfahani (2009) [2]. The airfoil, with a chord length C, is shown in the context of the uniform flow with a velocity U_∞.

These vortices are organized into two distinct rows separated by a distance S_y and aligned with the flow direction. Within each row, the vortices are spaced apart by S_x. The core coordinates of the vortices are labeled, and a core boundary radius (r_v) is defined.

Additionally, the drift velocity, $\vec{v}_{drift} = \left(v_{drift,x}, v_{drift,y}\right)$, is represented by an instantaneous velocity vector at the grid point coinciding with the vortex core. Furthermore, the vortices are characterized by their peak vorticity (ω_{peak}) and circulation (Γ). The circulation is approximated by summing the vorticity at each velocity measurement location determined by PIV within a radius r_v of a vortex as follows:

$$\Gamma = \iint_S \left(\vec{\nabla} \times \vec{v}\right) ds = A_{pixel} \sum_{i=1}^{N_i} \sum_{j=1}^{N_j} \omega_{i,j} \quad (25)$$

Here, *s* refers to the surface integral over area *ds*, and A_{pixel} is the surface area of a rectangle formed from the coordinates of 4 adjacent velocity vectors.

By manipulating the parameters θ_A and *f*, various wake configurations can be achieved. Common wake patterns for sinusoidal pitching symmetric airfoils include the following [57–60]:

1. von Karman Wake: This configuration, illustrated in Figure 15a, resembles the wake pattern typically associated with vortex shedding from a cylinder.
2. Aligned Wake: In this arrangement, the transverse separation distance S_y between vortices approaches zero.

3. Inverted von Karman Wake: This wake pattern, illustrated in Figure 15b, differs from the von Karman wake by reversing the rotation direction of each row of vortices, as described by Godoy-Diana et al. (2009) [10].

At an airfoil pitch angle of 0°, instantaneous velocity vector fields are computed for 100 wake datasets using PIV. These fields are then averaged to generate a phase-averaged representation since all the data are collected at a fixed oscillation phase. Subsequently, WA and CSL methods are applied to the phase-averaged fields, and the results are depicted in Figure 15. Within Figure 15, the CSL method identifies vortex core locations, represented as white crosses, and vortex boundary radii, denoted by black circles, superimposed on the vorticity field.

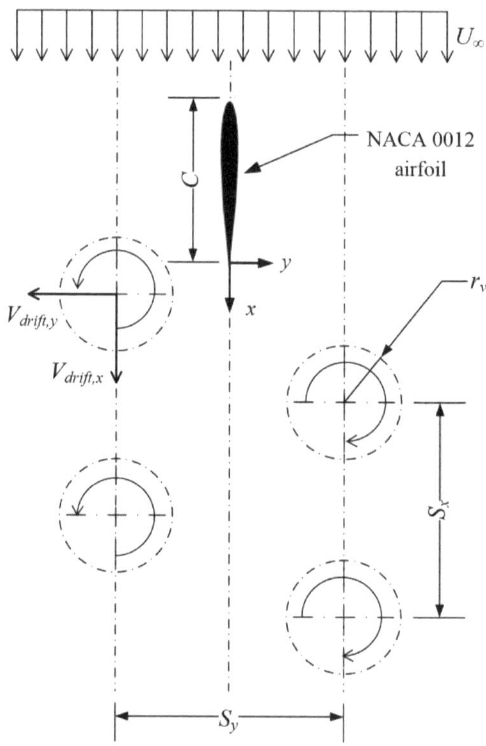

Figure 14. Schematic representation of the oscillating airfoil and the associated wake flow features. The airfoil, with chord length C, interacts with a uniform flow (U_∞), generating organized vortices. These vortices form two distinct rows, separated by S_y and spaced apart by S_x, aligned with the flow direction. Core coordinates and the core boundary radius (r_v) are identified. The drift velocity is represented by an instantaneous vector at the vortex core grid point.

Figure 15a illustrates the von Karman wake pattern, resulting from sinusoidal pitching at $f = 1.6$ rad/s and $\theta_A = 8°$, while Figure 15b showcases the inverted von Karman wake pattern generated by sinusoidal pitching at $f = 3.4$ rad/s and $\theta_A = 8°$. The von Karman wake typically displays weak vortices characterized by low peak vorticity and circulation. It exhibits minimal vortex decay over downstream distances and features larger streamwise spacing (S_x). Conversely, the inverted von Karman wake shown in Figure 15b exhibits vortices with a higher peak vorticity and circulation. These vortices also experience the most rapid decay as they progress downstream and display relatively smaller streamwise spacing (S_x). These two wake configurations at $Re = 146$ exemplify the extreme cases of wake vorticity for $\theta_A = 8°$.

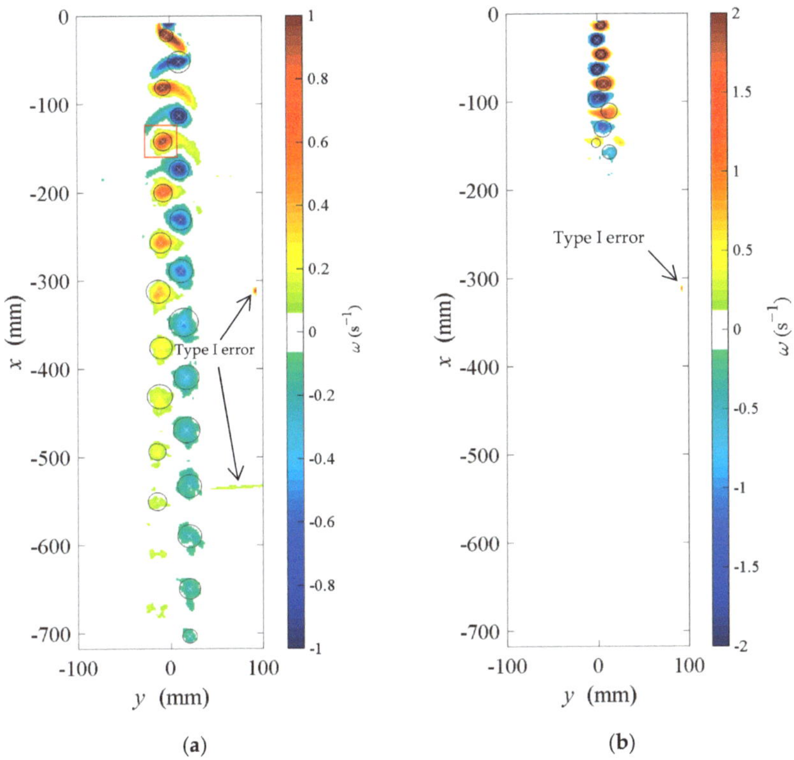

Figure 15. Phase-averaged results obtained at an airfoil pitch angle of 0 degrees using PIV data. The cross-section lines (CSL) method identifies vortex core locations (white crosses) and vortex boundary radii (black circles), superimposed on the vorticity field; (**a**) showcases the von Karman wake pattern generated by sinusoidal pitching at $f = 1.6$ rad/s and $\theta_A = 8°$, while (**b**) illustrates the inverted von Karman wake pattern resulting from sinusoidal pitching at $f = 3.4$ rad/s and $\theta_A = 8°$.

A false positive vortex diagnosis, or a Type I error, occurs when a vortex is identified where none exists [35], potentially due to local shear, boundary influences, or image stitching inaccuracies. Figure 15 highlights this issue with the CSL method, which may incorrectly confirm a ROI containing no vortex. However, when the CSL data are processed through the WA algorithm, these false positives are eliminated, as it checks for streamlines forming semi-closed, semi-elliptical paths, which are absent in the false detections.

The Supplementary Material includes animations (Supplemental Files S1 and S2) featuring instantaneous velocity vector fields for the von Karman Wake and inverted von Karman Wake cases. These animations dynamically illustrate vortex core locations and vortex boundary radii superimposed on the vorticity field for each wake configuration.

5.2.1. Sample Experimental Vortex for WA and CSL Verification

Figure 16 provides a detailed view of a vortex field sample from Figure 15a, highlighted within a red square, and includes a locally calculated velocity field centered on the vortex core. This calculation employs a moving reference frame that matches the drift velocity of the sample vortex, as described in the pre-WA section of the combinatorial algorithm.

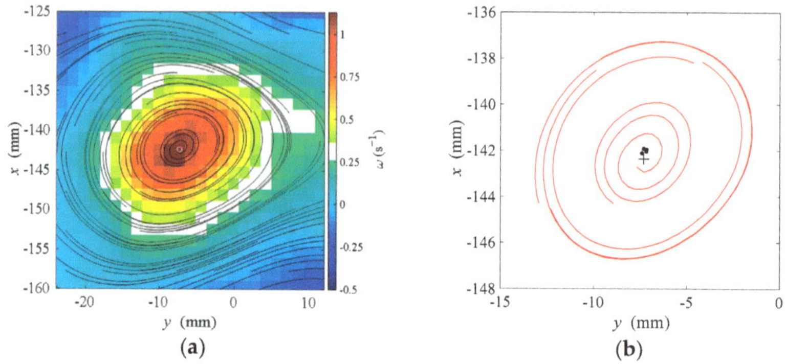

Figure 16. Illustration of a sample vortex along with a locally computed velocity field centered on the vortex core. The velocity field is calculated in a moving reference frame matching the drift velocity of the sample vortex. In (**a**), streamlines generated from the velocity field overlay the local vorticity field, and (**b**) highlights five streamlines meeting WA criteria, along with cluster points (black), determined via the CSL method, denoted by a black '+' symbol.

In Figure 16a, the streamlines generated from this velocity field are plotted over the local vorticity field. Figure 16b highlights five streamlines that meet the criteria set by the WA method. Each of these compliant streamlines is represented by a point (depicted as black dots), and the vortex core, previously determined using the CSL method, is indicated with a black '+' symbol. The WA algorithm serves to confirm the presence of a vortex within the sample ROI, consistent with the vortex definition of Robinson (1991) [36]. It also verifies the existence of a single vortex within the sample ROI and corroborates that the core location, computed by CSL, closely aligns with the geometric center of all the complying streamlines. Should the WA algorithm identify multiple vortices within a single ROI, it signals the possibility of Type II errors. However, it is crucial to note that the WA algorithm does not correct these errors directly. Instead, it acts as an indicator, and addressing Type II errors requires additional processing, such as using clustering algorithms. The primary vortex is determined by the largest number of complying streamlines, while other streamline clusters are marked as potential Type II errors. This underscores that while the WA algorithm detects deviations, further analyses and processing are necessary to correct and characterize potentially missed vortices within the ROI.

A false negative, or a Type II error, occurs when a vortex exists in the flow but remains undetected by the algorithm. Unfortunately, due to the a priori knowledge required by the WA algorithm, it is inevitable to encounter Type II errors for vortices that fall outside the predefined ROI. However, when multiple vortices coexist within a single ROI, the WA algorithm has the capability to identify a Type II error, but only if the drift velocity of the primary vortex closely matches that of any secondary vortices represented by separate streamline clusters.

The proposed CVD algorithm offers ways to minimize Type II errors through two key mechanisms:

- Wider Threshold Bounds: By expanding the threshold bounds in the threshold intensity vector (*TIV*) defined in Equation (1) for the MV algorithm, the ROI can encompass weaker vortices, including those with lower vorticity. This adjustment ensures that such vortices are considered by both the CSL and WA algorithms.
- Erosion Process Optimization: Employing two smaller sizes of image morphology (IM) structuring elements in the erosion process preserves smaller vortices, those with small values of (r_v), allowing them to remain within the ROI. These vortices can then be assessed by the CSL and WA algorithms.

It is important to note that implementing these modifications significantly increases the computational time. Therefore, it is recommended to define a minimum vortex size $r_{v,min}$

and strength $\omega_{peak,min}$ before running the algorithm. Weak and/or small vortices falling below the specified threshold are intentionally disregarded, reducing the algorithm's computational demands. The selection of both the threshold defined in Equation (1) and the sizes of IM structuring elements (S_{e1}, S_{e2}, and S_{e3}) are based on the goal of minimizing Type II errors while maintaining efficient computation.

5.2.2. Vorticity Distribution Profiles

With the vortex identified and core location determined, the distribution of vorticity within the vortex can be examined. The vorticity profiles of these vortices play a crucial role in assessing various aspects of vortex characteristics, including the accuracy of the vortex radius (r_v), vortex shape and symmetry and their suitability for fitting analytical models.

In Figure 17, the vorticity distribution of the sample vortex depicted in Figure 16 is plotted against the dimensionless radius $(y - y_c)/r_v$ along a constant y line traversing the vortex core. The vorticity profile indicates a Gaussian distribution with slight asymmetry, which may be attributed to interactions with nearby counter-rotating vortices.

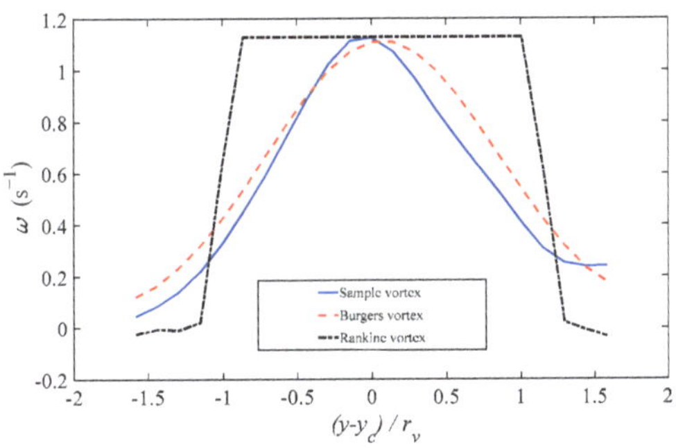

Figure 17. Vorticity distribution profiles of an identified vortex, showing the vorticity plotted against the dimensionless radius along a constant y line traversing the vortex core. Additionally, vorticity profiles for ideal Burgers and Rankine vortices are included, sharing the same peak vorticity and radii as computed by the CSL algorithm.

Figure 17 also includes vorticity profiles for ideal Burgers and Rankine vortices. These profiles have the same peak vorticity and radii as those computed by the CSL algorithm. The Burgers vortex was described in Equations (17) and (18) and the Rankine vortex represents a simplified model that attempts to mimic real vortices by dividing them into two regions: the inner core with uniform vorticity, resembling a forced vortex, and the outer core, which lacks vorticity, simulating an irrotational or free vortex [61]. In contrast, the Burgers vortex solution is a more intricate model used to illustrate fundamental elements of modern turbulence theory, providing an exact solution to the cylindrical Navier–Stokes equations that accounts for flow on a cylindrical vortex core inducing circulation (Γ_∞) at large distances [51].

The vorticity distribution plot in Figure 17 reveals that the sample vortex profile closely resembles a Burgers vortex, sharing the same radius and peak vorticity values computed by the CSL algorithm. This correspondence validates the CSL algorithm's effectiveness in predicting vortex radius and core coordinates from a PIV-generated velocity vector field. A slight spatial offset between the experimental vortex and the Burgers vortex is attributed to the spatial resolution of the experimental PIV data. This discrepancy arises from the CSL algorithm's sequential reading of $v_p(\varsigma, \psi)$ in Equation (4), from left to right. When the true core coordinate falls between two velocity vectors, the algorithm consistently selects the leftmost value.

5.2.3. Circumferential Velocity Profiles

In addition to vorticity, circumferential velocity profiles are valuable for characterizing vortical flow structures. The circumferential, or azimuthal, velocity represents the velocity component perpendicular to any straight line passing through the vortex core.

Figure 18 provides a comparison of the circumferential velocity profile for the sample vortex from Figure 16 against theoretical profiles for a Rankine vortex and Burgers vortex. The sample vortex profile is plotted alongside the analytical curves. As observed in Figure 17, defining the vortex boundary radius based solely on vorticity profiles requires an arbitrary cutoff [1]. This highlights the limitations of relying solely on the MV method for radius determination. However, circumferential velocity profiles offer a distinct advantage for defining the boundary. Within the vortex radius, the circumferential velocity magnitude generally increases with radial distance, reaching a maximum value precisely at the boundary (r_v) [1]. This characteristic peak at r_v is clearly illustrated for the sample vortex in Figure 18. Using the peak velocity avoids ambiguity and arbitrary cutoffs. Additionally, circumferential velocity can be measured farther from the high shear at the core, where PIV more accurately represents curvature. Thus, circumferential profiles provide an unambiguous vortex boundary definition without relying on vorticity gradients alone.

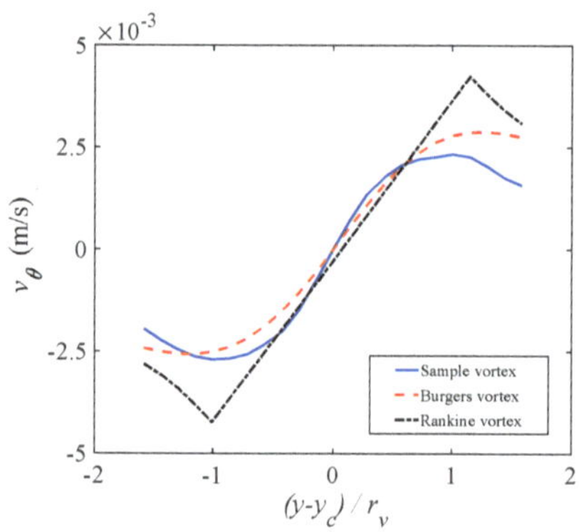

Figure 18. Comparison of circumferential velocity profiles for the sample vortex, a Rankine vortex, and a Burgers vortex.

Circumferential velocity profiles effectively showcase the CSL algorithm's ability to distinctly identify the boundary of a vortex core by pinpointing the radial location where the absolute circumferential velocity is maximized. This method of defining the core boundary radius is less prone to ambiguity compared to one derived from vorticity profiles, making it a superior approach for computing vortex boundary radii.

5.2.4. Circulation Analysis in the Von Karman Wake

With the CVD approach now in place, it can be used to analyze a large number of PIV datasets without the need for setting limits to individual vector sets to detect vortices. The flow field in Figure 15a has some unique characteristics. In the near wake, vortices exhibit more consistent spatial locations compared to those further downstream. In other words, vortex cores close to the trailing edge of the flapping wing consistently appear in the same positions across oscillation periods. However, structures farther downstream demonstrate greater spatial variation. Phase averaging multiple vector fields reduces the magnitudes of spatially varying vortices while preserving more consistent structures. The CVD method enables quantifying key wake parameters after computing vorticity distributions and defining vortex boundaries and cores. As highlighted before, these parameters include circulation (Γ), peak vorticity (ω_{peak}), vortex radii (r_v), vortex drift velocity (v_{drift}), streamwise spacing (S_x), and transverse spacing (S_y).

Figure 19 plots the circulation of 25 instantaneous vector fields from the case illustrated in Figure 15a against dimensionless downstream distance when the airfoil angle is $\theta_{af} = 0°$. Individual field circulations from the instantaneous images appear as black dots, while phase-averaged circulation is shown in red. This illustrates the increasing variability of the vortex core streamwise coordinate x_C with downstream distance. In the near field $x_C = x/C < 2$, tight vertical data groups indicate consistent vortex locations. However, groups become indistinguishable farther downstream as locations vary. Consequently, phase averaging underestimates circulation versus individual fields.

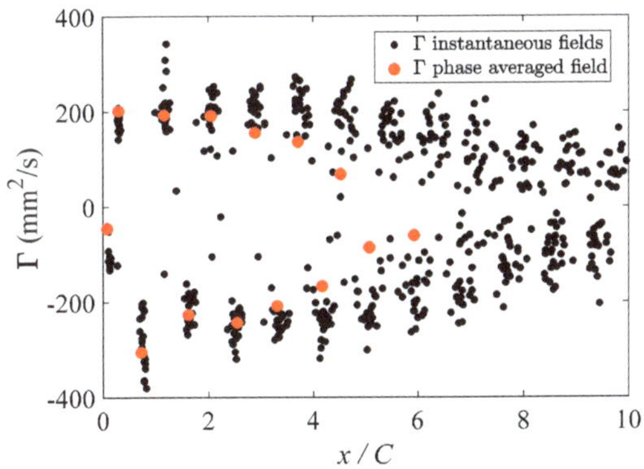

Figure 19. Circulation vs. dimensionless downstream distance for 25 instantaneous fields (black dots) and phase-averaged data (red circles) when the airfoil is at $\theta_{af} = 0°$. Near field ($x_C < 2$) shows tight vertical data grouping, indicating consistent vortex positions.

Circulation provides critical insights into vortex strength and influence within the wake. Examining circulation across instantaneous fields reveals vortex evolution and interactions over time, granting a deeper understanding of von Karman wake dynamics.

5.3. Considerations for Higher Reynolds Number Applications

At the demonstrated Reynolds number of 146 based on airfoil thickness and freestream velocity, the flow remains laminar and ordered, facilitating structured vortex shedding amenable to characterization with the combinatorial vortex detection (CVD) technique. However, applying the algorithm to scenarios with significantly higher Re introduces additional complexities that must be considered.

As Re increases in the transitional flow regime, turbulence levels intensify. This places greater demand on spatial resolution to fully resolve steeper velocity gradients and smaller-scale flow structures. If spatial resolution becomes insufficient, discretization errors may prevent accurate calculation of vorticity and vortex parameters. Additionally, heightened turbulence increases variability in vortex properties such as circulation, boundary radius, and peak vorticity. To achieve a statistically reliable description of vortex characteristics, an increased number of velocity field samples is necessary. Lastly, elevated turbulence can cause vortex core boundaries to become more diffuse [62,63], complicating boundary identification. Addressing these challenges would necessitate modifications such as improving localized velocity calculations and using an adaptive ROI threshold approach.

Proper validation of the CVD algorithm at higher Re requires a more controlled experimental facility and PIV setup meeting minimum resolution requirements. While not explored presently with the available resources, the proposed method shows promise for characterization of transitional flows. Future efforts should focus on method refinements targeting higher Re applications.

6. Conclusions

In this study, a robust vortex detection and characterization algorithm has been developed and rigorously tested on 2C2D velocity vector fields obtained from the wake of an oscillating NACA 0012 airfoil within a uniform flow. Existing vortex detection methods face major limitations in handling experimental data, including sensitivity to measurement noise and substantial velocity gradients near vortex cores that complicate particle seeding and the performance of the PIV algorithm [12,39].

The proposed combinatorial vortex detection (CVD) algorithm aimed to address these challenges and accurately identify wake vortices and their attributes. It harnesses the strengths of three distinct vortex detection methods, ensuring reliable vortex detection solutions. The central technique was the cross-section lines method, which played a key role in determining vortex core coordinates, boundary radii, and drift velocities. Evaluation of vorticity and velocity profiles verified the algorithm's precision in locating cores, defining boundaries, and calculating drift vectors, thus precisely enabling definition of the vortex swirling strength.

In the pursuit of comprehensive verification, the winding angle method emerged as a valuable tool for the verification of vortex detection, facilitating the reliable identification and subsequent elimination of false positives within the dataset. Such precise measurements have significant implications for comprehending and characterizing vortical structures with accuracy.

A limitation of the current study is that rigorous validation was restricted to a single low Reynolds number airfoil case. As turbulence levels intensify with increasing Reynolds numbers, factors such as sufficient spatial resolution, statistical convergence of vortex parameters, and diffuse vortex boundaries would need to be addressed. However, the CVD methodology shows promise for adaptation to characterize transitional flows at higher Reynolds numbers with proper experimental design. Refinements to the algorithm could incorporate spatial and temporal mesh refinement within vortex cores and adaptive ROI thresholds tuned to local vorticity gradient thicknesses. Further assessment across a range of higher turbulence flows would aid in refinement and demonstrate expanded applicability. Additionally, future work could explore combining this vortex analysis approach with machine learning for automated parameter tuning at varying flow conditions.

In summary, the CVD method could significantly contribute to the advancement of vortex detection and characterization techniques in the realm of fluid dynamics research. The demonstrated efficacy of the algorithm in addressing the complexities associated with experimental data reaffirms its potential as a valuable asset in the study of vortical flows. By mitigating challenges related to measurement uncertainties and the intricate dynamics of vortex structures, this approach facilitates enhanced understanding and modeling of vortical phenomena, ultimately advancing the comprehension of fluid mechanics and its manifold of applications.

Supplementary Materials: Figure S1 illustrates the von Karman Wake, and Figure S2 showcases the Inverted von Karman Wake. The Figure S1, Figure S2 and CVD algorithm's MATLAB codes are available on GitHub at https://github.com/dsnobes/Combinatorial-Vortex-Detection-Algorithm.

Author Contributions: The following outlines the contributions of each author. Conceptualization, M.B., C.R.K. and D.S.N.; methodology, M.B., G.M.B., C.R.K. and D.S.N.; software, M.B., G.M.B. and D.S.N.; validation, M.B., G.M.B., C.R.K. and D.S.N.; formal analysis, M.B., G.M.B., C.R.K. and D.S.N.; investigation, M.B., G.M.B., C.R.K. and D.S.N.; resources, C.R.K. and D.S.N.; data curation, M.B. and D.S.N.; writing—original draft preparation, M.B.; writing—review and editing, M.B., G.M.B., C.R.K. and D.S.N.; visualization, M.B., C.R.K. and D.S.N.; supervision, C.R.K. and D.S.N.; project administration, C.R.K. and D.S.N.; funding acquisition, C.R.K. and D.S.N. All authors have read and agreed to the published version of the manuscript.

Funding: This research was funded by the Natural Sciences and Engineering Research Council of Canada from Discovery Grants held by C.R.K. and D.S.N.

Data Availability Statement: Data are available on request.

Acknowledgments: Writing assistance from the AI tool ChatGPT is acknowledged. The ChatGPT writing assistant was used to review and improve the readability of the text. Feedback from ChatGPT was manually reviewed before integrating the suggested sentences into the manuscript.

Conflicts of Interest: The authors declare no conflicts of interest.

References

1. Jeong, J.; Hussain, F. On the identification of a vortex. *J. Fluid Mech.* **1995**, *285*, 69–94. [CrossRef]
2. Bohl, D.G.; Koochesfahani, M.M. MTV measurements of the vortical field in the wake of an airfoil oscillating at high reduced frequency. *J. Fluid Mech.* **2009**, *620*, 63–88. [CrossRef]
3. Andersen, A.; Bohr, T.; Schnipper, T.; Walther, J.H. Wake structure and thrust generation of a flapping foil in two-dimensional flow. *J. Fluid Mech.* **2017**, *812*, R4. [CrossRef]
4. Wu, X.; Zhang, X.; Tian, X.; Li, X. A review on fluid dynamics of flapping foils. *Ocean. Eng.* **2020**, *195*, 106712. [CrossRef]
5. Visbal, M.R.; Garmann, D.J. Analysis of Dynamic Stall on a Pitching Airfoil Using High-Fidelity Large-Eddy Simulations. *AIAA J.* **2018**, *56*, 46–63. [CrossRef]
6. Seshadri, P.K.; Aravind, A.; De, A. Leading edge vortex dynamics in airfoils: Effect of pitching motion at large amplitudes. *J. Fluids Struct.* **2023**, *116*, 103796. [CrossRef]
7. Chatelain, P.; Duponcheel, M.; Caprace, D.-G.; Marichal, Y.; Winckelmans, G. Vortex particle-mesh simulations of vertical axis wind turbine flows: From the airfoil performance to the very far wake. *Wind. Energy Sci.* **2017**, *2*, 317–328. [CrossRef]
8. Lai, J.C.S.; Platzer, M.F. Jet Characteristics of a Plunging Airfoil. *AIAA J.* **1999**, *37*, 1529–1537. [CrossRef]
9. Ribeiro, J.H.M.; Wolf, W.R. Identification of coherent structures in the flow past a NACA0012 airfoil via proper orthogonal decomposition. *Phys. Fluids* **2017**, *29*, 085104. [CrossRef]
10. Godoy-Diana, R.; Marais, C.; Aider, J.-L.; Wesfreid, J.E. A model for the symmetry breaking of the reverse Bénard–von Kármán vortex street produced by a flapping foil. *J. Fluid Mech.* **2009**, *622*, 23–32. [CrossRef]
11. Gupta, S.; Zhao, J.; Sharma, A.; Agrawal, A.; Hourigan, K.; Thompson, M.C. Two- and three-dimensional wake transitions of a NACA0012 airfoil. *J. Fluid Mech.* **2023**, *954*, A26. [CrossRef]
12. Vollmers, H. Detection of vortices and quantitative evaluation of their main parameters from experimental velocity data. *Meas. Sci. Technol.* **2001**, *12*, 1199. [CrossRef]
13. Hunt, J.C.R.; Wray, A.A.; Moin, P. Eddies, Streams, and Convergence Zones in Turbulent Flows. 1988. Available online: https://ntrs.nasa.gov/citations/19890015184 (accessed on 22 September 2023).
14. Dallmann, U. Topological structures of three-dimensional vortex flow separation. In Proceedings of the 16th Fluid and Plasmadynamics Conference, American Institute of Aeronautics and Astronautics, Danvers, MA, USA, 12–14 July 1983. [CrossRef]
15. Vollmers, H.; Kreplin, H.P.; Meier, H.U. Separation and Vortical-Type Flow Around a Prolate Spheroid—Evaluation of Relevant Parameters. In Proceedings of the AGARD Symposium on Aerodynamics of Vortical Type Flows in Three Dimensions, AGARD-CP-342, Rotterdam, The Netherlands, 1 July 1983; Available online: https://www.semanticscholar.org/paper/Separation-and-Vortical-Type-Flow-Around-a-Prolate-Vollmers-Kreplin/f58135b4da512b8c577589ac3ecf0bdf923284b2 (accessed on 13 October 2023).
16. Chong, M.S.; Perry, A.E.; Cantwell, B.J. A general classification of three-dimensional flow fields. *Phys. Fluids A Fluid Dyn.* **1990**, *2*, 765–777. [CrossRef]
17. Graftieaux, L.; Michard, M.; Grosjean, N. Combining PIV, POD and vortex identification algorithms for the study of unsteady turbulent swirling flows. *Meas. Sci. Technol.* **2001**, *12*, 1422. [CrossRef]
18. Kolář, V. Vortex identification: New requirements and limitations. *Int. J. Heat Fluid Flow* **2007**, *28*, 638–652. [CrossRef]
19. Zhang, Y.; Liu, K.; Xian, H.; Du, X. A review of methods for vortex identification in hydroturbines. *Renew. Sustain. Energy Rev.* **2018**, *81*, 1269–1285. [CrossRef]
20. Epps, B. Review of Vortex Identification Methods. In *55th AIAA Aerospace Sciences Meeting*; American Institute of Aeronautics and Astronautics: Grapevine, TX, USA, 2017. [CrossRef]
21. Elsas, J.H.; Moriconi, L. Vortex identification from local properties of the vorticity field. *Phys. Fluids* **2017**, *29*, 015101. [CrossRef]
22. Gampert, M.; Boschung, J.; Hennig, F.; Gauding, M.; Peters, N. The vorticity versus the scalar criterion for the detection of the turbulent/non-turbulent interface. *J. Fluid Mech.* **2014**, *750*, 578–596. [CrossRef]
23. Fraňa, K.; Stiller, J.; Grundmann, R. Taylor-Görtler vortices in the flow driven by a rotating magnetic field in a cylindrical container. *J. Vis.* **2005**, *8*, 323–330. [CrossRef]
24. Liu, C.; Wang, Y.; Yang, Y.; Duan, Z. New omega vortex identification method. *Sci. China Phys. Mech. Astron.* **2016**, *59*, 684711. [CrossRef]
25. Beach, M.J.S.; Golubeva, A.; Melko, R.G. Machine learning vortices at the Kosterlitz-Thouless transition. *Phys. Rev. B* **2018**, *97*, 045207. [CrossRef]
26. Stevens, P.R.R.J.; Sciacchitano, A. Application of clustering and the Hungarian algorithm to the problem of consistent vortex tracking in incompressible flowfields. *Exp. Fluids* **2021**, *62*, 173. [CrossRef]
27. Wang, J.; Guo, L.; Wang, Y.; Deng, L.; Wang, F.; Li, T. A Vortex Identification Method Based on Extreme Learning Machine. *Int. J. Aerosp. Eng.* **2020**, *2020*, e8865001. [CrossRef]

28. Liu, C.; Gao, Y.; Dong, X.; Wang, Y.; Liu, J.; Zhang, Y.; Cai, X.; Gui, N. Third generation of vortex identification methods: Omega and Liutex/Rortex based systems. *J. Hydrodyn.* **2019**, *31*, 205–223. [CrossRef]
29. Van Waeyenberge, B.; Puzic, A.; Stoll, H.; Chou, K.W.; Tyliszczak, T.; Hertel, R.; Fähnle, M.; Brückl, H.; Rott, K.; Reiss, G.; et al. Magnetic vortex core reversal by excitation with short bursts of an alternating field. *Nature* **2006**, *444*, 461–464. [CrossRef] [PubMed]
30. Tasci, M.O.; Tumse, S.; Sahin, B. Vortical flow characteristics of a slender delta wing in ground effect. *Ocean. Eng.* **2022**, *261*, 112120. [CrossRef]
31. Liu, C.; Gao, Y.; Tian, S.; Dong, X. Rortex—A new vortex vector definition and vorticity tensor and vector decompositions. *Phys. Fluids* **2018**, *30*, 035103. [CrossRef]
32. Gao, Y.; Liu, C. Rortex and comparison with eigenvalue-based vortex identification criteria. *Phys. Fluids* **2018**, *30*, 085107. [CrossRef]
33. Dong, X.; Wang, Y.; Chen, X.; Dong, Y.; Zhang, Y.; Liu, C. Determination of epsilon for Omega vortex identification method. *J. Hydrodyn.* **2018**, *30*, 541–548. [CrossRef]
34. Gao, Y.; Liu, J.; Yu, Y.; Liu, C. A Liutex based definition and identification of vortex core center lines. *J. Hydrodyn.* **2019**, *31*, 445–454. [CrossRef]
35. Jiang, M.; Machiraju, R.; Thompson, D. *The Visualization Handbook*; Elsevier Butterworth-Heinemann: Burlington, MA, USA, 2005; pp. 295–309. [CrossRef]
36. Robinson, S.K. Coherent Motions in the Turbulent Boundary Layer. *Annu. Rev. Fluid Mech.* **1991**, *23*, 601–639. [CrossRef]
37. Chakraborty, P.; Balachandar, S.; Adrian, R.J. On the relationships between local vortex identification schemes. *J. Fluid Mech.* **2005**, *535*, 189–214. [CrossRef]
38. Pierce, B.; Moin, P.; Sayadi, T. Application of vortex identification schemes to direct numerical simulation data of a transitional boundary layer. *Phys. Fluids* **2013**, *25*, 015102. [CrossRef]
39. Saikrishnan, N.; Marusic, I.; Longmire, E.K. Assessment of dual plane PIV measurements in wall turbulence using DNS data. *Exp. Fluids* **2006**, *41*, 265–278. [CrossRef]
40. Moin, P.; Mahesh, K. Direct numerical simulation: A Tool in Turbulence Research. *Annu. Rev. Fluid Mech.* **1998**, *30*, 539–578. [CrossRef]
41. Lugt, H.J. The Dilemma of Defining a Vortex. In *Recent Developments in Theoretical and Experimental Fluid Mechanics: Compressible and Incompressible Flows*; Müller, U., Roesner, K.G., Schmidt, B., Eds.; Springer: Berlin/Heidelberg, Germany, 1979; pp. 309–321. [CrossRef]
42. Sadarjoen, I.A.; Post, F.H. Geometric Methods for Vortex Extraction. In *Data Visualization '99*; Gröller, E., Löffelmann, H., Ribarsky, W., Eds.; Springer: Vienna, Austria, 1999; pp. 53–62. [CrossRef]
43. Strawn, R.C.; Kenwright, D.N.; Ahmad, J. Computer Visualization of Vortex Wake Systems. *AIAA J.* **1999**, *37*, 511–512. [CrossRef]
44. Portela, L.M. Identification and Characterization of Vortices in the Turbulent Boundary Layer. Ph.D. Thesis, Stanford University, Stanford, CA, USA, 1998. Available online: https://www.proquest.com/docview/304454244/abstract/B9604AD78D0B41E2PQ/1 (accessed on 1 September 2023).
45. Arora, S.; Acharya, J.; Verma, A.; Panigrahi, P.K. Multilevel Thresholding for Image Segmentation through a Fast Statistical Recursive Algorithm. *Pattern Recognit. Lett.* **2008**, *29*, 119–125. [CrossRef]
46. Sadarjoen, I.A.; Post, F.H. Detection, quantification, and tracking of vortices using streamline geometry. *Comput. Graph.* **2000**, *24*, 333–341. [CrossRef]
47. Pagendarm, H.-G.; Henne, B.; Rutten, M. Detecting vortical phenomena in vector data by medium-scale correlation. In Proceedings of the Visualization '99 (Cat. No.99CB37067), San Francisco, CA, USA, 24–29 October 1999; pp. 409–552. [CrossRef]
48. Perry, A.E.; Chong, M.S. Topology of flow patterns in vortex motions and turbulence. *Appl. Sci. Res.* **1994**, *53*, 357–374. [CrossRef]
49. Varun, A.V.; Balasubramanian, K.; Sujith, R.I. An automated vortex detection scheme using the wavelet transform of the d2 field. *Exp. Fluids* **2008**, *45*, 857–868. [CrossRef]
50. Gibbon, J.D.; Fokas, A.S.; Doering, C.R. Dynamically stretched vortices as solutions of the 3D Navier–Stokes equations. *Phys. D Nonlinear Phenom.* **1999**, *132*, 497–510. [CrossRef]
51. Drazin, P.G.; Riley, N. *The Navier-Stokes Equations: A Classification of Flows and Exact Solutions*; Cambridge University Press: Cambridge, UK , 2006.
52. Prochazka, A.; Pullin, D.I. Structure and stability of non-symmetric Burgers vortices. *J. Fluid Mech.* **1998**, *363*, 199–228. [CrossRef]
53. Ponta, F.L. Vortex decay in the Kármán eddy street. *Phys. Fluids* **2010**, *22*, 093601. [CrossRef]
54. Hilderman, T.L. Measurement, Modelling, and Stochastic Simulation of Concentration Fluctuations in a Shear Flow. Ph.D. Thesis, University of Alberta, Edmonton, AB, Canada, 2004. [CrossRef]
55. Raffel, M.; Willert, C.E.; Wereley, S.T.; Kompenhans, J. *Particle Image Velocimetry: A Practical Guide*; Springer: Berlin/Heidelberg, Germany, 2007. [CrossRef]
56. Adrian, R.J.; Westerweel, J. *Particle Image Velocimetry*; Cambridge University Press: Cambridge, UK, 2011.
57. Anderson, J.M.; Streitlien, K.; Barrett, D.S.; Triantafyllou, M.S. Oscillating foils of high propulsive efficiency. *J. Fluid Mech.* **1998**, *360*, 41–72. [CrossRef]
58. Lua, K.B.; Lim, T.T.; Yeo, K.S.; Oo, G.Y. Wake-Structure Formation of a Heaving Two-Dimensional Elliptic Airfoil. *AIAA J.* **2007**, *45*, 1571–1583. [CrossRef]

59. Platzer, M.F.; Jones, K.D.; Young, J.; Lai, J.C.S. Flapping Wing Aerodynamics: Progress and Challenges. *AIAA J.* **2008**, *46*, 2136–2149. [CrossRef]
60. Koochesfahani, M.M. Vortical patterns in the wake of an oscillating airfoil. *AIAA J.* **1989**, *27*, 1200–1205. [CrossRef]
61. Giaiotti, D.B.; Stel, F. The Rankine Vortex Model By. 2006. Available online: https://www.semanticscholar.org/paper/The-Rankine-Vortex-Model-By-Giaiotti-Stel/b1346a6f4cf55af5decd2689a794745db3d38449 (accessed on 21 September 2023).
62. JMarshall, S.; Beninati, M.L. External turbulence interaction with a columnar vortex. *J. Fluid Mech.* **2005**, *540*, 221–245. [CrossRef]
63. Ghimire, H.C.; Bailey, S.C.C. Experimental examination of vorticity stripping from a wing-tip vortex in free-stream turbulence. *Phys. Rev. Fluids* **2018**, *3*, 034702. [CrossRef]

fluids MDPI

Article

Quantitative Color Schlieren for an H_2–O_2 Exhaust Jet Developing in Air

Emilia-Georgiana Prisăcariu [1,*] and Tudor Prisecaru [2]

1 Romanian Research and Development Institute for Gas Turbines COMOTI, 061126 Bucharest, Romania
2 Faculty of Mechanical Engineering, Politehnica University of Bucharest, 060042 Bucharest, Romania; tudor.prisecaru@upb.ro
* Correspondence: emilia.prisacariu@comoti.ro; Tel.: +40-764060265

Abstract: Throughout many decades, the Schlieren visualization method has been mainly used as means to visualize transparent flows in a qualitative manner. The images recorded provide data regarding the existence of the flow, or illustrate predicted flow geometries and details. The colored Schlieren method has been developed in the late 1890s and has always had the intent to provide quantitative data rather than qualitative pictures of the studied phenomena. This paper centers on applying a quantitative color Schlieren method to help determine the gasodynamic parameters of an H_2–O_2 exhaust jet, developing in air. A comparison between the parameters obtained through calibrating the color filter for the Schlieren method and the results from a CFD simulation is performed to assess the range of the CS (color Schlieren) measurement. This paper's findings address the issues of calibrated color filter Schlieren encounter during its implementation and discusses possible errors appearing when the method is applied to a 3D flow. While the qualitative Schlieren images are still impressive to observe, the quantitative Schlieren presents challenges and a low measurement accuracy (75%) when applied to 3D flows and compared to 2D cases found in the literature (97–98%).

Keywords: rainbow Schlieren; post processing; color filter; calibration

Citation: Prisăcariu, E.-G.; Prisecaru, T. Quantitative Color Schlieren for an H_2–O_2 Exhaust Jet Developing in Air. *Fluids* **2024**, *9*, 19. https://doi.org/10.3390/fluids9010019

Academic Editors: Ashwin Vaidya, Guangjian Zhang, Mingming Ge and Xin-Lei Zhang

Received: 2 October 2023
Revised: 29 December 2023
Accepted: 3 January 2024
Published: 8 January 2024

1. Introduction

The color Schlieren method is a quantitative visualization method developed to allow data extraction regarding physical parameters of the imaged flow. Color Schlieren can also be found in the literature under the name of 'rainbow Schlieren', due to the rainbow segmented filters used in some applications, in which the filter's colors change abruptly. It has a similar working principle to the classic Schlieren, with a slight difference regarding the knife edge. In a classic Schlieren setup, density gradients can be observed by deflecting the light rays in the Fourier plane of the light source image by introducing a sharp knife edge to partially block the rays, either horizontally, causing the vertical gradients to be visible, or vertically, revealing the horizontal gradients. A circular knife edge can also be applied, showing both density gradients. Color Schlieren uses a color filter to replace the circular knife edge.

Color Schlieren is used in most papers to provide different color ratios for each density gradient, due to the fact that the deflection takes place on the filter, but if the color filter is not calibrated beforehand, the result will only consist out of qualitative pictures. Most papers apply the quantitative rainbow Schlieren to 2D flows, in which case the Schlieren technique is more prone to have an increased accuracy, given the optical path-integrative character of the method.

The concept of color Schlieren dates back to 1892, and it is attributed to Rheinberg [1], who was inspired to use color filters in his microscope by the chemical staining method, applied at that time to differentiate elements of transparent flow. Using the color contrast technique to account for different structures of his biological subject analyzed through a microscope, he achieved a technique which is characterized by Settles as "analog to Toepler's black and white microscopical Schlieren method" [2].

The quantitative Schlieren method has been used to determine thermodynamic properties, species concentrations and velocity of different oil and flows, aerodynamic research and shock waves by [3], using optical tomography to reconstruct density information.

A short review is presented by [4], whose objective was creating a color shock tube, allowing a quantitative analysis.

A very representative study was conducted by Elsinga et al. [5], explaining the process of applying different types of quantitative Schlieren to obtain different series of Schlieren images containing a 2D Prandtl–Meyer expansion fan. It concludes that both calibrated color filter Schlieren (CCS), as well as the BOS (background-oriented Schlieren) method, are capable of returning the light deflection angle in two spatial directions, projecting the density gradient vector.

Another color-based analysis can be found in the works of Greenberg [6], where the spectral color used was transformed from RGB into HIS (hue–saturation–intensity), which in that case proved to be more efficient regarding the intensity fluctuation of the light source, pixel to pixel variation in gain within a given detector array, absorption or scattering of light within the test section, etc. The rainbow Schlieren used by Greenberg demonstrated a system sensitivity comparable with that of a normal interferometer, while being easily implemented and less sensitive to mechanical misalignment.

The quantitative color Schlieren technique used in this paper relies on the calibration of a colored filter. The calibration of the filter is made possible given the special configuration of the filter, which has been designed to vary color ratios over its entire area, providing unique values for each pixel. This filter is integrated into a Z-type Schlieren system, used for recording the images of a turbulent exhaust jet, generated by a micro-thruster that develops a 1N force while using a H_2–O_2 mixture as propellant and has been designed for small satellites' altitude control [7]. The quantitative Schlieren method is applied with the goal of extracting the turbulent H_2–O_2 jet's gasodynamic parameters (such as density and temperature gradient maps). A short discussion is made in comparison to the CFD model obtained in Ansys CFX of the jet, in the same working conditions.

2. Experimental Setup

As presented by [8], where the velocity profiles of the turbulent jet were obtained by applying a series of Schlieren image velocimetry methods, the turbulent jet analyzed is encapsulated in a vacuum chamber. Although the present analysis presents the parameters of the jet dispersing in free air, the influence of the vacuum chamber must be considered, given the high temperatures developing into the jet, which is destined to return at a very high rate because of the vacuum chamber's walls. However, in spite of the analyzed jet presenting the same parameters and being developed in the same way, the first images analyzed by [8] were recorded with a U-turn circular knife edge Schlieren system, while this study uses the equipment presented in Table 1 and the configuration illustrated in Figure 1 to record the color Schlieren images which are further analyzed. The study presented in [8] investigates the possibility of applying SIV (Schlieren image velocimetry) methods to a 3D turbulent flow, and improving them by creating algorithms for digital post processing automatization. This study investigates the possibility of retrieving other quantitative jet parameters, while applying a different type of system calibration.

Table 1. Schlieren equipment used to record images of the exhaust jet.

Type of the Equipment	Name and Manufacturer	Important Features
Two twin parabolic mirrors	Al-plated parabolic mirrors (Edmund Optics, Barrington, NJ, USA)	Effective focal length (EFL) = 1524 mm
light source	LS-W1 (Lightsource Tech, Göttingen, Germany)	Laser-pumped white light source Diameter: 1.5 mm Optical output power: 500 mW
Small biconvex lens	(Edmund Optics, Barrington, NJ, USA)	Focal length (f_{lens}) = 250 mm
Square diaphragm	(Edmund Optics, Barrington, NJ, USA)	3D printed, 1 mm × 1 mm
Color filter	Square	CCS method
High-speed camera (CMOS)	Phantom Veo 1310L (Vision Research Inc., Charlottetown, PE, Canada)	Recording speed: 30.000 fps Spatial resolution (pixels): 600 × 480 24-bit depth Exposure time: 0.93 µs
Camera lens	Nikon Nikkor (Nikon Corporation, Minato, Japan)	50 mm 1:1.8 G
Filter	Custom color filter image	Exposed Fujifilm Superia Extra 400 Photographic, 35 mm fine grain film

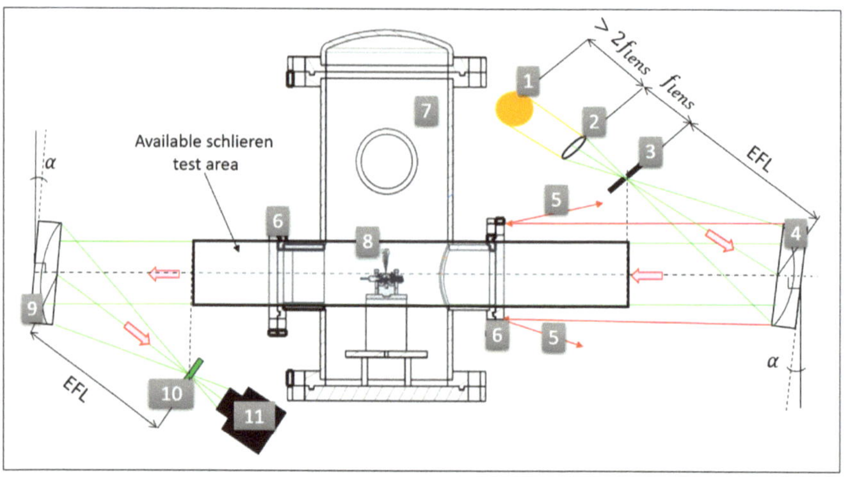

Figure 1. The Z-type color Schlieren setup: 1—Ls-W1 light source; 2—biconvex lens; 3—square diaphragm; 4— parabolic mirror; 5—(red) rays reflected back by the vacuum chamber's outer casing; 6—circular glass windows; 7—vacuum chamber assembly; 8—investigated exhaust jet; 9—second parabolic mirror; 10—color filter calibration mechanism; 11—high-speed camera.

The Z-type configuration used for the imaging of the analyzed jet is presented in Figure 1, showing the optical path and the placement of the exhaust jet into the vacuum chamber. There are several rays reflected back when hitting the opaque surface of the vacuum chamber, given the fact that the diameter of the twin parabolic mirrors is larger than the visualization windows with which the vacuum chamber is equipped. The light travels from the square light source to a biconvex lens with the focal length $f_{lens} = 250$ mm, placed at a distance longer than he double of the focal length. Afterwards, the rays are cut in the focal plane by a square 3D printed diaphragm, which allows the passing of the light through a 1mm square area. The light from the diaphragm placed at a distance equal to the effective focal length (EFF = 1524 mm) of the parabolic mirror, reaches the first parabolic mirror, and is collimated throughout the testing area. The second mirror images the light source in its focal point, creating the knife edge plane. In the plane of the knife edge, an optical transparent filter with a color gradient distribution is framed into a calibration mechanism. The camera is placed behind it, at the point where the mirror is seen as fully illuminated. The angle α represents the off-axis rotation of the mirrors. In this study, $\alpha = 10^{\circ}$, which is more than the recommended amount ($\alpha = 3^{\circ}$) found in the literature [2], given the special geometrical constraints. However, the optical errors resulting from the system are visually negligible, the image of the source does not present any astigmatism or coma effects. The distance between the two parabolic mirrors is larger than 3 times their EFL, resulting in a generous testing area, while introducing the disadvantage of reducing the resolution of the Schlieren image, given the type of the camera lens used and other alignment difficulties.

The optical filter is designed according to the next equations for the RGB channels of the image. These equations are the same as the ones used by Elsinga [5], with the difference that the filter is manufactured in a different manner and the experiment positioning regarding the filter is considered to be x_f and y_f representing the x and y as generated by Matlab 2020a in the process of creating the image of the filter (rows and columns).

$$R = \frac{255}{1 + x_f + y_f} \tag{1}$$

$$G = \frac{255 \cdot x_f}{1 + x_f + y_f} \tag{2}$$

$$B = \frac{255 \cdot y_f}{1 + x_f + y_f} \tag{3}$$

The filter was manufactured by exposing a fine grain photographic film to the image obtained in Matlab displayed on a computer screen. This process is very similar to the one explained in [5], but the filter applied in the current analysis was used in its raw form. [5] uses a different type of technology which allows the film to become completely transparent after exposure.

Given the effect of the film on the light that passes through it, the color gradients change, and the filter's response is no longer linear, as previously intended.

The photographic film's structure is presented in Figure 2.

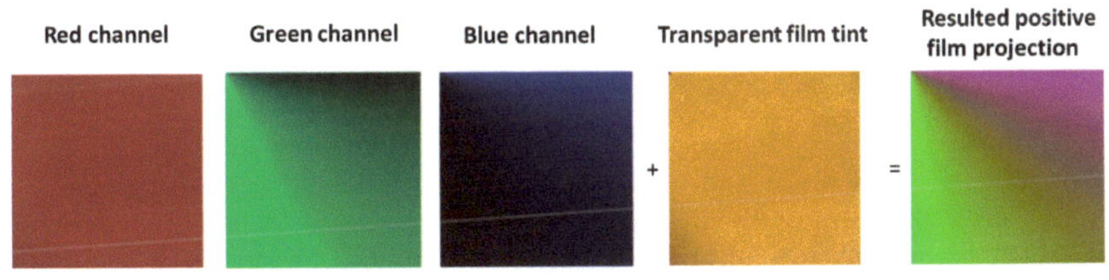

Figure 2. Decomposition of the colored filter and digital addition of the transparent film tint.

The photographic film is calibrated without removing the tint effect. The Phantom Veo 1310L CMOS camera produced by Vision Research Inc. (Wayne, NJ, USA) can record 24-bit depth images, which means it is capable of illustrating $256 \times 256 \times 256$ possible color combinations. The structure of the colored filter and its individual color channels as shown by Figure 2 with the film tint applied is as follows: the red channel's values vary very little causing it to look rather constant without any variation, the green gradient intensifies from left to right, and the blue channel intensifies from the bottom to the top.

The film tint effect is replicated by considering the image of a blank portion of the film, placed in the knife edge plane. An average of the color is performed and displayed with a 60% transparency over the image of the filter. The transparent film tint is represented in

Figure 2 by the real image of the blank photographic film introduced into the knife edge plane. The projection of the film is considered to be the real image transferred from the photographic film onto the camera.

3. Calibration of the Filter for Color-Calibrated Schlieren

The calibration process is achieved by translating the film with regard to the fixed light source image. The photographic film dimensions are 5 mm × 5 mm.

Figure 3a presents the calibration mechanism used for translating the filter. The calibration mechanism is composed of a fixed part, and the displacement plate which contains the optical filter. This mechanism allows the displacement of the filter in 25 positions; therefore, the coverage of the entire filter can be assumed. For each position, the system will acquire information about the color ratio of each pixel. Figure 3b represents the color filter mounted into the calibration plate, while also displaying the calibration grid. Each node of the grid marked by a red cross represents the position of the light source image center, following displacement, and every white square represents the position of the light source related to the new position of the filter.

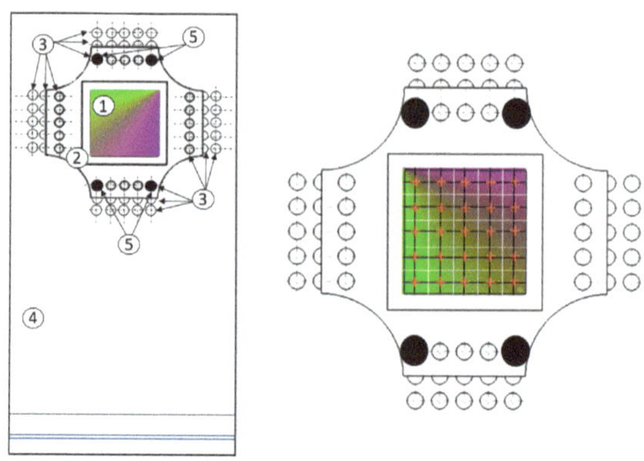

(a) (b)

Figure 3. (**a**) Calibrating mechanism with 1—color-graded filter, 2—calibration plate with filter frame, 3—positioning rows of holes in the fixed plate for the displacement of the calibration plate, 4—fixed plate, 5—locking pins. (**b**) Color filter calibration grid, where the red crosses represent the center of the light source during calibration, the white rectangles represent its 25 calibration positions, and the dark circles represent the fixing pins which are designed to position the calibration plate onto the fixed plate.

Figure 4 depicts the device placement diagram of the color-graded filter. In this case, the horizontal arrows represent a one-hole shift in the indicated direction and the vertical arrows represent the lowering of the calibration plate by one hole. The diagram starts and ends with the filter positioned in the center of the light source which can be used to detect displacement errors appearing during the calibration process. The fixing pins are represented by dark circles in order to confirm the overlap of the holes found in the calibration plate and the fixed plate.

The calibrating procedure is very similar to the one found in [5]. The displacement is achieved by starting with the colored filter in the center position and afterwards displacing it into the other 24 positions, recording different color gradients for each pixel in the 25 points mentioned. The analyzed images are recorded with the image of the light source placed in the center of the graded filter, which can be considered to be the initial position of the filter.

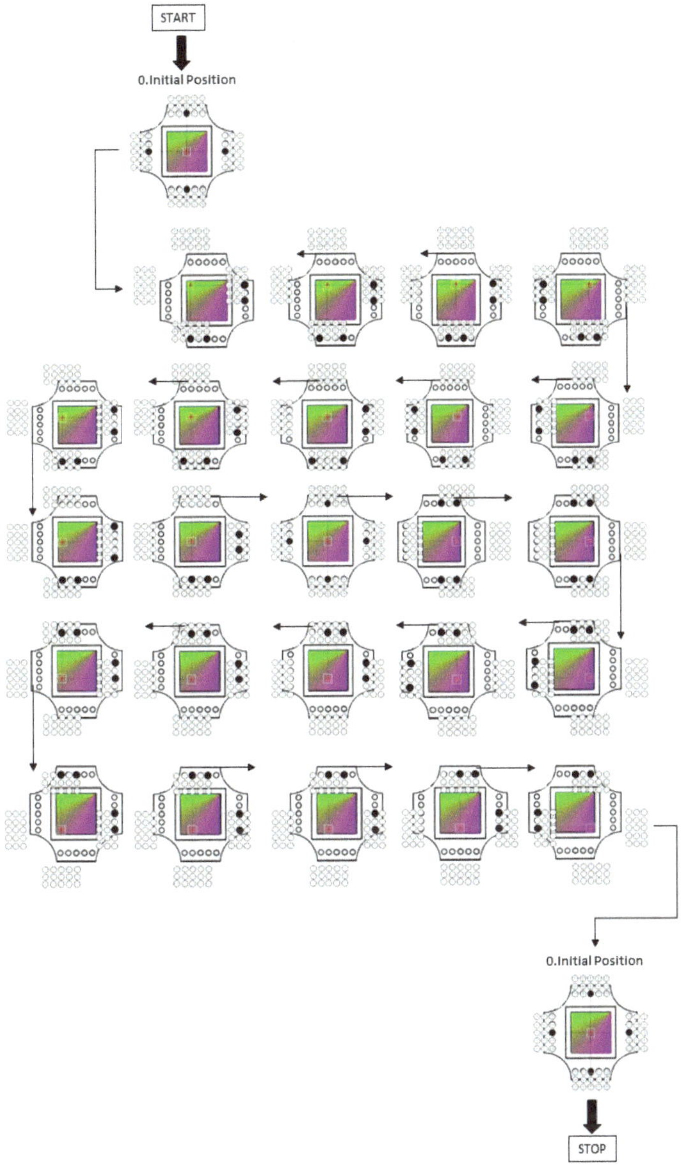

Figure 4. Device placement diagram of the calibration process.

Figure 5 presents the initial moments of the jet development. This sequence is presented here for two important reasons. The first one is to observe the influence of the vacuum chamber on the jet's development, illustrating the velocity with which the pressure wave returns, which can be similar to the velocity of the jet returning into the main visualized area as background noise, and the second one is to underline that the black gradient pictured into the first frame containing the jet is caused by the system recording a ray deflection angle which surpasses the system's capabilities.

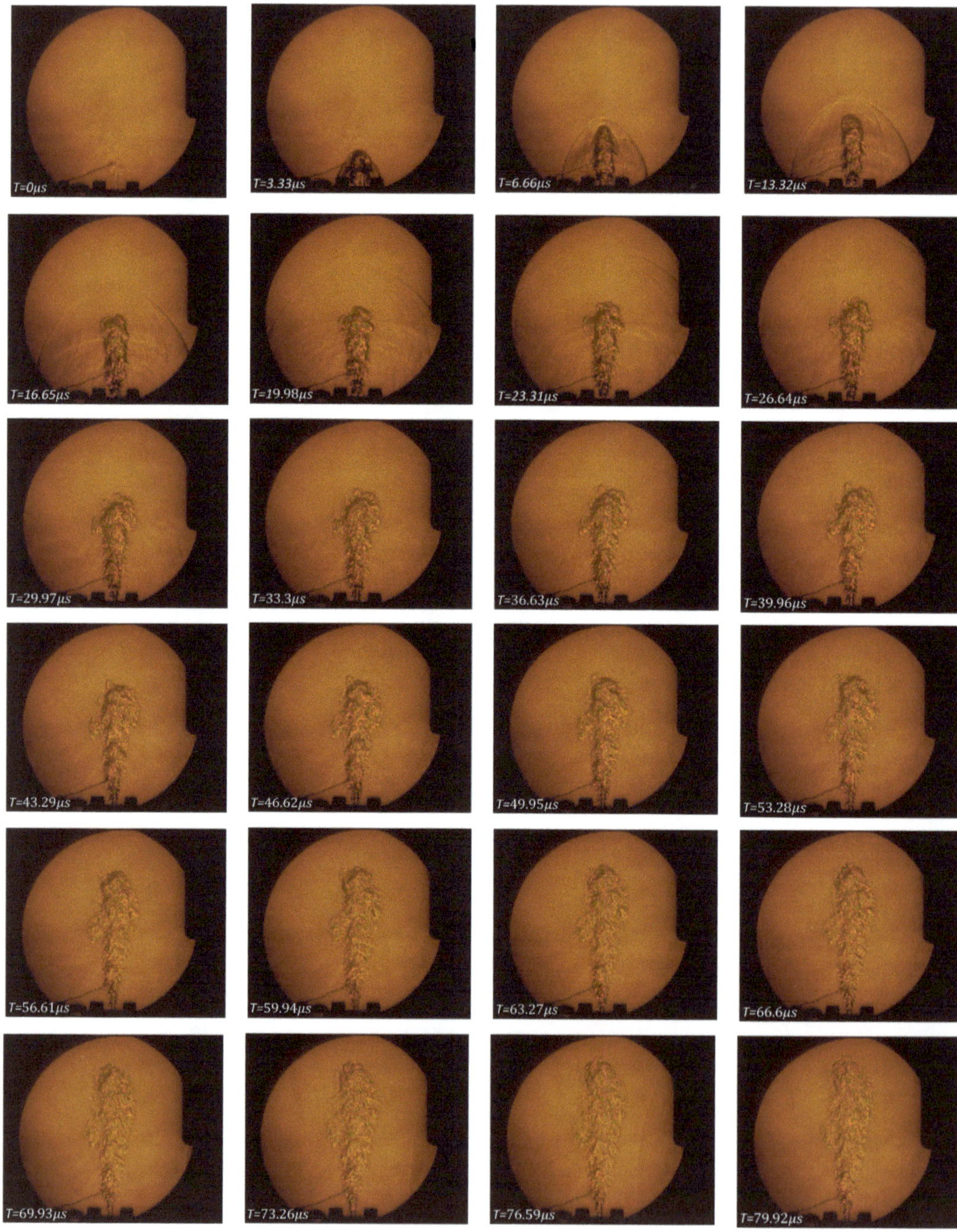

Figure 5. Color Schlieren image sequence (without post processing) of the turbulent jet, underlining the sensitivity of the colored Schlieren system by observing the formation of the pressure shock waves present at in the jet's initial frames and their return (in the last frames of the sequence). The frame markings represent the frame's position in the sequence and the time (relative to t = 0) at which the frame was taken.

For the reason described above, the analyzed pair of images will be chosen to contain the fully developed exhaust jet in air, where the density gradients are in good correlation with the system's sensitivity.

The system's theoretical sensitivity in this case can be calculated using the equations used by [5], and applying it to the current case. The first step in calculating the theoretical sensitivity is to calculate the range of the displacement angle which can be recorded by the system, ϵ_{range}. This is defined as the ratio from Equation (4) [5], where b_f represents the dimension of the filter, b_s represents the dimension of the source, and f represents the focal length of the second parabolic mirror.

$$\varepsilon_{range} = \frac{b_f - b_s}{f} \quad (4)$$

The resulting ratio value is 2.6 mrad, but the ε_{range} is equal in each direction, given the square light source and filter. The displacement range is, therefore, [−1.3 mrad, 1.3 mrad]. Equation (4) confirms the hypothesis that if one uses a filter with a larger dimension, the overall detectable displacement range increases. For example, for the 12 mm × 12 filter, $\epsilon_{range} = [-3.6,\ 3.6]$ mrad. The theoretical sensitivity is related to the camera's dynamic range, is 0.80% for the 5 mm × 5 mm filter and 0.35% for the 12 mm × 12 mm filter. A visual comparison of the 5 mm × 5 mm filter and the 12 mm × 12 mm is illustrated in Figure 6.

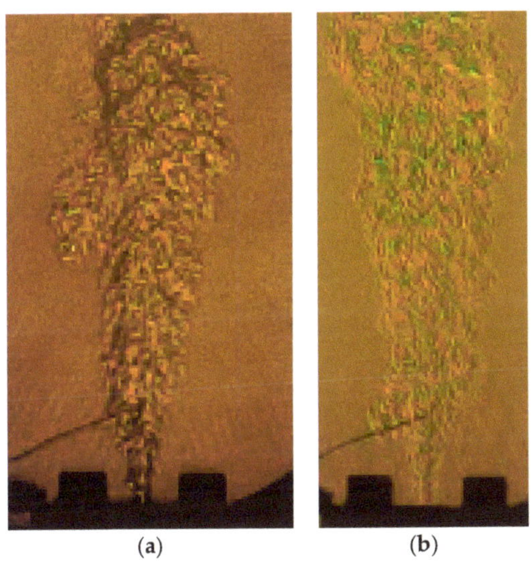

Figure 6. (**a**) Raw image of the jet with the 5 mm × 5 mm filter; (**b**) raw image of the jet with the 12 mm × 12 mm filter.

The low resolution of the Schlieren images is caused by the issues arising from the alignment of the system. Other filter manufacturing techniques with using more modern technology such as printing the filter on a transparent laser printer sheet were discarded, as it did not allow enough light to properly pass through it because of the ink opacity, despite it seeming perfectly transparent when held against natural light.

This paper describes the calibration method and the study resulting from using the 5 mm × 5 mm Schlieren filter. Despite the initial images presenting dark areas, the images of the analyzed jet contain pixels with color ratios which could be matched to the ones found on the above-mentioned filter. The calibration method yields for each pixel a calibration curve. These calibration curves are exemplified in Figure 7 for 3 pixels of the test image, found at the following locations: [320, 294], [321, 295], and [322, 296]. These can be described as being in the Schlieren area, in the vicinity of the thruster's exit. Each calibration curve can be described in relation to two different color ratios, the indicated ones in this case being R/G and B/R. These pixel color ratio values exemplified in Figure 7 are obtained from the filter calibration, without the image of the studied jet. The calibration curves present five inflexion points due to the calibration mechanism and filter structure which cause an abrupt change in color ratios five times, as the filter is lowered relatively to the light source.

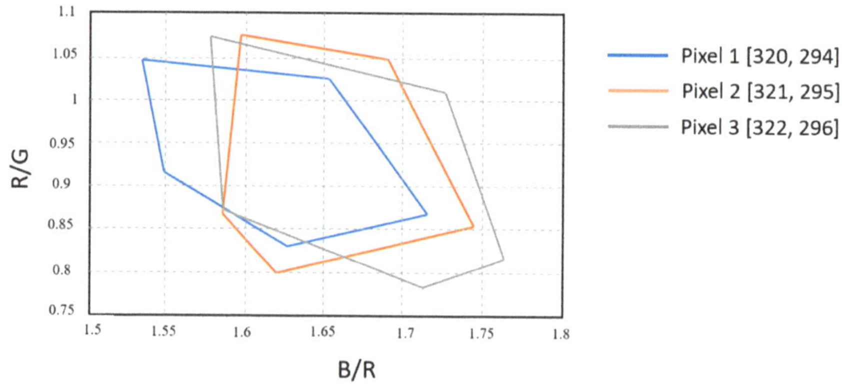

Figure 7. Calibration curves for three different pixels, represented by the R/G and B/R color ratios.

The deflection of the rays on the calibration filter results in a deflection angle which can be calculated for the reference image, as well as for the image containing the studied jet. The graph pictured in Figure 7 provides a glimpse into the differences between the initial design of the filter, which is assumed to only project linear changes over a single-color ration, and the color ratio distribution obtained from the physical filter obtained. The explanation of the unpredicted behavior of the filter color ratios is that the image of the filter is polluted with the film tint and there are differences between the image created in Matlab and the one displayed by the computer screen. This translates into a variation of the B/R and R/G curves, different than expected.

The deflection angle of the rays in the source image plane is considered to be split in the two directions of the Schlieren image, with ε_x and ε_y corresponding to the x and y directions, respectively. The deflection angle results from the calibration function, which in this case is written in the form presented by Equation (5).

$$\varepsilon = c_1 + c_2\frac{B}{R} + c_3\frac{G}{R} + c_4\frac{B^2}{R^2} + c_5\frac{B\cdot G}{R} + c_6\frac{R^2}{G^2} + c_7\frac{B^3}{R^3} + c_8\frac{B^2R}{G^3} + c_9\frac{BR^2}{G^3} + c_{10}\frac{R^3}{G^3} \tag{5}$$

The coefficients of the calibration function are calculated, as suggested in [5], at each pixel. The calibration function and the calibration coefficients provide the means needed to extract the deflection angle.

This paper uses the principles applied by [5] and finds that, in a similar manner, a linear function is not enough to describe the color ratio progression. The calibration curves for the color ratios displayed in Figure 7 present similar characteristics, but do not perfectly overlap.

4. CFD Simulation

The CFD simulation of the jet was conducted for the velocimetry study found in [8], performed on the jet provided in the same functioning conditions. Due to the path integrated character of the Schlieren technique, the results of the CFD analysis can only be used for a comparison achieved in a wide range of density values. The error calculation is therefore performed only as a reference for the CS technique, for a more precise result, a DNS or LES simulation would be required.

5. Results

After the calibration of the filter is achieved and the deflection angle at each pixel is found, the density map is retrieved by applying Equation (6), where W is the length of the optical path and K is the Gladstone–Dale constant. The deflection angles expected to be obtained can be considered as sufficiently small for the approximation made for Equation (6). Equation (6) is used by Elsinga [5] to determine the density gradients by assuming that the gradient along the integrated path of the investigated 2D flow is constant. This equation leads the study found in [5] to a final density gradient obtained by means of a numerically solved Equation (7), where ρ is the array containing the density values in each pixel and g is the array containing the x and y elements of the density at each pixel, while D is a sparse matrix which contains a second order central differencing scheme.

$$\varepsilon = KW\nabla\rho \tag{6}$$

$$D_{ij}\rho_i = g_j \tag{7}$$

This, for a non 2D flow, introduces a considerable error. For this reason, the studied jet will be divided into five different sections. The Gladstone–Dale equation applied (8) in this study considers the change in K, which varies slightly with the wavelength of the light.

$$n = 1 + K\rho \tag{8}$$

A comparison of the averaged result obtained per section by CCS and the result yielded by the CFD simulation will also be further conducted.

Figure 8 presents the lines from which the date is extracted in order to compare the color schlieren results to those offered by the CFD analysis.

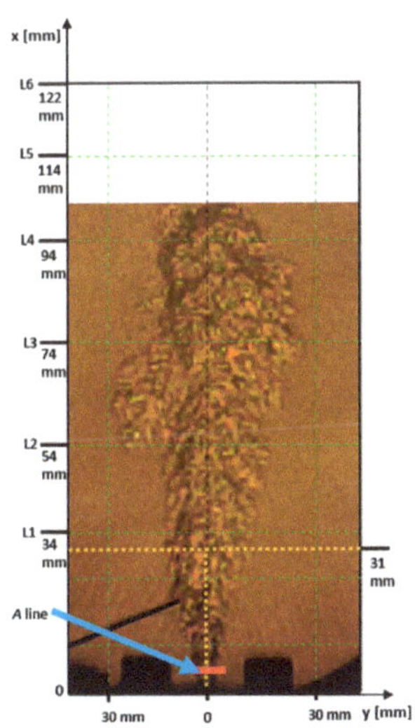

Figure 8. Test image with axes and the lines of interest: A (blue) and the vertical centerline stopping at 31 mm after the nozzle's exit.

The density map of values at every pixel location, as resulted from the CCS analysis, is found in Figures 9 and 10 (both curves in black). It presents the data obtained by calculating the density for each resulted ε.

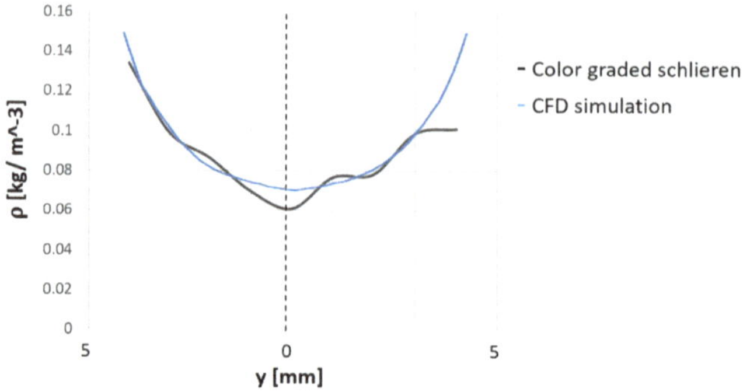

Figure 9. Density profiles along the A-line pictured in Figure 8, as obtained from the color-graded Schlieren analysis (black curve) and the CFD results (blue curve).

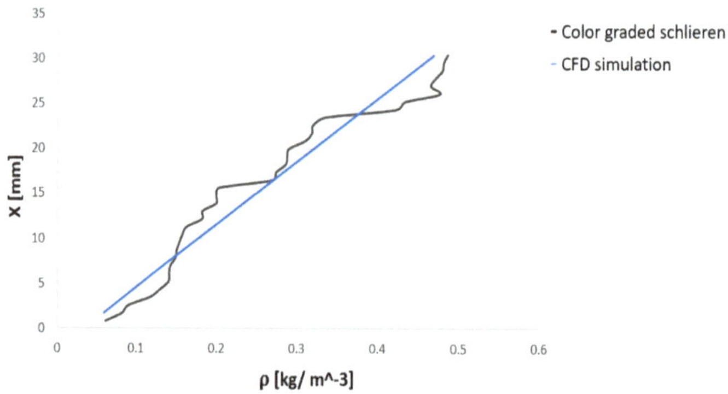

Figure 10. Density profiles on central axis density profile from origin to 31 mm on the x axis, pictured in Figure 8a, as obtained as obtained from the color-graded Schlieren analysis (black curve) and the CFD results (blue curve).

The results obtained through Schlieren CCS present a generally good accuracy in the 0–31 mm area, between 2% and 10%, given the position of the filter and the displacement recorded in this point. There are some pixels which present a dark character, and the explanation for those are that the filter's dimensions were too small, the deflection angle created by the system was too big, causing the rays to land outside the filter, in a dark area. For these pixels, no density value could be retrieved. The measurement provides a good coherency with the CFD simulation in the area starting at the origin and ending at 31 mm on the x axis. After the 31 mm, the error increases dramatically and the measurement becomes irrelevant. Between line 1 and line 3, the general accuracy of the measurement is 75, when reported to the CFD result, and over line 4, the ratio of quantifiable/unusable pixels is too low to properly describe the flow. However, central axis measurements prove to be less accurate when related to the CFD results than the horizontal samples. This can be caused by many experimental factors, like the one previously mentioned (the 3D character of the flow, filter limitations, etc.).

The CFD simulation parameters are described in [8]. The functioning conditions of the installation are the thruster has been designed to provide a nominal value of 1 N in vacuum conditions, with an expansion rate of 50 and a nominal mass flow of 0.3 g/s. The thruster's geometry consists of a convergent–divergent nozzle. Its minimal section is less than 1 mm.

ANSYS CFX was used to estimate the thruster's performance and determine the magnitude of the flow field at the thruster's outlet. A 3D model was created to replicate the combustion chamber's outlet and the convergent–divergent nozzle. Figure 11 presents the conic wall used for lowering the computational demand, as the flow around the external geometry of the thruster is not considered of importance. The computational model's dimensions are determined by the experimental chamber dimensions, which has a cylindrical geometry [8].

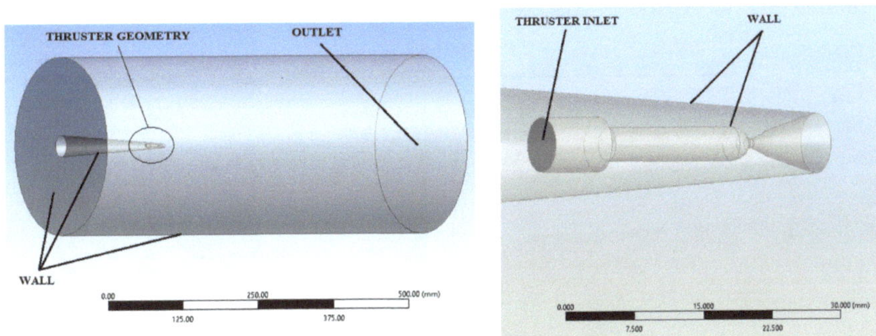

(a) (b)

Figure 11. (**a**) Fluid volume; (**b**) thruster's interior geometry [8].

ICEM was used for designing the numerical grid, which has a hexahedral structure with a smaller dimension in the thruster's inside volume and the vicinity of the thruster outlet, as can be observed in Figure 12. The grid independence was checked in terms of resulted y+ values, which were in the suggested intervals for the turbulence model used [8].

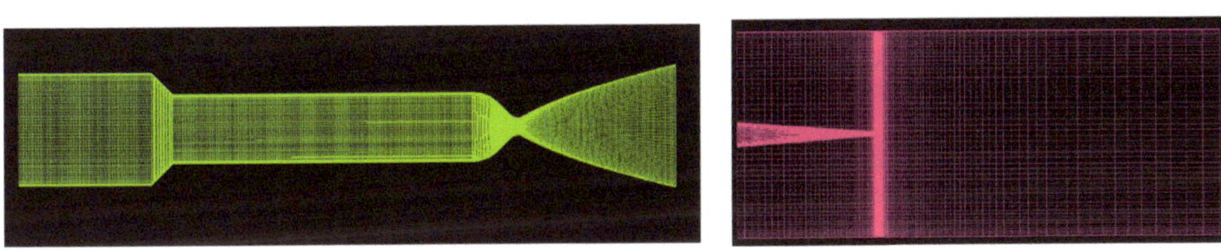

(a) (b)

Figure 12. (**a**) Numerical grid for thruster internal flow; (**b**) numerical grid for fluid volume [8].

The numerical case was set based on measurements for the respective test, and thus the total mass flow was set at 0.27 g/s. In order to reduce the complexity of the numerical simulation and lower the computational demand, the combustion process was not modeled. The combustion process was considered complete at the volume inlet, with the working fluid composed of water vapor forming the complete reaction of gaseous H_2 and O_2. The temperature was determined using CEARun software (last updated in January 2023) for a stoichiometric reaction of gaseous hydrogen and oxygen resulting in a value of 3450 K. The fluid used for this simulation was a mixture of H_2O, H_2, O_2, and N_2 in order to simulate the interaction between the water vapors and ambient air [8].

The boundary conditions for this study are presented in Table 2. Inlet boundary conditions were set for the thruster inlet (fixing the mass flow, total temperature and chemical components of the fluid, water vapor in this case), opening conditions for the fluid volume outlet (fixing the temperature, pressure, and chemical components). For the thruster and experimental chamber walls, adiabatic no slip conditions were used. Reynolds-averaged Navier–Stokes (RANS) simulations were conducted using the K-ε turbulence model due to its known performance for high-turbulent flows [8]. Figure 13 presents the density and temperature maps obtained from the CFD simulation.

Table 2. Boundary conditions [8].

Type	Location	Details
Inlet	Thruster inlet	Mass flow = 0.27 g/s Total temperature = 3450 K
Wall	Thruster walls, testing chamber walls	Adiabatic walls with no slip conditions
Opening	Testing chamber outlet	Temperature = 25 °C Pressure = 1 bar

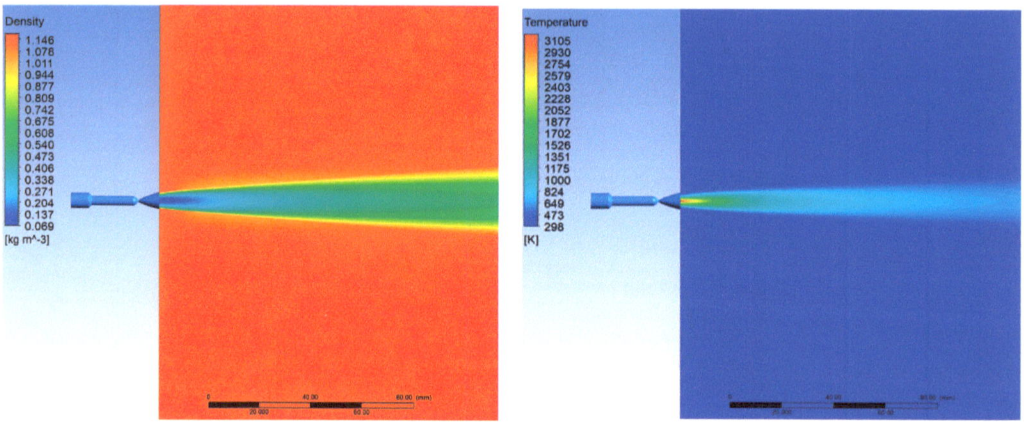

(**a**) (**b**)

Figure 13. (**a**) The exhaust jet's CFD obtained density profile; (**b**) the exhaust jet's CFD obtained total temperature profile.

To obtain a temperature map from the density map obtained by Schlieren, the equation of state can be applied to each value, individually. However, this will most likely present a greater error than the one obtained for the density map, as the calculation basis will already be flooded by errors, caused by the shortcomings of the method, such as the path integrated character of the turbulent axisymmetric jet, the filter's reduced dimensions which reduces the filter's sensitivity, and the 1pixel error introduced by the spatial resolution limitations. The temperature profile presents a 12% error in the 0–31 area and is inconclusive in the upper stages, which renders the temperature measurement by CS in this case futile.

6. Conclusions

The color-calibrated filter Schlieren (CS) method has been previously used for obtaining density maps and other parameters of 2D flows, representing a very powerful tool, registering errors from 2% to 3%, according to [5]. This paper addresses its use on 3D flows, and exemplifies a simple and low-cost method to manufacture and implement such a filter. Although the idea of using a photographic film has been around for years, the calibration mechanism has never been exemplified. The errors introduced by the filter are separated from the images the moment the filter is calibrated experimentally. For the horizontally place line A, depicted in Figure 8, the density graph reveals the issue of the jet's symmetry, which is affected by one of the vacuum chamber's windows, which causes the flow to propagate a little to the left of its central axis. Apart from the slight asymmetric density profile of the jet, the density curve on line A, the experimental density curve follows closely the CFD curve, with a clearer gap between the two curves values as it comes closer to the center line. This could be caused by the 3D character of the flow, which integrates the path and can cause an imprecise reading. Unfortunately, the experimental data are also scarce due to the above-mentioned pixels, where the deflection angle is large and information could not be extracted from. The measurement error was calculated by relating each experimental value to its CFD correspondent and eliminating the highest and lowest resulted error. This yielded a percentage of 2–3% on the A line.

The method can be improved with regards to the 3D flow by using a better fitted color filter, with a better color distribution. The easier way to achieve a better-known color ratio distribution would be to use a transparent step filter manufactured from gelatine sheets, like the ones produces by COMAR [9]. If the filer is cut into small sheets with a well-known length, the calibration process can be very similar to the one proposed by [5].

Corrections regarding the path-integrative character of the flow can be achieved by performing an inverse Abel transformation if the study can be considered to be propagating in a symmetrical manner.

A study in progress at this moment includes using a step color filter and applying an inverse Abel transformation, while relating the obtained measurements to the ones recorded by physical temperature sensors introduced into the flow.

Author Contributions: Conceptualization, E.-G.P. and T.P.; methodology, E.-G.P.; software, E.-G.P.; validation, E.-G.P.; resources, E.-G.P.; data curation, E.-G.P.; writing—original draft preparation, E.-G.P.; writing—review and editing, E.-G.P. and T.P.; visualization, E.-G.P.; supervision, T.P. All authors have read and agreed to the published version of the manuscript.

Funding: This work was carried out through the "Nucleu" Program, within the framework of the National Plan for Research, Development and Innovation 2023–2026, supported by the Romanian Ministry of Research, Innovation and Development, project number PN23.12.01.02.

Data Availability Statement: Data are contained within the article.

Acknowledgments: This work was carried out on the installation resulted from the "GREENTH" project (contract number 4000117301/16/NL/CBi), developed at the Romanian Institute for Gas Turbines COMOTI. Special thanks to the project manager, Eng. Jeni A. Vilag, for providing the optimal framework for the application of the CCS method in turbulent flows.

Conflicts of Interest: The authors declare no conflicts of interest.

References

1. Ting, C.C.; Cheng, C.C. Detection of gas leakage using microcolor schlieren technique. *Measurement* **2013**, *46*, 2467–2472. [CrossRef]
2. Settles, G.S. *Schlieren and Shadowgraph Techniques*; Springer: Berlin, Germany, 2012.
3. Iffa, D.E. A Method of Measuring Velocity Field. Patent Nb. PI 2010001427.
4. Stevenson, D. Quantitative Colour Schlieren Techniques. Master's Thesis, University of the Witwatersrand, Johannesburg, South Africa, 2011.
5. Elsinga, G.E.; van Oudheusden, B.W.; Scarano, F.; Watt, D.W. Assessment and application of quantitative schlieren methods: Calibrated color schlieren and background oriented schlieren. *Exp. Fluids* **2003**, *36*, 309–325. [CrossRef]
6. Greenberg, P.S.; Kimek, R.B.; Buchele, D.R. Quantitative rainbow schlieren deflectometry. *App. Opt.* **1995**, *34*, 3810–3825. [CrossRef] [PubMed]
7. Nicoară, R.E.; Vilag, J.; Suciu, C.; Vilag, V. Experimental testing of small H_2/O_2 thrusters in atmospheric and vacuum condition. In Proceedings of the 7th Edition of the Space Propulsion Conference, Online, 15 March 2021.
8. Prisăcariu, E.G.; Nicoară, R.E.; Prisecaru, T.; Vilag, J.; Vilag, V. Velocity Mapping of an H_2–O_2 Exhaust Jet in Air by Means of Schlieren Image Velocimetry (SIV). *Aerospace* **2023**, *10*, 576. [CrossRef]
9. Available online: https://www.comaroptics.com/components/filters/gelatin-and-polyester-colour-filters (accessed on 27 July 2023).

fluids MDPI

Article

Volumetric Flow Field inside a Gas Stirred Cylindrical Water Tank

Yasmeen Jojo-Cunningham [1], Xipeng Guo [2], Chenn Zhou [2] and Yun Liu [1,*]

[1] Department of Mechanical and Civil Engineering, Purdue University Northwest, Hammond, IN 46323, USA; yjojocun@pnw.edu
[2] Center for Innovation through Visualization and Simulation, POWERS Building 123 Purdue University Northwest, Hammond, IN 46323, USA; guo468@pnw.edu (X.G.); czhou@pnw.edu (C.Z.)
* Correspondence: liu739@pnw.edu; Tel.: +1-219-9895-381

Abstract: Ladle metallurgy serves as a crucial component of the steelmaking industry, where it plays a pivotal role in manipulating the molten steel to exercise precise control over its composition and properties. Turbulence in ladle metallurgy influences various important aspects of the steelmaking process, including mixing and distribution of additives, alongside the transport and removal of inclusions within the ladle. Consequently, gaining a clear understanding of the stirred flow field holds the potential of optimizing ladle design, improving control strategies, and enhancing the overall efficiency and steel quality. In this project, an advanced Particle-Tracking-Velocimetry system known as "Shake-the-Box" is implemented on a cylindrical water ladle model while compressed air injections through two circular plugs positioned at the bottom of the model are employed to actively stir the flow. To mitigate the particle images distortion caused by the cylindrical plexi-glass walls, the method of refractive matching is utilized with an outer polygon tank filled with a sodium iodide solution. The volumetric flow measurement is achieved on a 6 × 6 × 2 cm domain between the two plugs inside the cylindrical container while the flow rate of gas injection is set from 0.1 to 0.4 L per minute. The volumetric flow field result suggests double gas injection at low flow rate (0.1 L per minute) produce the least disturbed flow while highly disturbed and turbulent flow can be created at higher flow rate of gas injection.

Keywords: ladle metallurgy; water ladle model; volumetric flow field; Shake-the-Box system

Citation: Jojo-Cunningham, Y.; Guo, X.; Zhou, C.; Liu, Y. Volumetric Flow Field inside a Gas Stirred Cylindrical Water Tank. *Fluids* **2024**, *9*, 11. https://doi.org/10.3390/fluids9010011

Academic Editors: Guangjian Zhang and D. Andrew S. Rees

Received: 27 October 2023
Revised: 3 December 2023
Accepted: 6 December 2023
Published: 28 December 2023

1. Introduction

The global steel industry stands as a cornerstone of modern civilization, serving as the backbone of infrastructure, manufacturing, and technological advancement. The rising demand for high-quality steel products with precise characteristics has prompted ongoing innovations and advancements in steelmaking industry. Secondary steelmaking processes, a vital stage following primary steel production, have emerged as a critical avenue for refining and enhancing steel properties [1]. In particular, the intrinsic part of gas stirring has garnered significant attention in recent years. It entails injecting an inert gas into the molten metal to achieve uniform mixing to facilitate the homogenization of the chemical composition of different alloy elements and the removal of inclusions from the molten steel [2]. The implementation of stirring aids elevates the caliber and purity of the steel, enhancing its mechanical attributes and minimizing defects in the final product. Turbulence, driven by the injection of inert gas, stirs the molten metal vigorously, facilitating efficient mass transfer and promoting uniform distribution of alloying elements. This dynamic mixing action intensifies the interaction between the molten steel and the slag, a process can further amplified by the distinctive depression of the slag eye. This intensified interaction zone becomes a hotspot for chemical reactions, promoting the removal of undesired impurities and inclusions from the steel, which are buoyed to the surface and subsequently removed through slag formation [1,3]. Therefore, the flow turbulence can aids in achieving the desired steel characteristics though the formation of the slag eye, enhances its purity, consistency of mechanical attributes, and overall quality in the final product. Over the past few decades, researchers have extensively studied the slag eye formation under laboratory conditions. And dynamically- scaled water model apparatus have been applied, which involves using water at room temperature instead of molten steel. Water at room temperature has the similar kinetic viscosity to molten steel [4], and thus the properly scaled water ladle apparatus is a suitable physical model to be studied for understanding the related flow process inside steel ladles.

Szekely et al. [5] first proposed a simplified water model to study the flow characteristics of a ladle by injecting gas from the bottom, assuming a constant bubble size, and using Spalding's k-ω model to solve the Navier-Stokes equations for velocity and turbulence predictions. The model was developed to simulate the thermal and flow behavior of molten steel during the pouring process using a water ladle, offering a simple and effective way to predict the behavior of molten steel during pouring process. Debroy et al. [6] improved Szekely's model by refining the bubble model and accounting for the effects of turbulence and bubble coalescence. These added terms helped to predict the behavior of bubbles more accurately in the liquid metal and their impact on the mixing and refining of the steel. Johansen et al. [7] furthered the research by discovering that bubbles can create turbulence and affect flow velocity in the bubble plume region thus a bottom injection water model was adapted. Peranandhantan et al. [8] conducted an experiment to study the behavior of the slag eye in a ladle during steelmaking process by injecting air into a ladle filled with water and various fluids to simulate the injection of gas into molten steel. The size and shape of the slag eye were then measured using a high-speed photography and image analysis techniques. Based on their observation from tested multiple variables, including gas flow rate, slag thickness and liquid depth, an empirical expression was derived to formulate the slag eye size in terms of gas flow rate, the ladle dimension, gravity, surface tension, and the momentum of the gas bubbles. Mazumdar et al. [9] reviewed several studies related to the physical modeling and empirical correlation of gas-stirred ladle, highlighting the impact of various variables involved in the gas-stirring process, specifically the plug positions.

Several researchers experimented with changing the plug positions in their water ladle model to achieve a better mixing and wall shear stress distribution [9–11]. It is argued that plug design can change the bubble size distribution close to the plug, but not the average size and distribution in the whole ladle [8,12–14]. The study conducted by Gajjar et al. [15] delve into the influence of injector design on turbulence within ladle metallurgy processes, investigating how different injector designs affect the level and characteristics of turbulence in the flow. In 2019, Owusu et al. [16] used Particle Image Velocimetry (PIV) to investigate the behavior of bubbles and their effect on turbulence kinetic energy (TKE) on the cross-sectional plane of the water ladle. So far, many researchers have used PIV to study the flow field in water ladle models, but only the flow field on two-dimension planes were resolved.

Therefore, in this study, to further investigate the three-dimensional internal flow field in a water ladle model of a cylindrical container, a Particle Tracking Velocimetry system with refractive index matching is implemented to quantify the unsteady/three-dimensional flow field while eliminating the particle imaging distortion. Such study cannot only improve our understanding of the complex flow behavior in gas-stirred ladles but also can be used to validate the CFD simulation models.

2. Materials and Methods

Based on an industrial steel refining ladle, a cylindrical water ladle model is designed and built. The dimensions and parameters of the industrial prototype and the downscaled water ladle are summarized in Table 1. Four typical flow rates of gas injection on the industrial ladle is also presented. To maintain a dynamic similarity on the Froude number, the flow rates of gas injection on water ladle model are calculated and determined, ranging from 0.1 L per minute to 0.4 L per minute. Although, we were able to keep a very similar Froude number between the industrial prototype and water ladle model, the Reynolds number are different remaining a considerable challenge of achieving a complete dynamic similarity. Equation (1) is utilized for estimating the Froude number, where U_p is the velocity of the plume and H is the height of the molten steel/solution. Plume velocity U_p is calculated with the Equation (2), where Q is the volumetric flow rate of the gas injection, R is the radius of the ladle/model [17].

$$Fr = \frac{{U_p}^2}{gH} \tag{1}$$

$$U_P = 3.1Q^{\frac{1}{3}}H^{\frac{1}{4}}R^{-0.58} \tag{2}$$

Table 1. Parameter of prototype and downscaled water ladle.

Prototype—Steel Ladle		
Ladle Diameter		**2.5 m**
Molten steel height		3 m
Molten steel density		6795 kg/m^3
Molten steel dynamic viscosity		0.006 pa·s [18]
Surface tension—molten steel		1.82 n/m [18]
Flow rate	Reynolds number	Froude number
206.97 NL/min	2,775,000	0.0227
413.95 NL/min	3,496,000	0.036
620.92 NL/min	4,002,000	0.0471
827.90 NL/min	4,405,000	0.0571
Model—Water Ladle		
Tank Diameter		**0.070 m (2.75 in)**
NaI solution height		0.084 m (3.30 in)
NaI solution density		1793 kg/m^3 [19]
Solution dynamic viscosity		0.002 pa·s [19]
Surface tension—NaI solution		0.073 n/m [18]
Flow rate (20 °C)	Reynolds number	Froude number
0.10 L/min	12,000	0.0232
0.20 L/min	15,000	0.0368
0.30 L/min	17,000	0.0483
0.40 L/min	18,000	0.0585

Consequently, in this experiment, eight conditions of gas stirring/injection with a volumetric flow rate ranging from 0.1 L per minute (LPM) to 0.4 LPM were adopted. An overview of the experimental conditions is presented in the Table 2. Since two plugs are geometrically identical and symmetric to each other, therefore for single gas injection the same plug (plug 2) was used to inject the gas for the cases 5–8.

Table 2. Eight Test Conditions of gas injection flow rate in liter per minute.

Condition	1	2	3	4	5	6	7	8
Plug 1	0.1	0.2	0.3	0.4	0.1	0.2	0.3	0.4
Plug 2	0.1	0.2	0.3	0.4	0	0	0	0

2.1. Cylindrical Water Ladle Model and Refractive Index Matching

In this experiment, a cylindrical container with an inner diameter of 70 mm is implemented to simulate and replicate the characteristics of a cylindrical ladle (To avoid NaI solution spill onto the high-speed cameras and Laser a tall cylindrical wall of 178 mm is implemented). However, due to the curved surface of the cylindrical container that would introduce significant particle image distortion from refraction, the conventional flow field measurement methods, such as Particle Image Velocimetry and Particle Tracking Velocimetry, cannot be directly implemented to measure the flow field. Therefore, this study adopted the refractive index matching method to counteract the effects of image distortion due to refraction. Additionally, to capture and film the particle images from four different perspectives, a four-camera particle tracking velocimetry system was strategically configured on a larger hexagon tank with six flat walls meticulously designed to accommodate the angle of imagining. This larger hexagon tank was designed to allow the cameras to film the flow inside the cylindrical tank, which was positioned at the center of the larger hexagon tank, with the cameras filming perpendicular to the flat walls. Both tanks were fabricated from plexi-glass. To ensure optimal refractive index matching and thereby minimize image distortion, a Sodium Iodide solution was prepared and used to fill the tanks to a height of 84 mm. Importantly, the refractive index of the Sodium Iodide solution closely aligns with that of plexi-glass, ensuring minimum light refraction as the scatted light from the seeding particles traverses the curved plexiglass walls and the solution, eliminating particle image distortion when the Particle Tracking Velocimetry is employed. During the experiment, the compressed gas is introduced to the cylindrical tank through two 4.7 mm circular plugs at the bottom of the tank. The plugs sat on the centerline of tank with a distance of 40 mm. Two mechanical flow meters with an accuracy of 0.02 L per minute (Brooke Instrument, Hatfield, UK) were utilized to measure and monitor the flow rate of the compressed gas during the experiments.

2.2. Particle Tracking Velocimetry System

The state-of-the-art: Shake-the-Box system (Lavision, Gottingen) was implemented on the cylindrical water ladle model, which inject compressed air to stir the flow. To accurately track the intricate gas-stirred water flow within the water ladle model, hollow glass spheres with a diameter range of 8 to 12 μm were used as seeding particles. To capture the seeding particles in the flow field and film the particle images, a high repetition rate laser (Nd: YLF single cavity, Photonics DM-30-527) and four high-speed cameras (Phantom VEO 640), were strategically positioned on two sides of the experimental setup (See Figure 1a,b). Lenses (Tokina Macro) with a focal length of 100 mms and aperture size of f/4.5 and f/11 were incorporated into the imaging system to facilitate capturing particle images within the flow field. To ensure optimal illumination and imaging, cylindrical optical lenses were added to the laser head, generating a 20 mm thick laser light that penetrated the walls and illuminated the inner cylindrical tank from the side. The sampling frequency of the images/laser was set at 100 Hz, with an image resolution of 1024 × 1024 pixels maintained across all four high-speed cameras, enabling a flow measurement volume of 57 mm × 62 mm × 19 mm that locates between the plugs and 18 mm above the bottom of the tank (See Figure 1c). For each testing condition (see Table 1), on each high-speed camera, the maximum number of 2000 image/data samples were collected over a time duration of 20 s. During the experiment, the gas injection was started at least 2 min before the data collection (image collection with laser illumination) with a 5 min break between each case of data collection.

The software Davis 10 (LaVision, Gottingen, Germany) was employed for calibration, data collection, and velocity field constructions. To achieve accurate calibration, a 55 × 55 mm calibrate target was positioned within the Sodium Iodide solution, allowing for the acquisition of calibration images. Calibration was performed in the Davis 10 using four images of the target. On each data set of collected particle images, for enhanced accuracy in sub-pixel measurements and to facilitate volumetric flow field measurement, volume self-calibration was incorporated during the final calibration. The Shake-the-Box algorithm (LaVision, Gottingen, Germany) was then used to carry out the particle reconstruction/tracking on each data set by shaking the particle position by 1 voxel during the iterations. The culmination of these process involves the reconstruction of the instantaneous volumetric velocity field through post-processing within Davis 10. This resulted in three

13 × 14 × 5 matrices, each representing volumetric velocities along the three dimensions with a spatial resolution of 4.7 mm/velocity vector. For a more in-depth understanding of Shake-the-Box algorithm and its principle, please refer to Schanz et al. [20].

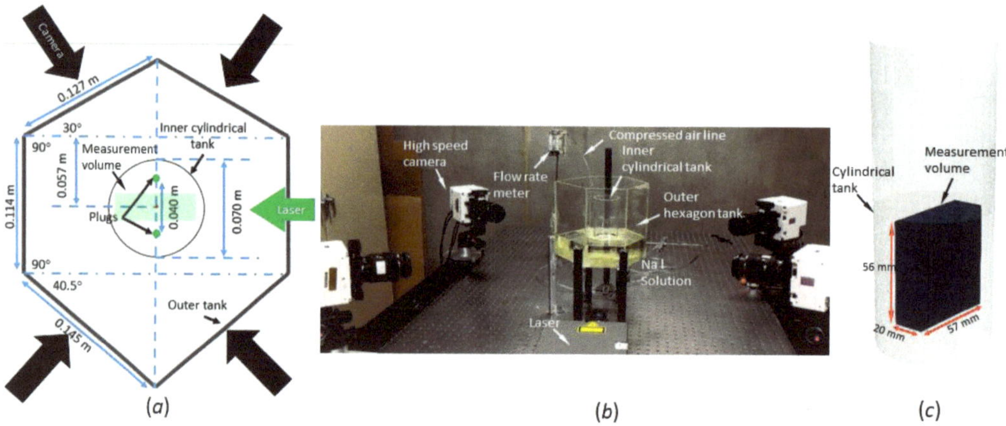

Figure 1. (**a**) Top view sketch of the setup. (**b**) Setup. (**c**) Measurement Volume.

3. Results

The primary focus of this study revolves around the aggregated flow field behavior represented by the mean velocity field and turbulent kinetic energy. The experimental results of each condition of gas flow injection will be discussed with a deeper focus on conditions 1, 3, 5 and 8 to further detail the flow behavior of the water ladle model through the plots of volumetric mean velocity field, streamlines, and turbulent kinetic energy.

3.1. Mean Velocity Field

Figures 2 and 3 present the volumetric contour plots of mean velocity in the x (transverse direction), y (vertical direction), and z (through thickness direction) directions, as well as the corresponding velocity magnitude, for both double and single gas injection cases. In Figure 2, representing the double gas injection, the distinct flow field behavior at case 1 (double gas injection at a flow rate of 0.1 LPM) with significantly lower mean velocity field stands out as compared to the flow field at other gas injection conditions. To accommodate this significantly lower velocity flow field at condition 1, a different scale is implemented (from −0.05 to 0.05 m/s for mean velocity plot in x, y, z directions and 0 to 0.1 m/s for velocity magnitude). In the x-direction (transverse direction) velocity plot, the flow domain splits into two at the top with negative x velocity flow (blue) on the left-side and positive x velocity flow (red) on the right-side. This flow pattern at case 1 suggests the stirred flow divided into two branches heading to opposite directions of x (transverse direction) at the top of the flow field. For the vertical velocity (y-direction velocity), a high velocity concentration is observed at the center of the top region. Moving away from the center, the vertical velocity decreases and creates a lower vertical velocity flow, resulting in a concave bowl shape from the center line and extends downwards. For the through thickness direction velocity (z-direction velocity), high velocity region moves away from the top and concentrates at the center of the flow domain, forming round contours with lower velocity away from the center.

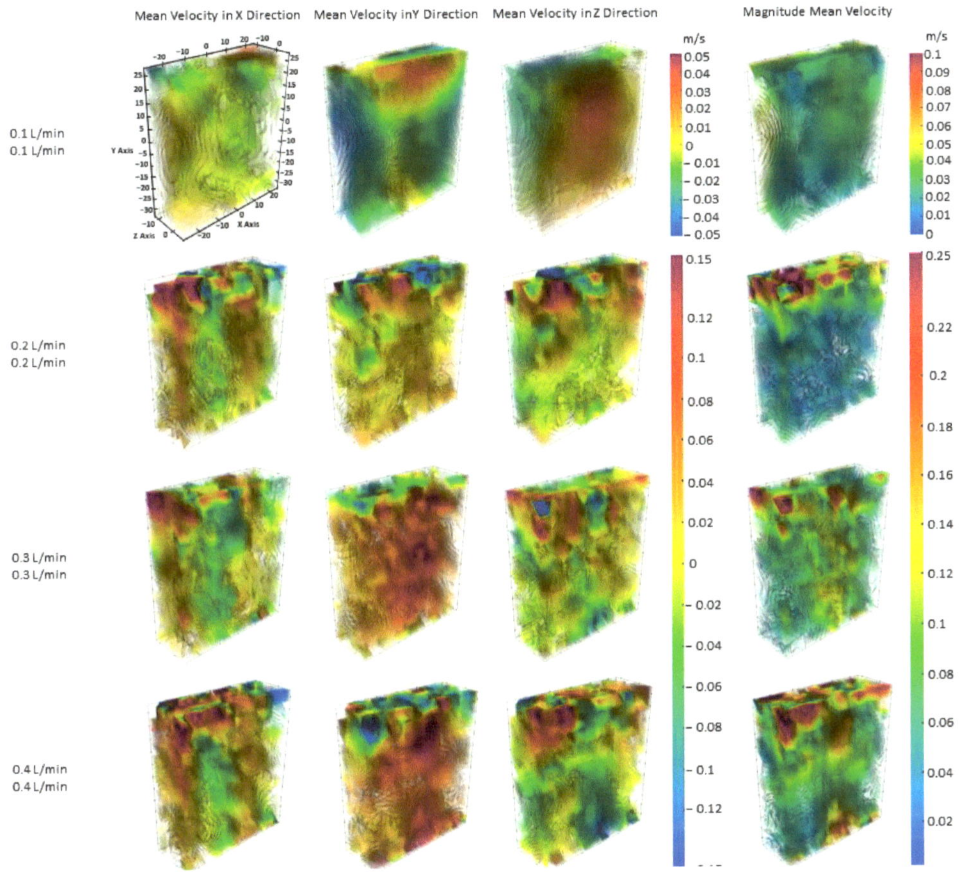

Figure 2. Mean velocity field in X, Y, Z directions with double gas injections.

At higher flow rates of gas injection, as more vertical momentum is added to the flow field, flow field quickly become random and lack a clear structure or pattern to be characterized. With flow rate of double gas injection goes above 0.2 L per minutes, instead of enhanced flow velocity, no clear separation of positive and negative transverse velocity can be observed at the top of the flow field. Meanwhile, instead at the top of the flow domain, the upward velocity (Y direction velocity) is mostly intensified at the mid-height of the flow field. Yet, for the through thickness velocity plot, no clear characteristics was observed. As for the mean velocity magnitude results, with increasing flow rates, the high-velocity region becomes more pronounced and extends from the top towards the bottom of the ladle model. Conversely, with single gas injection (Figure 3), the mean velocity field plots start with random distributions from the beginning at 0.1 L per minute and with less distinct flow patterns. As the flow rate increases, the high velocity region experiences subtle changes with no distinguishing variations, except the upward velocity gets more noteworthy at the center of the flow domain. Regarding to the magnitude of mean velocity, as the flow rate increases, the flow contour pattern remains consistent and becomes more distinct, intensifying at the concentrated regions at the corners. While Figures 2 and 3 presenting the volumetric velocity distribution, to gain a better understanding of the underlying physical process inside the ladle model, further investigation into the complex flow field is warranted.

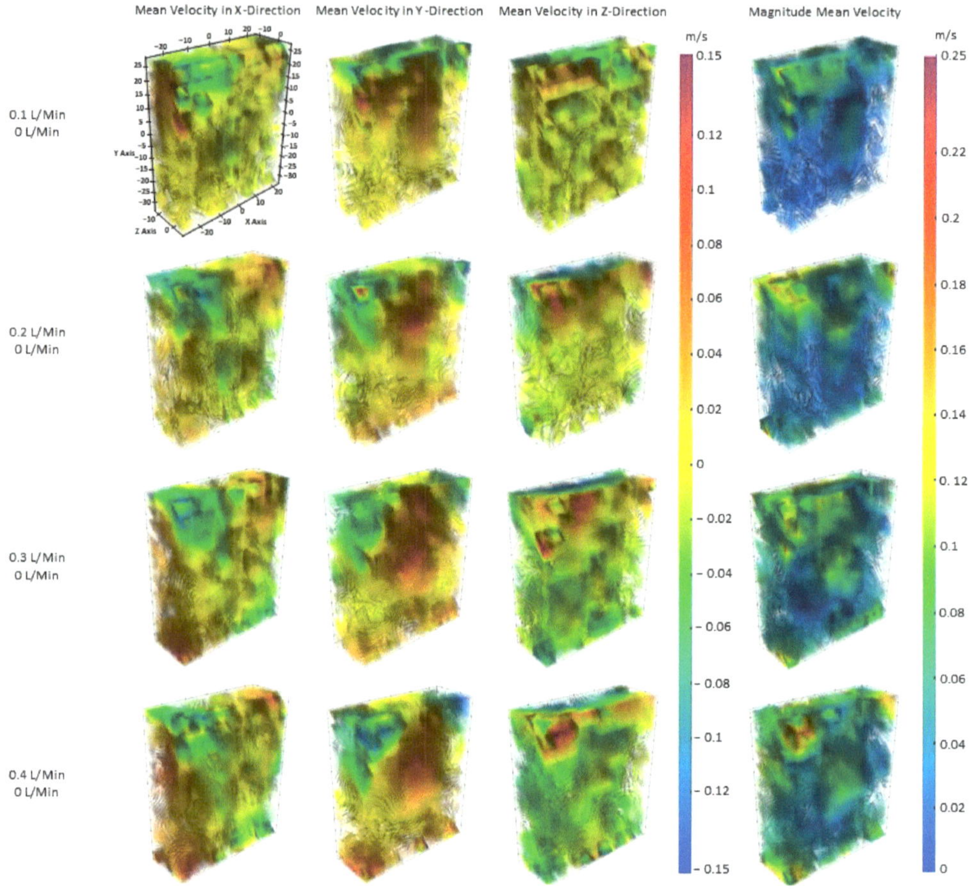

Figure 3. Mean velocity field in X, Y, Z directions with single gas injections.

3.2. Mean Velocity Streamlines

The derivation of three-dimensional streamlines from the mean velocity field provides a valuable tool for visualizing fluid flow and comprehending intricate flow behaviors. Figure 4 shows the three-dimensional streamlines that are color coded with velocity magnitude. For double gas injection cases, at the low flow rate of 0.1 L per minute, the three-dimensional streamlines exhibit a high degree of structure and organization. The streamlines adheres to a well-defined, coherent pattern within the X-Y planes (the planes that are parallel to the frontal planes). Also these streamlines symmetrically distribute about the sagittal plane. However, as the gas injection flow rate increases, the streamlines adopt a more intricate and less organized character with the velocity magnitude intensified. Conversely, streamline patterns in the case of single gas injection mark disorderliness and a lack of clear patterns, even at low gas injection rates of 0.1 L per minute. The streamlines with single gas injection evade any discernible pattern while clusters or groups of streamlines exist within the flow field, regardless of the injection rate. Notably, the velocity magnitude of the streamlines is significantly lower under the single gas injection as compared to the double gas injection cases as much less momentum is added in the flow for single gas injection cases.

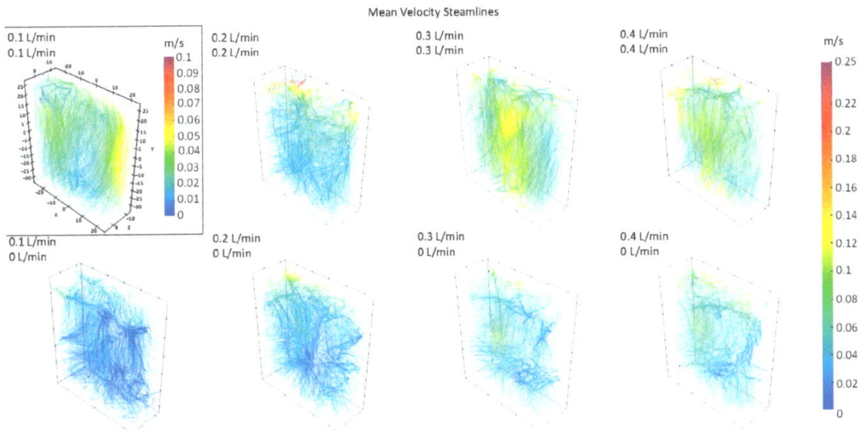

Figure 4. 3D streamlines for magnitude mean velocity field.

Furthermore, the mean velocity field results are analyzed on two dimensional planes that can detail distinct flow features such as nodal points of separation/attachment, focus points of separation/attachment, and saddle points [21]. A nodal point of attachment occurs when velocity lines radiate outward from the node, while nodal points of separation involve velocity lines converging inward towards the node. Contrasting with nodal points, focuses exhibit a spiraling movement around a singular point—either spiraling away from it (a focus of attachment) or spiraling into it (a focus of separation). Saddle points present a unique scenario where streamlines move in differing directions across two perpendicular planes. As fluid moves away from this saddle point, it diverges in one direction and converges in another. To further examine the gas stirred flow field of the water ladle model, Figures 5–8 depict two dimensional streamlines and vectors with velocity magnitude contour on three planes on x = −5 mm (Sagittal plane), y = −5 mm (Transverse plane), and z = −5 mm (Frontal plane). Similar to the observations made on Figures 2 and 4, streamlines exhibit an organized and orderly pattern for double gas injection at low flow rate of 0.1 L per minute (condition 1). In contrast, higher gas injection flow rates and single gas injection induce greater disorder and more flow features (Nodal, Focus and Saddle points) on the streamlines. Figure 5 distinctly portrays a well-structured flow pattern in the case of double gas injection at a rate of 0.1 L per minute (condition 1). In Figure 5b, on the Transverse plane a saddle point is observed near the center, where the flow bifurcates into opposing directions. Surrounding this saddle point, a focus point is captured on the upper left corner where a symmetry flow field can be observed in the plot. Figure 5c demonstrates the organized streamlines with a slight deviation from the perfect symmetry on the frontal plane. Counteracting the injected upward momentum, a prevailing downward flow is observed on the sagittal plane, leading to a concentrated region of low velocity on the plane near the saddle point.

Figure 6 presents the streamlines plot with the double gas injection at a flow rate of 0.3 LPM, marking the shift from a downward dominated to an upward dominated flow field, as previously noted. Additionally, high velocity becomes more pervasive throughout each plane, displaying a more random flow pattern compared to Figure 5. Transitioning to a single gas injection scenario at a flow rate of 0.1 L per minute is presented in Figure 7. The flow field lacks a single predominant momentum direction, exhibiting the presence of multiple flow directions. Also, low velocity concentration prevails across the flow field, while high velocity zones emerge mainly at the tips of the top flow domain, suggesting no distinct flow patterns. Finally, Figure 8 showcases the outcomes for a single gas injection at the highest rate of 0.4 L per minute. The flow remains disrupted and irregular, lacking any dominant momentum. Concentrated high velocity zones intensify at the boundaries, devoid of any discernible or consistent pattern. In summary, the two-dimensional streamline visualizations provide insightful understanding about the mean velocity field inside the model. The findings underscore the considerable influence of injection rate and gas injection type on flow behavior and resulting streamlines. Lower gas injection rates, particularly in double injections, yield more structured and organized flow patterns. In contrast, higher gas injection rates and single gas injection can lead to increased randomness and disorganized flow behavior. This randomness and disorganized flow behavior shall be attributed to the elevated turbulence in the gas stirred flow field.

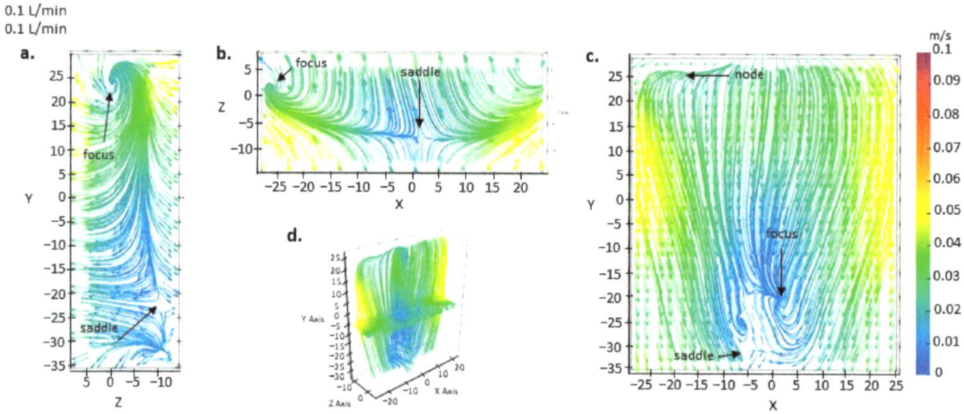

Figure 5. Streamlines and velocity magnitude on different planes with double gas injection at 0.1 L/Min. (**a**) streamline plot on X = −5 mm plane. (**b**) streamline plot on Y = −5 mm plane. (**c**) streamline plot on Z = −5 mm plane.

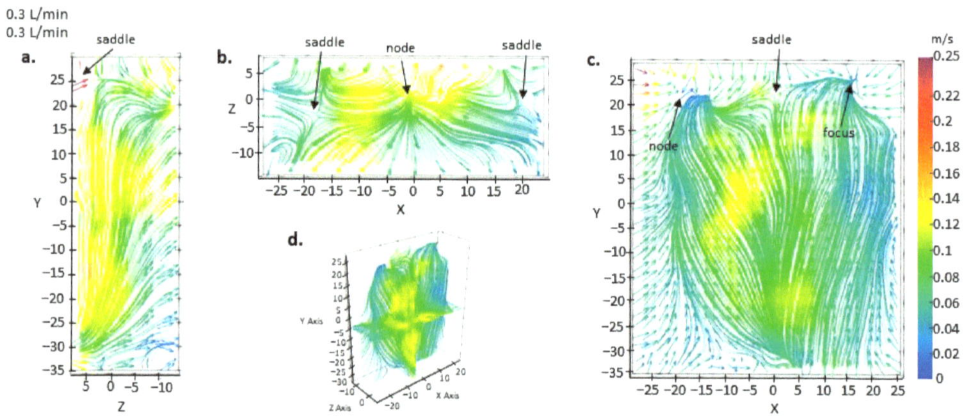

Figure 6. Streamlines and velocity magnitude on different planes with double gas injection at 0.3 L/Min. (**a**) streamline plot on X = −5 mm plane. (**b**) streamline plot on Y = −5 mm plane. (**c**) streamline plot on Z = −5 mm plane.

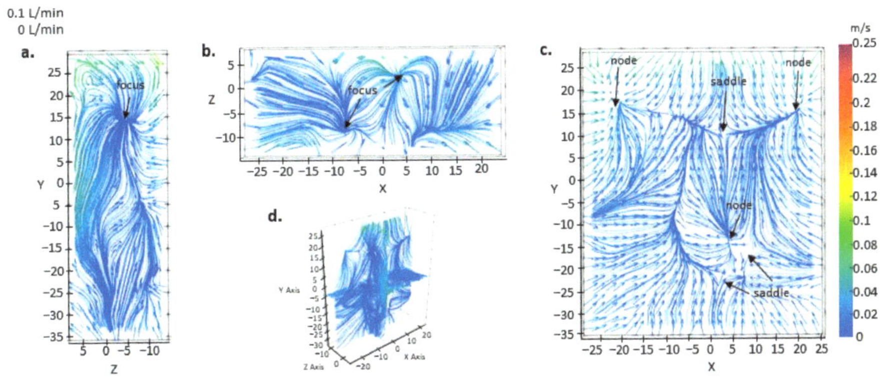

Figure 7. Streamlines and velocity magnitude on different planes with single gas injection at 0.1 L/Min. (**a**) streamline plot on X = −5 mm plane. (**b**) streamline plot on Y = −5 mm plane. (**c**) streamline plot on Z = −5 mm plane.

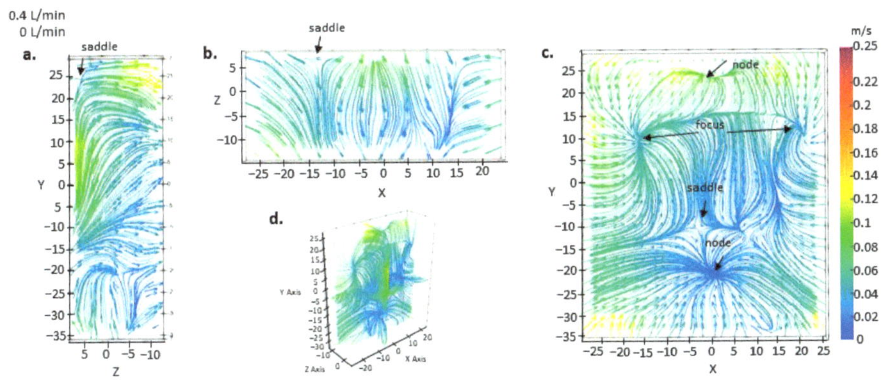

Figure 8. Streamlines and velocity magnitude on different planes with single gas injection at 0.4 L/Min. (**a**) streamline plot on X = −5 mm plane. (**b**) streamline plot on Y = −5 mm plane. (**c**) streamline plot on Z = −5 mm plane.

3.3. Turbulent Kinetic Energy

Turbulence holds paramount importance in the realm of ladle metallurgy, where its influence resonates through the entire process, borne from the vigorous and intricate motion of molten metal, underpins the homogenization of temperature, composition, and additives within the metal. In this study, the turbulent kinetic energy is calculated from the resolved volumetric flow field to understand the turbulent behavior of the stirred flow field inside the water ladle model, using the equation below [22], where u', v' and w' represent the fluctuating components of velocity in the x, y, z directions, respectively.

$$TKE = \frac{1}{2}\left(u'^2 + v'^2 + w'^2\right) \tag{3}$$

Figure 9 illustrates the volumetric distribution of turbulent kinetic energy for both single and double injections. In the case of double injection at 0.1 LPM, the value registered is a hundred times lower than the turbulent kinetic energy observed with cases 2–8. If the same color bar scale of cases 2–8, 0 to 0.3 m^2/s^2, is applied on condition 1, the depiction of the turbulent kinetic energy volumetric flow would have appeared uniformly blue, signifying minimal turbulence under condition 1. Conversely, for a single injection at

0.1 LPM, a distant area of turbulence is evident, presenting a notable contrast. Observable in both single and double injections, an increase in the gas rate corresponds to a considerable escalation in turbulence, primarily concentrated within the uppermost portion of the flow domain which is consistent with other studies on the gas stirred steel refining ladle [16,23,24]. However, in the case of double injection, the turbulence exhibits greater turbulence strength and is more consistently concentrated throughout as more momentum is added to the flow, comparing to single gas injection. This observation aligns with the findings of with many other studies [15,25,26] that demonstrated the outward dispersion of turbulence from the plume region near the top.

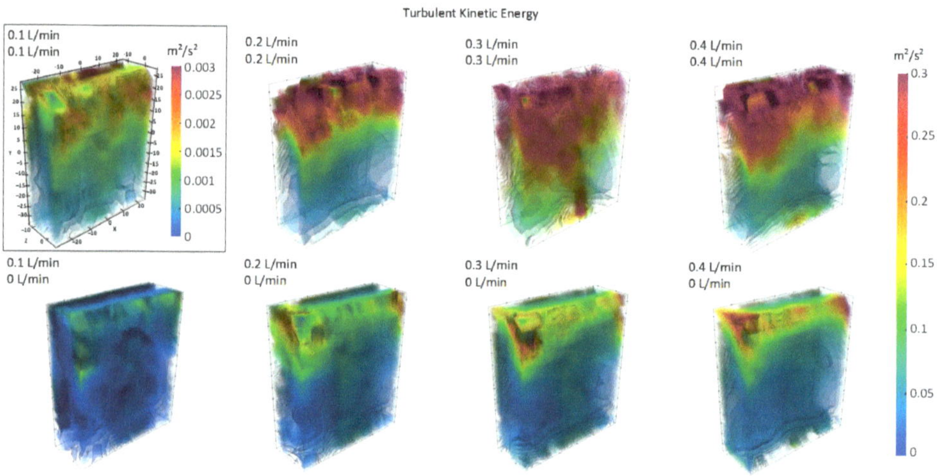

Figure 9. Turbulent kinetic energy for single and double injections.

4. Summary and Discussion

Through the analysis of the velocity field within the volumetric flow of the water ladle model using mean velocity magnitude (as discussed in Section 3.1), streamlines (as outlined in Section 3.2), and turbulent kinetic energy (as outlined in Section 3.3), a consistent trend has emerged and can be confirmed across multiple figures. In the context of double gas injection, a clear flow trend emerges as the gas injection rate increases. This flow trend is characterized by the extension of high velocity concentration from the top to the bottom in the vertical direction and higher fluid velocity magnitudes throughout the entire volumetric region. Particularly, in considering the vertical velocity and velocity magnitude, this trend is more prominent, suggesting the uneven fluid motion at the top under the stirring of gas injections. This has exceptional importance as critical processes, such as desulfurization, takes place at the interface between the molten steel and slag. Furthermore, when the flow rate is increased from 0.1 to 0.2 L per minute, the flow field shifts from an organized and structured mean velocity field pattern to a disorganized one. This transition is clearly illustrated in the volumetric distribution of vertical velocity plot in the Figure 2. As the flow rate continues to rise, the disorderly flow pattern persists, accompanied by a notable elevation in velocity magnitude across flow region. This progression of irregular and asymmetrical flow patterns with increasing gas flow rates is further underscored by a particularly vivid visual representation of Figure 4 of the mean 3D streamline plots.

With the gas injection flow rate elevated from 0.1 to 0.2 L per minute, the previously symmetrical, organized, and predominantly vertical streamlines evolve into irregular and predominantly horizontal flow pattern. While the effect is less pronounced, another transition occurs as the flow rate is further increased to 0.3 L per minute. This shift is evident in the changing orientations of the streamlines, which regain a more vertical alignment. Upon examination of the mean velocity streamlines at various planes, additional confirmation of the observed trends is confirmed in Figures 5 and 6. Figure 5 solidifies the depiction of a highly organized and structured flow pattern at the low double gas injection rate of 0.1 L per minute. Intriguingly, it becomes apparent that despite gas injection aiming to introduce upward momentum, downward flow dominates the measured flow field at the low gas injection rate of 0.1 L per minute. A focused analysis of the streamlines and mean velocity across stream-slices reveals the concentration of low velocity around the focal point on frontal plane (Figure 5c). The presence of this low velocity spot in the mean velocity field is unwanted, as it leads to fluid accumulation, hindering the efficient fluid mixing process. Raising the gas flow rate to 0.3 L per minute, the dominant flow momentum alters from downward to upward while exhibiting reduced outward curvature (Figure 6). Furthermore, high velocity concentrations become more prevalent across all planes. These trends suggest that higher gas flow rates could be advantageous in attaining the desired flow uniformity and mitigating the accumulation of stagnant fluid.

For single gas injection in Figure 3, however, increasing the gas flow rate does not introduce changes in location of contour concentration. Instead, it augments overall magnitude of velocity while maintaining a consistent contour pattern. As flow rate increases, the contour patterns of high velocity regions become sharper and more pronounced, suggesting the mean velocity field becomes more concentrated. Additionally, in Figure 4, streamlines under single gas injection consistently lack orderly structure across all flow rates. They display a disorganized and irregular flow pattern marked by random flow directions and clusters of streamline concentration. This observation is further verified through the examination of the 2D mean velocity streamlines plots in Figures 7 and 8. Figure 7 confirms the presence of a disorganized flow pattern, lacking a dominant flow momentum direction. This complexity could be attributed to the existence of multiple nodal, focal, and saddle points. Moreover, the flow field is primarily characterized by low velocity concentration, indicating the potential for stationary fluid within the volume under single gas injection with low flow rate. With the gas flow rate increased to 0.4 L per minute, as demonstrated in Figure 8, the flow maintains its disrupted nature without a prevailing momentum direction. However, the concentration of high velocity becomes more intense and extended, encompassing the outer boundaries and enhancing fluid bulk motion. By contrasting the effects of double and single gas injections, distinct flow differences become evident. Figure 4 illuminates the evolution of the streamlines for double and single gas injections.

As the flow rate increases, double gas injections tend to align the streamlines more within the vertical orientation, whereas single gas injections continue to exhibit fluid motion in multiple directions. Additionally, single gas injections manifest more concentrated streamline groups, while double gas injections maintain a more uniform distribution throughout the region. The volumetric contour plots of turbulent kinetic energy further support our observation from the mean velocity field plots. At condition 1, with double gas injection at a low flow rate of 0.1 L per minute, the turbulent kinetic energy distribution is significantly low at the range of 0~0.003 m^2/s^2 while the turbulent kinetic energy plot for other conditions rise to much higher value in the range of 0~0.3 m^2/s^2 with higher TKE value distribution in other double gas injection cases and lower TKE value distribution in single gas injection cases, implying a fundamental flow behavior difference between the condition 1 and other conditions which can be correlated to the laminar and turbulent flow. With double gas injections at low flow rate, the fluid inside the cylindrical container is stirred but still able to maintain at laminar flow regime. However, at higher flow rate or single gas injection, the fluid inside the container is turned into turbulent when more momentum and asymmetry are added to the flow.

Therefore, to achieve the homogenization of the molten steel inside a real gas stirred steel refining ladle, a high flow rate and single (asymmetrical) gas injection should be implemented to introduce desired turbulent flow inside the ladle. In gas stirred refining ladles, the process of injected argon bubbles rising stirs the molten steel, promoting uniform composition and temperature, reducing segregation, and facilitating the removal of inclusions. In essence, ladle argon blowing serves to clean the molten steel, decrease hydrogen, oxygen, and nitrogen levels, enhance temperature and composition uniformity. In this study, the implementation of advanced Particle-Tracking-Velocimetry systems, particularly the Shake-the-Box method, on a cylindrical water ladle model, provided great insights into the three-dimensional flow field, illuminating the nuances of gas stirring, crucial for achieving desired steel quality.

Author Contributions: Y.J.-C. analyzed the data and wrote the manuscript. X.G. and Y.L. conducted the experiments. C.Z. and Y.L. conceptualize and supervise the work. All authors have read and agreed to the published version of the manuscript.

Funding: This work was supported in part by the National Science Foundation through its MRI program (CMMI-1919726) and GOALI program (CMMI-2113967). Support from all members of the Steel Manufacturing Simulation and Visualization Consortium (SMSVC), the SMSVC Ladle Project Technical Committee members, and Nucor Steel and the permission to publish this work are greatly acknowledged.

Data Availability Statement: The experimental data is available upon request.

Conflicts of Interest: The authors declare no conflict of interest.

References

1. Fruehan, R. *The Making, Shaping, and Treating of Steel*, 11th ed.; The AISE Steel Foundation: Pittsburgh, PA, USA, 1998.
2. Mishra, B. Steelmaking practices and their influence on properties. In *Metals Handbook: Desk Edition*; ASM International: Almere, The Netherlands, 1998; pp. 174–202.
3. Duan, H.; Zhang, L.; Thomas, B.; Conejo, A. Fluid flow, Dissolution, and Mixing Phenomena in Argon-Stirred Steel Ladles. *Metall. Mater. Trans. B* **2018**, *49*, 2722–2743. [CrossRef]
4. Mazumdar, D.; Guthrie, R. The physical and mathematical modelling of gas stirred ladle systems. *ISIJ Int.* **1995**, *35*, 1–20. [CrossRef]
5. Szekely, J.; Asai, S. The general mathematical statement of turbulent recirculatory flows. *Trans. ISIJ* **1975**, *15*, 270–275. [CrossRef]
6. Roy, T.D.; Majumdar, A.K.; Spalding, D.B. Numerical prediction of recirculation flows with free convection encountered in gas agitated reactors. *Appl. Math. Model.* **1978**, *2*, 146–150.
7. Johansen, S.T.; Robertson, D.G.C.; Woje, K.; Engh, T. Fluid dynamics in bubble stirred ladles: Part I. experiments. *Metall. Trans. B* **1988**, *19*, 745–754. [CrossRef]
8. Peranandhanthan, M.; Mazumdar, D. Modeling of slag eye area in argon stirred ladles. *ISIJ Int.* **2010**, *50*, 1622–1631. [CrossRef]
9. Mazumdar, D.; Roderick, G. Modeling energy dissipation in slag-covered steel baths in steelmaking ladles. *Metall. Mater. Trans. B* **2010**, *41*, 976–989. [CrossRef]
10. Nunes, R.; Pereira, J.; Vilela, A.; Der Laan, F. Visualisation and analysis of the fluid flow structure inside an elliptical steelmaking ladle through image processing techniques. *J. Eng. Sci. Technol.* **2007**, *2*, 139–150.
11. Li, L.; Liu, Z.; Li, B.; Matsuura, H.; Tsukihashi, F. Water model and CFD-PBM coupled model of gas-liquid-slag three-phase flow in ladle metallurgy. *ISIJ Int.* **2015**, *55*, 1337–1346. [CrossRef]
12. Vazquez, A.; Sanchez, R.; Salinas-Rodriguez, E.; Soria, A. A look at three measurement techniques for bubble size determination. *Exp. Therm. Fluid Sci.* **2005**, *30*, 49–57. [CrossRef]
13. Aoki, J.; Thomas, B.; Peter, J.; Peaslee, K. Experimental and theoretical investigation of mixing in a bottom gas-stirred ladle. *AISTech Iron Steel Technol. Conf. Proc.* **2004**, *1*, 1045–1056.
14. Fukuji, Y.; Mori, Y.; Fujita, S. Sizes and Size Distributions of Bubbles in a Bubble Column. *Chem. Eng. Jpn.* **2006**, *12*, 5–9.
15. Gajjar, P.; Haas, T.; Owusu, K.; Eickhoff, M.; Kowitwarangkul, P.; Pfeifer, H. Physical study of the impact of injector design on mixing, convection, and turbulence in ladle metallurgy. *Eng. Sci. Technol. Int. J.* **2019**, *22*, 538–547. [CrossRef]
16. Owusu, K.; Haas, T.; Gajjar, P.; Eickhoff, M.; Kowitwarangkul, P.; Pfeifer, H. Interaction of injector design, bubble size, flow structure, and turbulence in ladle metallurgy. *Steel Res. Int.* **2019**, *90*, 1800346. [CrossRef]
17. Conejo, A. *Fundamentals of Dimensional Analysis, Theory and Applications in Metallurgy*; Springer Nature: Singapore, 2021.
18. Liu, W.; Lee, J.; Silaen, A.; Zhou, C. Argon bubble coalescence and breakup in a steel ladle with bottom plugs. *Steel Res. Int.* **2019**, *90*, 1800396. [CrossRef]
19. Bai, K.; Katz, J. On the refractive index of sodium iodide solutions for index matching in PIV. *Exp. Fluids* **2014**, *55*, 1704. [CrossRef]
20. Schanz, D.; Gesemann, S.; Schröder, A. Shake-the-box: Lagrangian particle tracking at high particle image densities. *Exp. Fluids* **2016**, *57*, 70. [CrossRef]
21. Tobak, M.; Peake, D.J. Topology of three-dimensional separated flows. *Annu. Rev. Fluid Mech.* **1982**, *14*, 61–85. [CrossRef]
22. Pope, S. *Turbulent Flows*, 1st ed.; Cambridge University Press: Cambridge, UK, 2000.

23. Jardón-Pérez, L.; Gonalez-Rivera, C.; Trapaga-Martiez, G.; Amaro-Villeda, A.; Ramirez-Argaez, M. Experimental study of mass transfer mechanisms for solute mixing in a gas-stirred ladle using the particle image velocimetry and planar laser-induced fluorescence techniques. *Steel Res. Int.* **2021**, *92*, 2100241. [CrossRef]
24. Ramírez-Argáez, M.; Dutta, A.; Amaro-Villeda, A.; Gonzalez-Rivera, C.; Conejo, A. A novel multiphase methodology simulating three phase flows in a steel ladle. *Processes* **2019**, *7*, 175. [CrossRef]
25. Jardón-Pérez, L.; Amaro-Villeda, A.; Gonzalez-Rivera, C.; Trapaga, G.; Conejo, A.; Ramirez-Argaez, M. Introducing the Planar Laser-Induced Fluorescence technique (PLIF) to measure mixing time in gas-stirred ladles. *Metall. Mater. Trans. B* **2019**, *50*, 2121–2133. [CrossRef]
26. Jardón-Pérez, L.; Gonzalez-Morales, D.; Trapaga, G.; Gonzalez-Rivera, C.; Ramirez-Argaez, M. Effect of differentiated injection ratio, gas flow rate, and slag thickness on mixing time and open eye area in gas-stirred ladle assisted by physical modeling. *Metals* **2019**, *9*, 555. [CrossRef]

fluids

MDPI

Article

An Enhanced Python-Based Open-Source Particle Image Velocimetry Software for Use with Central Processing Units

Ali Shirinzad *, Khodr Jaber †, Kecheng Xu † and Pierre E. Sullivan †

Department of Mechanical and Industrial Engineering, University of Toronto, Toronto, ON M5S 3G8, Canada; khodr.jaber@mail.utoronto.ca (K.J.); kecheng.xu@mail.utoronto.ca (K.X.); sullivan@mie.utoronto.ca (P.E.S.)

* Correspondence: ali.shirinzad@mail.utoronto.ca

† These authors contributed equally to this work.

Abstract: Particle Image Velocimetry (PIV) is a widely used experimental technique for measuring flow. In recent years, open-source PIV software has become more popular as it offers researchers and practitioners enhanced computational capabilities. Software development for graphical processing unit (GPU) architectures requires careful algorithm design and data structure selection for optimal performance. PIV software, optimized for central processing units (CPUs), offer an alternative to specialized GPU software. In the present work, an improved algorithm for the OpenPIV–Python software (Version 0.25.1, OpenPIV, Tel Aviv-Yafo, Israel) is presented and implemented under a traditional CPU framework. The Python language was selected due to its versatility and widespread adoption. The algorithm was also tested on a supercomputing cluster, a workstation, and Google Colaboratory during the development phase. Using a known velocity field, the algorithm precisely captured the time-average flow, momentary velocity fields, and vortices.

Keywords: particle image velocimetry; OpenPIV; Python; image processing

Citation: Shirinzad, A.; Jaber, K.; Xu, K.; Sullivan, P.E. An Enhanced Python-Based Open-Source Particle Image Velocimetry Software for Use with Central Processing Units. *Fluids* **2023**, *8*, 285. https://doi.org/10.3390/fluids8110285

Academic Editors: D. Andrew S. Rees and Guangjian Zhang

Received: 29 September 2023
Revised: 20 October 2023
Accepted: 25 October 2023
Published: 27 October 2023

1. Introduction

Particle Image Velocimetry (PIV) is a non-intrusive experimental method that allows the measurement of fluid velocity vectors over a plane of interest [1]. PIV has been applied to a broad range of fluid flows, such as high–speed flows with shocks, laminar boundary layers, and near–wall flows, making it a widely used technique for velocity measurement [2]. Recently, PIV has also been used to look at cell motion [3], granular flows [4], ultrasonic images [5], and other fields where velocities or displacements need to be quantified. Cutting–edge hardware designed for PIV experiments can quickly capture and store thousands of image pairs. This capability stands as a critical factor in acquiring flow characteristics with sufficient statistics. Image processing techniques for PIV have also evolved in tandem with these hardware advancements, with the most notable of these methods being the window deformation iterative multigrid (WIDIM), which can significantly enhance the precision and spatial resolution of velocity fields in high–shear regions [6,7]. However, using the WIDIM approach for large datasets is computationally expensive and often limits the possible size of the datasets.

The growing significance of PIV over the past few decades has led to the emergence of numerous software packages. Among the available open–source packages are [8–10], e.g., PIVLab [11], OpenPIV [12], Fluere [13], Fluidimage [14], mpiv [15], JPIV [16], and UVMAT [17]. The advantages of open–source software development include the complete availability of algorithm details, which provides greater flexibility in future developments, especially in the context of community–driven collaboration, and compatibility with high–performance computing systems [8]. In this regard, PIVLab and OpenPIV have proven to be popular with the research community with the former being one of Matlab®'s most popular nonofficial free toolboxes [10].

Most open–source PIV algorithms are primarily developed for central processing units (CPUs) and leverage multiple cores and multiprocessing to accelerate calculations. In contrast, there are relatively fewer implementations optimized for GPUs [8,18]. GPU implementations have the potential to outpace their CPU counterparts and are essential for real–time PIV processing. For offline PIV analysis, deciding between GPU and CPU implementations is less straightforward. One crucial factor to consider is the hardware cost, as datacenter–grade GPUs are generally more expensive than CPUs. Additionally, the complexities of GPU programming, usually with regard to data structure selection and algorithm design suitable for the single–instruction multiple–data architecture, may deter some users from adopting GPU–based implementations due to the steeper learning curve compared to traditional CPU–based algorithms. Moreover, Python packages for CPUs tend to have greater stability compared to GPU packages, which undergo more frequent changes. Consequently, the CPU version of OpenPIV remains essential for maintaining stability and reliability.

Dallas et al. [8] developed a GPU–accelerated version of OpenPIV–Python that outperformed the CPU version of the software by a factor of 175. In their work, the GPU algorithms were executed on a supercomputing cluster while the CPU version was run on a standard workstation. Even though GPUs are expected to be faster than CPUs, the results of their work clearly showed that the CPU version of OpenPIV–Python suffers from poor performance, prompting the authors to conduct an in-depth analysis of both OpenPIV and PIVLab on all sub-levels. The current study presents an open-source CPU version for OpenPIV–Python, entitled OpenPIV–Python–CPU, combining essential features from both PIVLab and OpenPIV and aiming to improve processing time, accuracy, and spatial resolution of the velocity fields. The package is freely available on a GitHub repository, the link to which is provided in Appendix A. The remainder of the paper is organized as follows. The improved algorithms, implementation of the new software, and a description of the used datasets are presented in Section 2. The main results are presented in Section 3, followed by a more detailed discussion of the effect of the PIV parameters on the software performance in Section 4. The major findings and conclusions are summarized in Section 5.

2. Materials and Methods

2.1. Implementation and Architecture

OpenPIV–Python was initially developed as a Python package available on the Python Package Index (PyPI). Figure 1 illustrates a flowchart detailing the typical PIV process using the WIDIM approach in the current implementation. The readers are referred to Scarano [19] for a more detailed explanation of the WIDIM algorithm. The functionalities of the original software can be categorized into four main groups: correlation, validation, replacement, and smoothing. The following sections will elaborate on the detailed enhancements made to each of these tasks, excluding smoothing, as the algorithm remains consistent with the one utilized in PIVLab and previous versions of OpenPIV [8,20].

The numerical cross-correlation process begins by dividing a pair of images into square regions, known as interrogation windows, as shown in Figure 2a. In Python, striding is applied to create two 3D stacks of all the interrogation windows for an image pair, as working with 1D arrays is more efficient for future operations, such as removing masked windows. Similarly to PIVLab, a discrete Fourier transform (DFT) is used to perform a circular cross-correlation of the two 3D arrays. The mean intensity from each window may then be subtracted as typical image data include some noisy, non-zero background signal. An important assumption in circular cross-correlation is the periodicity of the signal (image data), which may introduce frequencies in the DFT spectrum that are non-existent [2]. To suppress this negative effect, the window stacks can be zero-padded, yielding an approximation of a linear non-periodic cross-correlation [10]:

$$W_{FFT} = n_{FFT} \times W \tag{1}$$

where W is the window size and W_{FFT} is the width of the Fourier transform. The cross-correlation is then calculated according to the cross-correlation theorem, using the *Fastest Fourier Transform in the West* (FFTW) available as pyFFTW on PyPI [21]. The Fourier transform width must be a power of two when using FFTW for performance reasons.

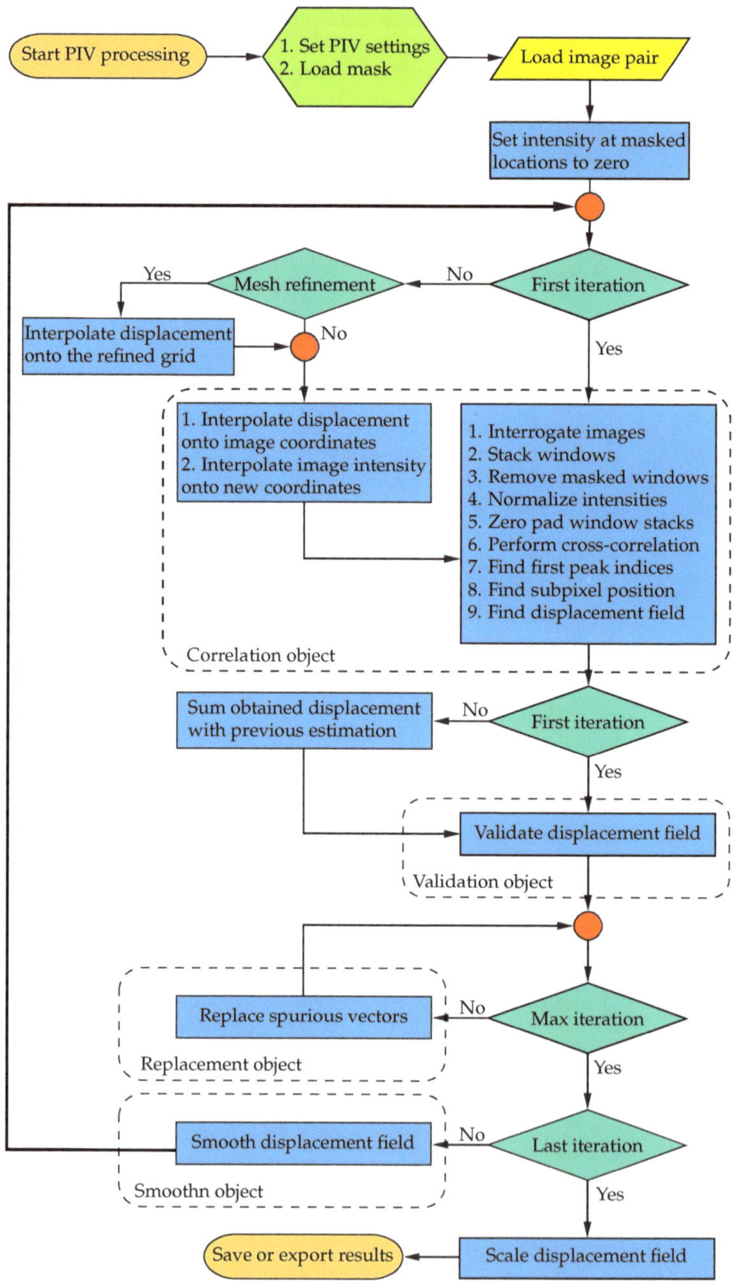

Figure 1. Flowchart of the architecture of OpenPIV–Python-CPU.

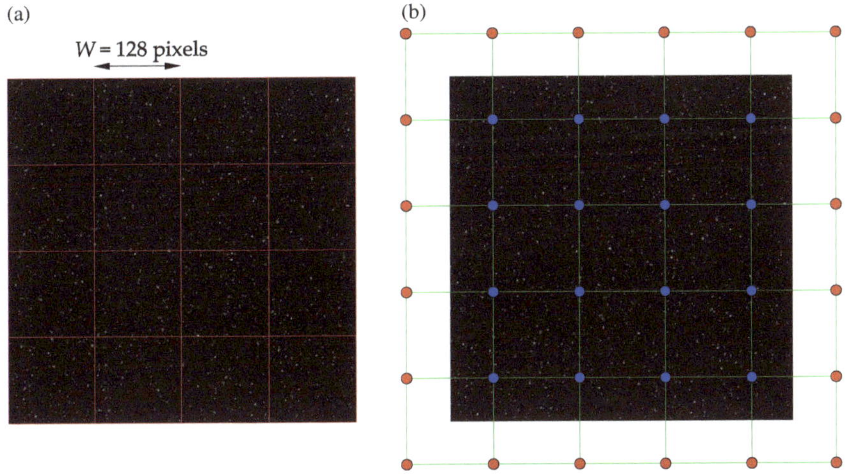

Figure 2. A 512 pixels × 512 pixels synthetic image of tracer particles in a Rankine vortex: (**a**) interrogation windows of size 128 pixels × 128 pixels with zero overlap; (**b**) measurement (blue) and padded (red) nodes are shown, respectively.

Once the cross-correlation map is calculated, the subpixel peak location may be approximated, typically using a Gaussian estimation:

$$i_{sp} = i + 0.5\frac{\log I_{i-1,j} - \log I_{i+1,j}}{\log I_{i-1,j} - 2\log I_{i,j} + \log I_{i+1,j}} \tag{2a}$$

$$j_{sp} = j + 0.5\frac{\log I_{i,j-1} - \log I_{i,j+1}}{\log I_{i,j-1} - 2\log I_{i,j} + \log I_{i,j+1}} \tag{2b}$$

where i and j are the peak location indices, $I_{i,j}$ denotes the value of the cross-correlation map corresponding to i and j indices, and i_{sp} and j_{sp} are the subpixel approximations. The displacement is then obtained by centering the subpixel peak locations using Equations (3a) and (3b) below. Note that from Equations (2a) and (2b) it is clear that no subpixel location may be estimated if the peak location is one of the borders of the correlation map. In such cases, the software uses the original peak indices to calculate the displacement.

$$u = j_{sp} - \frac{W_{FFT}}{2} \tag{3a}$$

$$v = i_{sp} - \frac{W_{FFT}}{2} \tag{3b}$$

After each iteration, the displacement field must be validated for spurious vectors to ensure the accuracy of the displacement field used as a predictor to shift and deform the interrogation windows in the next iteration. Various validation schemes have been implemented in OpenPIV [8]. The signal-to-noise ratio is a correlation-based validation method defined as the ratio of the first and second highest peaks (I_1 and I_2) in the correlation map, as shown in Equation (4) below:

$$\frac{I_1}{I_2} > \epsilon_{S2N} \tag{4}$$

where ϵ_{S2N} is the tolerance parameter, which is generally greater than 1.3 [8]. Note that it may not be feasible to use a single signal-to-noise tolerance value for all iterations as the signal-to-noise ratio decreases with every grid refinement.

Displacement-based validation methods provide a more robust alternative to the signal-to-noise ratio. Figures 2b and 3a show the measurement nodes and kernels used during a typical displacement-based validation process. As shown in Figure 2b, the displacement field is padded before being stridden to create a 3D array of all kernels. The center of every kernel is then evaluated against some statistics, such as mean, median, or median absolute deviations, of its neighbors:

$$u_0 - u_m > r_{m,u} \times \epsilon \tag{5a}$$

$$v_0 - v_m > r_{m,v} \times \epsilon \tag{5b}$$

In Equations (5a) and (5b), u_0 and v_0 are the displacements at the center of the kernel while ϵ is the tolerance parameter. For a simple median or mean validation, u_m and v_m are the median or mean of the neighbors, and $r_{m,u} = r_{m,v} = 1$. For the median-absolute-deviation test, however, u_m and v_m are the median while $r_{m,u}$ and $r_{m,v}$ are the median absolute deviation of the neighbors. The median and median-absolute-deviation validation methods have shown to be less sensitive to the variation in vectors during the PIV process. Westerweel and Scarano [22] argued that it is possible to apply a single tolerance value for all iterations in certain scenarios.

The outliers detected during the validation process must be replaced properly to ensure the accuracy of the final displacement field or to improve the displacement field used as a predictor for the next PIV iteration. The methods implemented in the previous versions of OpenPIV replace an outlier with the mean or median of its non-spurious neighbors iteratively until the maximum number of iterations is reached. The median and mean replacement methods fail to replace the vectors that are entirely surrounded by outliers. Such vectors may be replaced in the remaining replacement iterations if a large enough number of iterations are used for the information to spread to their neighbors. In the above-described method, it is also possible to revalidate the field after each replacement iteration to exclude the outliers from the next replacement iteration when they satisfy the validation criteria. The current software allows the mean and median methods to be used with or without the revalidation option.

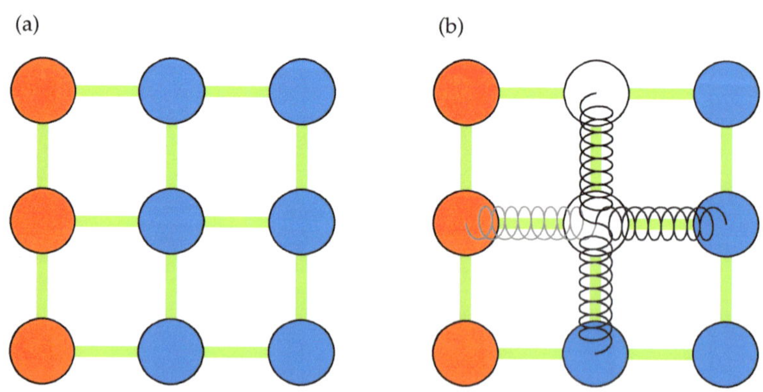

Figure 3. Schematics of a kernel of size three, showing center and neighboring nodes used during validation and replacement processes: (**a**) validation kernel; (**b**) spring analogy for outliers replacement. Measured nodes (blue), Padded nodes (red) and outliers (white) are shown.

PIVLab, on the other hand, uses a spring analogy to replace the outliers altogether. A schematic of a system of springs is shown in Figure 3b. In this method, all measurement nodes are treated as forces applied to the ends of a spring and the outliers are replaced by solving a system of inter-connected springs with zero net force. If there are no neighboring outliers, the outlier is simply replaced by the mean of four of its neighbors. Otherwise, the method uses all of the nodes surrounding a region of outliers to interpolate the nodes within. For a displacement field with n outliers, the force balance at each node may be expressed by Equations (6a) and (6b):

$$x_i - \frac{1}{a_i}\sum_{j=1}^{n}\kappa_{ij}x_j = \frac{1}{a_i}\sum_{k=1}^{c_i} b_{ik} \tag{6a}$$

$$c_i = a_i - \sum_{j=1}^{n}\kappa_{ij} \tag{6b}$$

where an outlier and a node connected to the outlier are denoted by x_i and b_{ik}, respectively. In Equation (6a), $\kappa_{ij} = 1$ if an outlier is connected to the other end of one of the four springs shown in Figure 3b and zero otherwise (note that by this definition $\kappa_{ii} = 0$ and $\kappa_{ij} = \kappa_{ji}$). The coefficient a_i is the number of available springs at a given node. For instance, there are only three free springs in Figure 3b since one is connected to a padded node. Generally, $a_i = 3$ if a node is at the edge, $a_i = 2$ if a node is in the corner, and $a_i = 4$ if a node is not on the borders.

Theoretically, in the spring analogy, all of the outliers need not be replaced at once, and only the linked nodes need to be solved together, simplifying the original problem to a set of smaller systems. Hence, Equations (6a) and (6b) form multiple linear equations systems, which may be solved either separately or simultaneously. For simplicity, however, the algorithm used in the present software constructs a system of all equations before solving the system using linear algebra. Equations (6a) and (6b) may be rewritten in matrix form suitable for numerical computations:

$$LX = B$$

$$L = \begin{bmatrix} 1 & \frac{-\kappa_{12}}{a_1} & \cdots & \frac{-\kappa_{1n}}{a_1} \\ \frac{-\kappa_{21}}{a_2} & 1 & & \frac{-\kappa_{2n}}{a_2} \\ \vdots & \vdots & \ddots & \vdots \\ \frac{-\kappa_{n1}}{a_n} & \frac{-\kappa_{n2}}{a_n} & \cdots & 1 \end{bmatrix} \tag{7}$$

In Equation (7), L and B are the linkage and coefficient matrices, respectively. Since every outlier is at most connected to four other outliers, the linkage matrix is sparse. The linkage matrix is constructed row by row using a row-based list of lists (LIL) sparse matrix before being converted into a compressed sparse row (CSR) format. In every loop, an array of zeros with the same shape as the padded displacement field is initialized, and the linkage kernel is placed at the outlier location, filling the coefficients of the connecting nodes. The row of the linkage matrix is then filled by selecting all elements corresponding to outlier locations from this array. This procedure is illustrated in Figure 4 for the first outlier in the corner ($a_1 = 2$) for the same field shown in Figure 2b.

The construction of the coefficient matrix in Equation (7) is straightforward. First, all elements corresponding to outlier locations in the padded field are replaced with zeros. The resulting field is stridden into a 3D array of kernels and the kernels centered around all outliers are selected from this array before calculating the mean of the nodes connected to the center of kernels. An example of this procedure for the field shown in Figure 4 is presented in Figure 5.

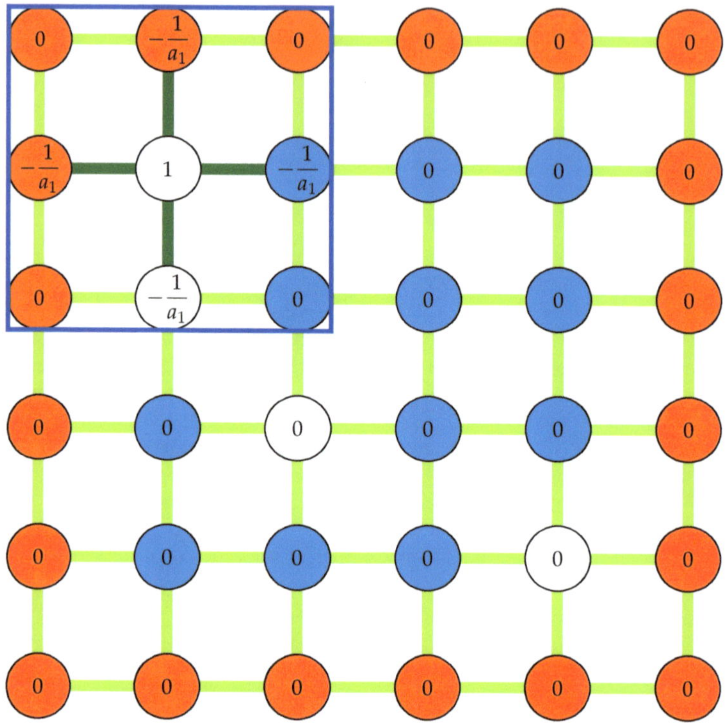

Figure 4. A schematic showing an array of zeros and a kernel of size three used to construct the linkage matrix. The white nodes are selected to fill the first row of the linkage matrix. Measured nodes (blue), Padded nodes (red), and outliers (white) are shown.

The remaining subroutines are described as follows. If the window size in the next PIV iteration is reduced, the filtered and smoothed displacement field is interpolated onto the new grid through bicubic interpolation. The displacement field is then interpolated onto all image coordinates using spline interpolation to build a predictor displacement field over all the image pixels. The two images are deformed according to the predictor spatial distribution. For performance reasons, the resampling of the pixel values at intermediate locations is only completed through bilinear interpolation. The deformed images are then used as input for the next iteration.

2.2. *Datasets and Image Processing*

Synthetic images generated from a known distribution are often used to assess the accuracy of PIV software. Synthetic images are also useful for analyzing the influence of image parameters, such as noise, particle size, or loss of particle pairs, on the performance of PIV software. Consider the turbulent flow through a rectangular channel slice of a half-height h. A Cartesian coordinate system is employed with the origin at the junction of the entrance cross-section, bottom wall, and the midspan plane, as sketched in Figure 6a.

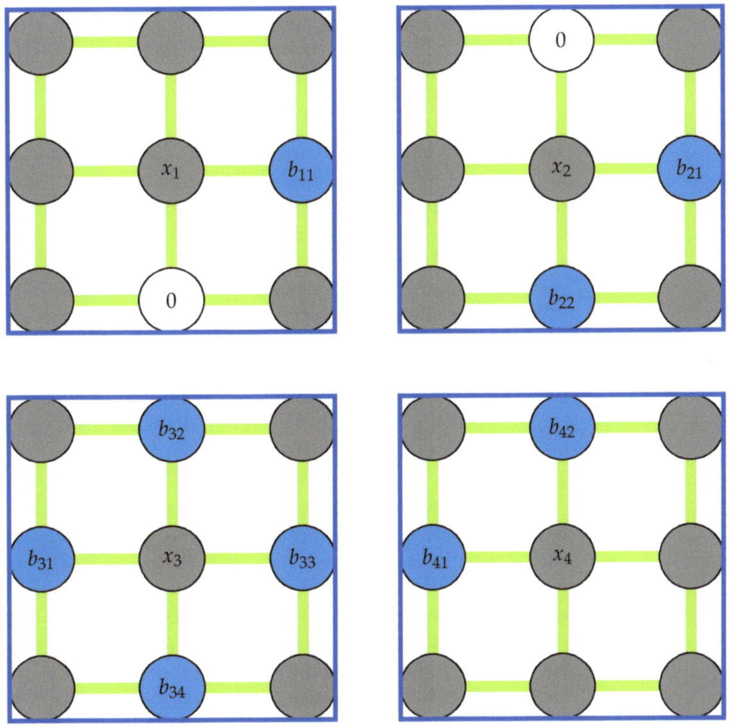

Figure 5. An example showing the kernels used for constructing the right-hand side of Equation (7). Note that the gray nodes will not be used when calculating the average in each kernel.

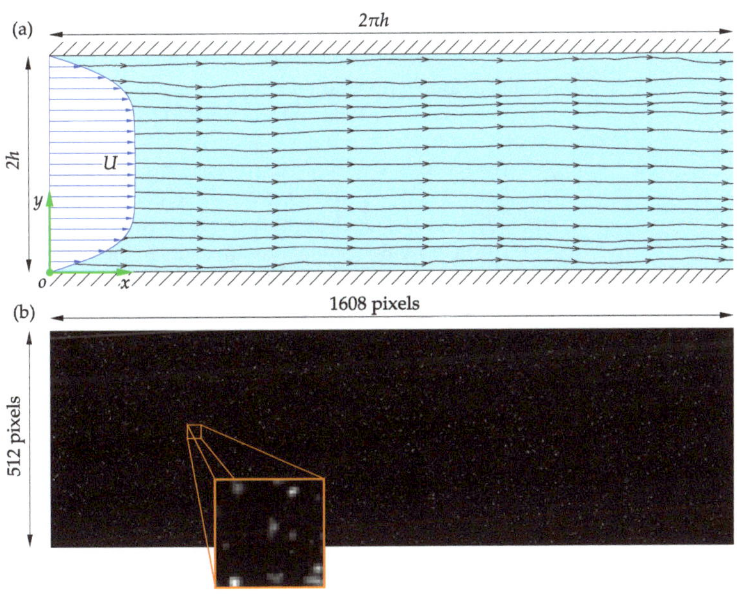

Figure 6. A schematic representation of the wall-bounded turbulent rectangular channel flow: (**a**) the adopted nomenclature; (**b**) a sample of the synthetic images used in the present study and tracer particles in a 32 pixels × 32 pixels window.

The velocities in the x and y coordinate directions are denoted by u and v, respectively. The velocity field in the x-y plane is denoted by $\vec{v}$ and its magnitude is given by Equation (8):

$$|\vec{v}| = \sqrt{u^2 + v^2} \tag{8}$$

since the z component of the velocity is not of interest. All time-averaged velocities are denoted by their corresponding capital letters. John Hopkins turbulent channel database contains the direct numerical simulation (DNS) results for wall-bounded turbulent flow in a rectangular channel [23]. The DNS domain size is $8\pi \times 2 \times 2\pi$ using $2048 \times 512 \times 1536$ nodes in the x, y, and z coordinate directions, respectively. The results were obtained by solving incompressible Navier–Stokes equations with periodic boundary conditions in the stream–wise and span–wise directions and a no-slip condition at the top and bottom boundaries. The DNS grid spacing was uniform in the stream–wise direction and non-uniform in the wall-normal direction with a higher density of nodes near the wall boundary. The database contains velocity fields for 4000 simulation time steps, corresponding to approximately one flow-through period of the channel. The fluid kinematic viscosity and the friction velocity were $\nu = 5 \times 10^{-5}$ and $u_\tau = 0.0499$, corresponding to a friction velocity Reynolds number of $Re_\tau = u_\tau \times h/\nu \approx 1000$.

In the present study, synthetic images generated from a $2\pi \times 2$ slice of the velocity fields at $z = \pi/10$ for 1000 time steps (amounting to 2000 images) were used to assess the performance of the software to realistic and complicated flow fields. The DNS velocity field was uniformly scaled for a maximum particle displacement of 8 pixels when generating the images, satisfying the one-quarter rule for a 32 pixel $\times$ 32 pixel initial interrogation window. This corresponds to adjusting the time delay between image pairs to best capture the flow dynamics when capturing real PIV images. The synthetic images are 1608 pixels $\times$ 512 pixels large and have the same aspect ratio as the channel slice. Samples of the synthetic images of the tracer particles in the channel slice and a 32 pixel $\times$ 32 pixel window are shown in Figure 6b.

The images were processed on a workstation and the Niagara computer cluster to ensure compatibility with both the Microsoft Windows and Linux platforms and evaluate computational performance. The workstation had a 12-core Intel® Core™ i5-10600K CPU (Intel Corporation, Santa Clara, CA, USA) at 4.1 GHz and 64 GB of RAM. On the Niagara cluster, a node with 40 Intel® Skylake cores at 2.4 GHz and 188 GB of RAM was used for PIV processing. All 1000 image pairs had to be transferred to the cluster using Globus. A list of the software input parameters and PIV settings is presented in Table 1.

The Rankine vortex is a useful case study to investigate the effects of the PIV parameters shown in Table 1 on the accuracy and performance of the software. The Rankine vortex and the adopted coordinate system are shown in Figure 7a.

The velocity field and its components in the x and y coordinate directions are denoted by $\vec{v}$, u, and v, respectively. The velocity magnitude and its distribution for the Rankine vortex are given by Equations (8) and (9):

$$\begin{cases} |\vec{v}| = \frac{r}{R}|\vec{v}|_R & r < R \\ \vec{v} = \frac{R}{r}|\vec{v}|_R & r \geqslant R \end{cases} \tag{9}$$

where r is the distance from the vortex center, R is the radius of the vortex core, and $|\vec{v}|_R$ is the velocity magnitude at $r = R$. A pair of synthetic images of the size 512 pixels $\times$ 512 pixels were generated from the Rankine vortex distribution using PIVLab. The vortex core was located at the center of the frame and had a radius of $R = 50$ pixels. The number of particles, particle diameter, and noise level were set to 50,000, 3 pixels, and 0.001, respectively. The maximum particle velocity was scaled to $|\vec{v}|_R = 16$ pixels/s to satisfy the one-quarter rule for a 64 pixels $\times$ 64 pixels initial interrogation window. A sample of these images is shown in Figure 7b.

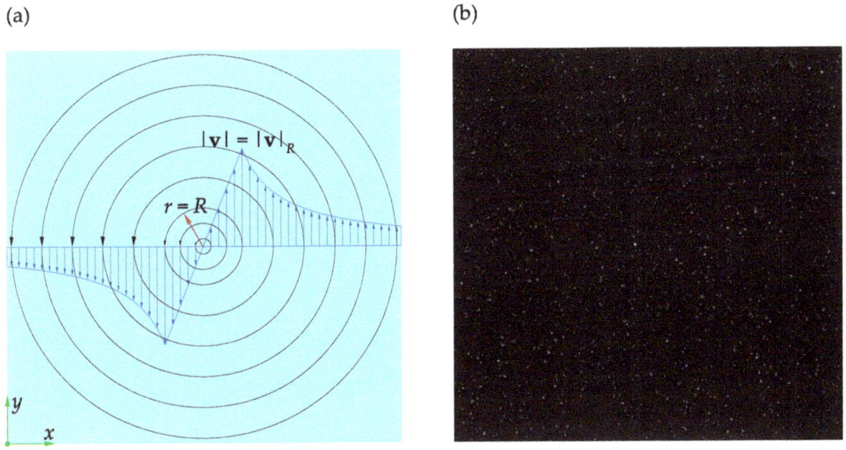

Figure 7. A schematic representation of Rankine vortex: (**a**) the adopted nomenclature; (**b**) a sample of the synthetic images used in the present study.

Table 1. Summary of PIV parameters for turbulent channel flow.

Settings	Variable	Description	Value
Masking	`mask`	2D array with non-zero values indicating the masked locations.	`None`
Data type	`dtype_f` [1]	Type of floating-point numbers.	"float32"
Geometry	`frame_shape`	Size of the images.	(512, 1608)
	`min_search_size`	Interrogation window size for the final iteration.	8
	`search_size_iters`	Number of iterations for each window size.	(1, 1, 2)
	`overlap_ratio`	Ratio of overlap for each window size.	0.5
	`shrink_ratio` [2]	Ratio to shrink the search size for the first iteration.	1
Correlation	`deforming_order`	Order of spline interpolation for window deformation.	2
	`normalize`	Normalize the window intensity by subtracting the mean value.	`True`
	`subpixel_method` [3]	Method to estimate subpixel location of the correlation peak.	"gaussian"
	`n_fft`	Size factor for the 2D FFT.	(1, 1, 2)
	`deforming_par` [4]	Ratio of the predictor used to deform each frame.	0.5
	`batch_size`	Batch size for calculating the cross-correlation.	1
Validation	`s2n_method` [5]	Method of signal-to-noise ratio measurement.	"peak2peak"
	`s2n_size`	Half-size of a square around the first peak ignored for second peak.	2
	`validation_size` [6]	Size parameter for validation kernel.	1
	`s2n_tol` [7]	Tolerance for signal-to-noise ratio validation.	`None`
	`median_tol` [7]	Tolerance for median validation.	2
	`mad_tol` [7]	Tolerance for median-absolute-deviation validation.	`None`
	`mean_tol` [7]	Tolerance for mean validation.	`None`
	`rms_tol` [7]	Tolerance for root mean squared validation.	`None`
Replacement	`num_replacing_iters`	Number of iterations per replacement cycle.	2
	`replacing_method` [8]	Method to use for outlier replacement.	"spring"
	`replacing_size` [6]	Size parameter for replacement kernel.	1
	`revalidate`	Revalidate the fields in between replacement iterations.	`True`

Table 1. *Cont.*

Settings	Variable	Description	Value
Smoothing	smooth [9]	Smooth the displacement fields.	True
	smoothing_par [10]	Smoothing parameter to apply to the velocity fields.	None
Scaling	dt [11]	Time delay separating the two images.	1
	scaling_par [11]	Scaling factor to apply to the velocity fields.	1

[1] The available types are *"float32"* and *"float64"* for single and double floating points, respectively. [2] Shrinking the window size allows for performing an extended search area PIV for the first iteration. [3] The available methods are *"gaussian"*, *"centroid"*, and *"parabolic"*. [4] A value of 0.5 corresponds to the central difference interrogation (CDI) scheme, minimizing the bias error [7]. [5] The available methods are *"peak2peak"*, *"peak2mean"*, and *"peak2energy"*. [6] The actual kernel size is obtained by kernel_size $= 2 \times$ size $+ 1$. [7] By selecting None, a validation method method may be ignored. [8] The available methods are *"spring"*, *"mean"*, and *"median"*. [9] No smoothing is applied at the end of the final iteration. [10] By selecting None, the generalized cross-validation (GCV) method is used to obtain the optimum value. [11] If no scaling is applied, the output values will be in units of pixels/s.

3. Results

In this section, contour plots, vector plots, and one-dimensional profiles are presented to evaluate the accuracy of the algorithms in resolving the instantaneous and time-averaged flow characteristics. Python was used to calculate the average of 1000 velocity fields for both the DNS and PIV results. The velocity and displacement throughout the rest of this paper are in units of pixels/s and pixels, respectively. All data visualizations were accomplished using the commercial software Origin®.

3.1. Mean Flow Field

Figure 8 compares the contour plots of the mean stream–wise velocity for the DNS data and PIV results. From Figure 8a,b, it can be seen that the OpenPIV results captured all of the major patterns found in the contour plot of the DNS data, and closely matched all the large-scale contour lines. The discrepancies found between Figure 8a,b are mostly concentrated near the bottom and top walls, near the entrance and exit cross-sections, and in high-velocity regions.

The one-dimensional profiles of the mean stream–wise velocity are provided in Figure 9 to better visualize the stream–wise evolution of the flow. The profiles were plotted at five successive stream–wise locations, starting at $x = 4$ and ending at $x = 1604$ with an increment of 400, allowing the DNS data and PIV results to be compared in both near boundary and internal regions. In the near-wall regions, the small-scale structures are averaged out from the DNS data during the generation of the images [8,24]. The wall coordinates are defined as follows:

$$y^+ = \frac{u_\tau \Delta y}{\nu} \tag{10}$$

The PIV measurement nodes closest to the bottom and top walls are located at $y = 4$ and $y = 508$, corresponding to $y^+ \approx 16$ (given that $Re_\tau \approx 1000$). Since these nodes are near the edge of the shear layers, their values are biased towards the larger velocities, which may also be observed in Figure 9.

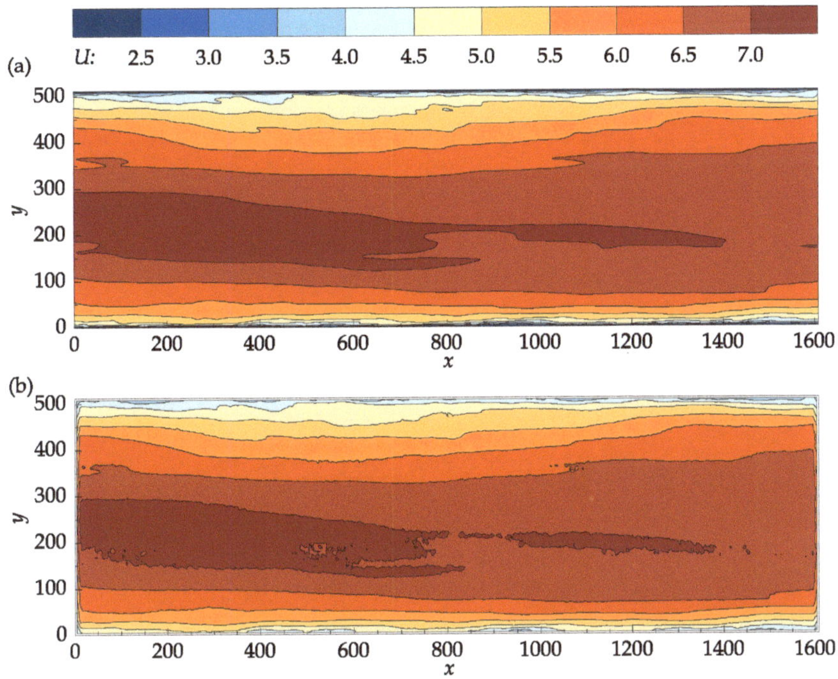

Figure 8. Contour plots of the mean stream–wise velocity: (**a**) John Hopkins DNS data; (**b**) results obtained with OpenPIV–Python.

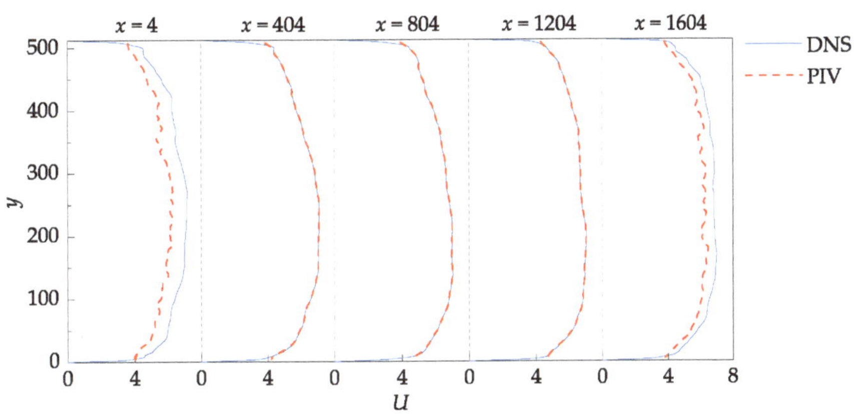

Figure 9. Stream–wise evolution of the mean stream–wise velocity profiles.

The disparity between the DNS data and the PIV results at the entrance and exit cross-sections ($x = 4$ and $x = 1604$) occurs due to the loss of particle pairs that cannot be redeemed by window translations and deformation. For a given image pair, the particles entering or leaving the frame are only present in one of the two images, leading to a significant loss of correlation. The current software tends to remedy this negative effect using bilinear extrapolation to resample the image intensity near the boundaries. In general, it is good practice to discard at least the layer immediately near the borders of the velocity fields.

Figure 10 shows the frequency distribution plots of the error (difference between the DNS and PIV results) for the mean velocity magnitude, mean stream–wise velocity, and the mean transverse velocity. The sample size, mean, and standard deviation of the velocity magnitude error were 50,927, 0.003, and 0.116, respectively. These results, as well as the 95 % confidence intervals shown in Figure 10a–c, indicate that the software is capable of resolving all time-averaged velocities with reasonable accuracy.

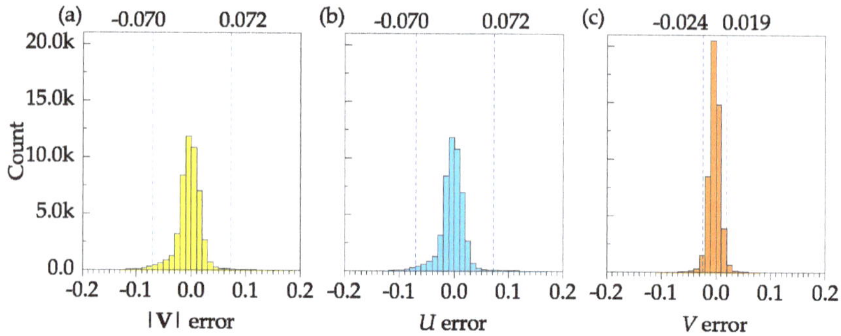

Figure 10. Frequency distribution plots of velocity error for the mean velocity fields: (**a**) velocity magnitude; (**b**) stream–wise velocity; (**c**) transverse velocity. The 95 % confidence interval is shown by dashed vertical lines on each plot.

3.2. Instantaneous Flow Field

The performance of the software in resolving the instantaneous flow field was evaluated by analyzing the fluctuating velocities, span–wise vorticity, and Q-criterion. The fluctuating velocities and the span–wise vorticity are given by Equations (11a) and (11b):

$$u' = u - U \quad (11a)$$

$$v' = v - V \quad (11b)$$

The coherent structures in the flow can be identified using the Q-criterion. The Q-criterion is a scalar field that defines structures as regions where the vorticity magnitude is greater than the magnitude of the strain rate. In a two-dimensional Cartesian coordinate system, the Q-criterion is given by Equation (12) below:

$$Q = -\frac{1}{2}\left(\left(\frac{\partial u}{\partial x}\right)^2 + 2\left(\frac{\partial u}{\partial y}\right)\left(\frac{\partial v}{\partial x}\right) + \left(\frac{\partial v}{\partial y}\right)^2\right) \quad (12)$$

In the present study, the instantaneous velocity gradients were calculated using second-order accurate central differences in the interior points and first-order accurate one-side differences at the boundaries to calculate Equation (12). The fluctuating velocity vectors superimposed on top of the contour plots of the Q-criterion for the DNS data and PIV results at an arbitrary time step are presented in Figure 11.

In both Figure 11a,b, the Q-criterion was normalized by its maximum value in the DNS data. The main structures in the DNS data are also captured in the PIV results. In the middle section of the channel where the velocity is large, small isolated regions of negative Q values are present in the PIV results, which may not be seen in the DNS data. Many of these regions, a few of which are marked with red circles in Figure 11b, corresponded to the vectors that were replaced during the last iteration, indicating the effectiveness of the median validation in detecting the outliers. It is important to note that the median validation did not classify these vectors as outliers at the end of the process. These vectors, nevertheless, are approximated from their neighbors and should be regarded as unreliable.

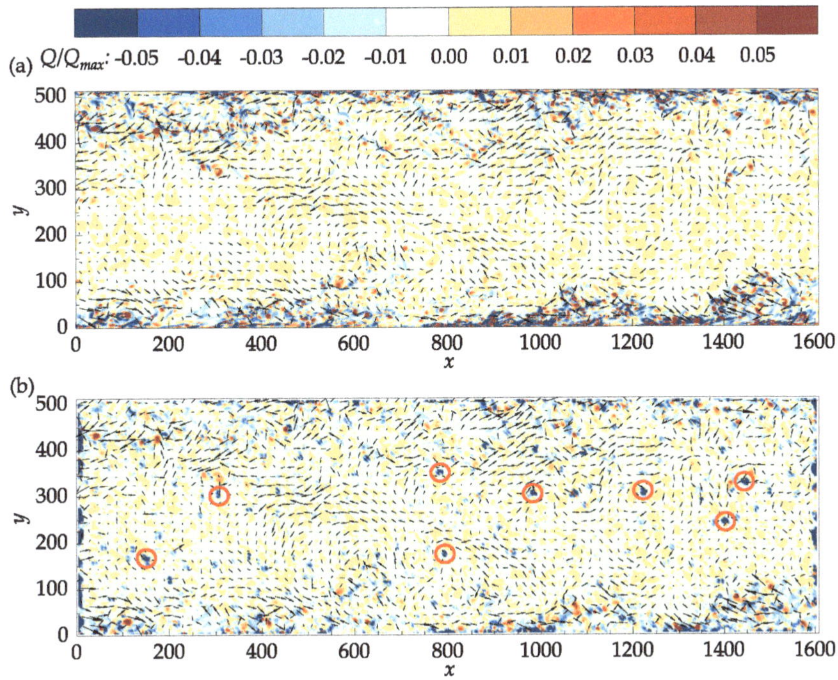

Figure 11. Vectors of the fluctuating velocities superimposed on the contour plots of the normalized Q-criterion: (**a**) John Hopkins DNS data; (**b**) results obtained with OpenPIV–Python.

The density of the vector field may be increased by means of decreasing the minimum window size or increasing the final overlap. The same vector spacing of 4 pixels may be achieved by setting only `min_search_size=16` and `overlap_ratio=(0.5, 0.5, 0.75)` from the parameters in Table 1. This will significantly improve the results away from the walls ($100 < y < 400$) and at the entrance and exit cross-sections. Obviously, the near-wall results will not be as detailed when a larger `min_search_size` is used. These results are provided as supplementary figures in Appendix B.

Figure 12 shows the frequency distribution plots of the error for the velocity magnitude, stream–wise velocity, and transverse velocity to further assess the accuracy of the instantaneous velocity fields. The sample size, mean, and standard deviation of the velocity magnitude error were 50,927,000, 0.005, and 0.208, respectively. In Figure 12a–c, the 95% confidence interval is specified by two dashed vertical lines. These confidence intervals may be significantly improved if the boundary nodes are discarded. The results show that the software is capable of resolving the instantaneous velocities with reasonable accuracy.

3.3. Computational Performance

Figure 13 demonstrates the effect of the number of processors on the computation time. Multiprocessing was employed to vary the number of utilized processors for both the workstation and Niagara cluster. Initially, for both platforms, the computation time decreases rapidly as more processors are utilized. Meanwhile, launching more processes above a certain threshold did not impact the calculation time. The time needed to transfer 2000 images to the Niagara cluster using Globus was 863 s at an effect speed of 3.81 MB/s. From Figure 13, it can be seen that the speedup achieved by using the Niagara cluster is not enough to compensate for the data transfer time. Using the cluster only becomes sensible when large images with computationally intensive settings are to be processed or when a dataset needs to be processed more than once.

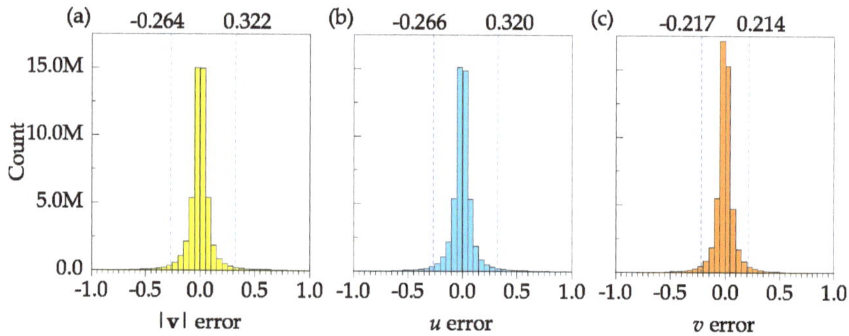

Figure 12. Frequency distribution plots of velocity error for 1000 instantaneous velocity fields: (**a**) velocity magnitude; (**b**) stream–wise velocity; (**c**) transverse velocity. The 95 % confidence interval is shown by dashed vertical lines on each plot.

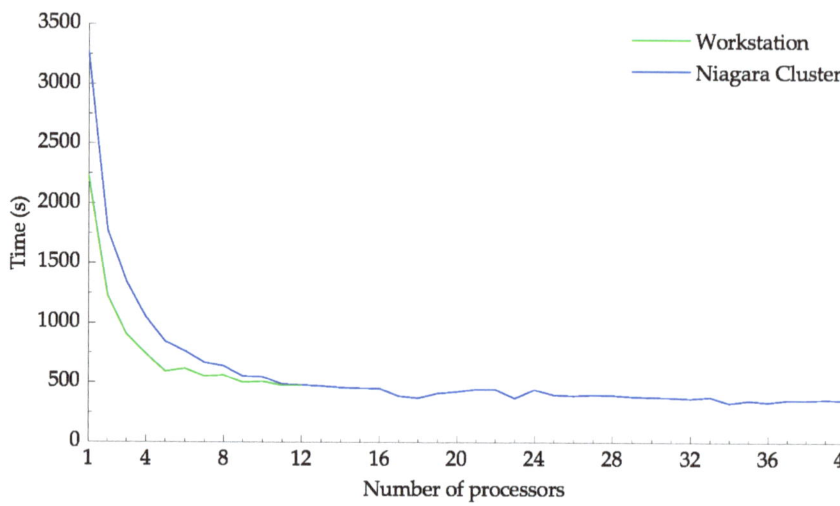

Figure 13. Plot of the computation time against the core number.

CPU–Based PIV software, unlike their GPU counterparts, generally should cycle through the windows rather than performing the cross-correlation simultaneously. Hence, it is recommended always to set `batch_size=1` for optimum performance. Regardless of whether the cross-correlation is parallelized or not, the most time-consuming calculation in a PIV algorithm is generally computing the cross-correlation between the interrogation windows. The relative computational time for most of the sub-routines is presented in Figure 1 is reported in Table 2. The relative computational time was calculated five times using the same parameters shown in Table 1 before being averaged. As expected, cross-correlation was the most time-consuming subroutine with nearly 65% of the total computation time. Although deformation is the second most time-consuming subroutine, it is worth mentioning that a larger portion of the time is spent on frame deformation. Since window deformation is not as time-consuming, using a `deforming_order` greater than one is recommended for better accuracy. Overall, 84.207% of the calculation time was spent in the correlation object, which handles the main PIV calculations, while only 7.449%

of the time was dedicated to validation, replacement, and smoothing altogether, further indicating that the software is implemented well.

Table 2. Performance profile of the CPU–Based software.

Subroutine			Computational Time per Run [1]			Average
Interrogation	0.893	0.894	0.833	0.867	0.791	0.855
Deformation [2]	14.596	14.533	14.626	14.308	4.511	14.515
Normalization	1.902	1.865	1.783	1.852	1.820	1.845
Cross-correlation	65.573	65.769	65.797	66.103	65.560	65.760
Peak estimation [3]	1.203	1.282	1.190	1.222	1.263	1.232
Validation	3.066	3.185	3.209	3.153	3.124	3.148
Replacement	3.493	3.615	3.727	3.587	3.639	3.612
Smoothing	0.658	0.698	0.672	0.748	0.671	0.689
Time per vector [4]	50.611	50.557	49.586	49.859	49.704	50.063

[1] For the subroutines, the reported values are the percentage of the relative computational time defined as the time required for the subroutine divided by the total processing time. [2] Deformation refers to the combination of window and frame deformation. [3] Peak estimation refers to the process of finding the first peak and estimating its subpixel location. [4] Computational time per vector is presented in units of μs and is obtained as the total calculation time divided by the number of vectors in the resulting vector field.

4. Discussion

The effect of several PIV parameters, including data type, number of PIV iterations, correlation width, number of replacement iterations, replacing method, and revalidation, on the Rankine vortex dataset described in Section 2 are discussed in this section. The readers are referred to Meunier and Leweke [7] for a detailed discussion on the effect of the deformation parameter on the bias error. For each case study, all parameters are the same as in Table 1 with the exception of the parameters being investigated, `frame_shape=(512, 512)` and `min_search_size=16`. After the effect of the PIV parameters on the resulting field is discussed, a comparison between the results derived using the best replacement method and PIVLab results obtained using the same number of iterations with a minimum window size of 16 pixels is presented.

The precision of the floating-point data used during the PIV process can significantly impact the computational performance. Although both single and double precisions are available options, the single precision data type is preferred as it can greatly reduce the cross-correlation computation time. Performing the PIV analysis for both double and single data types resulted in the same mean, standard deviation, and 95 % confidence interval of $-0.061, 0.374$, and $(-0.984, 0.204)$ for the velocity magnitude error. In general, using the single precision data type should not be a problem for most PIV applications.

The number of iterations used at each window size may remarkably improve the accuracy of the results. For instance, just performing one more iteration for the 32 pixels $\times$ 32 pixels window size, that is, changing `search_size_iters=(1, 1, 2)` to `(1, 2, 2)`, can change the 95 % confidence interval of the velocity magnitude error from $(-0.984, 0.204)$ to $(-0.657, 0.211)$.

As was discussed in Section 2, zero-padding the windows before taking the circular-cross correlation may improve the results. It is recommended to increase the correlation width for the minimum window size at least by a factor of two to strike a balance between accuracy and computation efficiency. Increasing the correlation width for other window sizes, although not as effective as the smallest window size, may also improve the accuracy. For example, changing `n_fft=(1, 1, 2)` to `(1, 2, 2)` reduces the 95 % confidence interval of the velocity magnitude error from $(-0.984, 0.204)$ to $(-0.805, 0.203)$.

The number of replacement iterations can become an important PIV parameter depending on the replacement method. For the spring replacement method, using one iteration is sufficient since this method replaces the outliers altogether. There is no rule to determine the number of iterations required for the convergence of the mean and median methods. It may take several iterations for these methods to converge if revalidation is not applied depending on the number of outliers and their connection. For this case study, the mean and median methods converged after three iterations when the fields were revalidated. Without revalidation, on the other hand, it took 16 and 34 iterations for the mean and median methods to converge, respectively.

The effect of the replacement method on the accuracy of the results is not straightforward. Replacing outliers with reasonable estimations increases the chance of obtaining a better correlation match in the subsequent PIV iterations. The outliers replaced during the last iteration, however, are not reliable since all of the replaced vectors are simply approximated from their neighbors. Figure 14 shows the exact velocity field and the results obtained with three replacement methods. The mean and median replacements were performed using five iterations to reach convergence. The green vectors indicate the outliers that were replaced during the last iteration and satisfied the validation criteria at the end of the process. The remaining outliers are designated in red color and amounted to three, one, and, two for spring, mean, and median methods, as can be seen in Figure 14a–c, respectively.

The mean, standard deviation (STD), and 95 % confidence interval (95 % CI) of the velocity components and velocity magnitude are reported in Table 3. It can be seen that all values in Table 3 are of the same order of magnitude. For this particular set of parameters, the mean method yielded the minimum mean absolute error (MAE) and root mean squared error (RMSE) of 0.303 and 0.459, respectively.

Table 3. Mean, standard deviation, and 95% confidence interval of velocity errors for different replacement methods.

Method	Error	Mean	STD [1]	95% CI [2]
Spring	u	−0.011	0.356	(−0.645, 0.480)
	v	0.004	0.419	(−0.593, 0.582)
	$\lvert\vec{v}\rvert$	−0.073	0.438	(−1.231, 0.217)
Mean	u	−0.005	0.345	(−0.569, 0.521)
	v	−0.007	0.303	(−0.544, 0.521)
	$\lvert\vec{v}\rvert$	−0.045	0.301	(−0.732, 0.196)
Median	u	−0.002	0.378	(−0.620, 0.504)
	v	0.001	0.374	(−0.541, 0.597)
	$\lvert\vec{v}\rvert$	−0.053	0.346	(−0.783, 0.210)

[1] Standard deviation. [2] 95% confidence interval.

The effect of revalidation on the results obtained with the mean and median replacement methods is shown in Figure 15. Although revalidation always reduces the number of iterations needed for convergence, it does not necessarily improve the accuracy of the results. By comparing Figure 15a with Figure 15b and Figure 15c with Figure 15d, it can be seen that revalidation has improved the results only for the mean method. The mean replacement with revalidation performed the best since fewer vectors were replaced during the last PIV iteration and only one outlier remains after the process. For the best practice, it is recommended to perform a few pilot runs before PIV analysis to decide on the replacement and validation options, aiming to minimize the replaced vectors during the last PIV iteration. Note that revalidation is not applicable to the signal-to-noise ratio validation method since the signal-to-noise ratio is correlation-based and is not affected by the replaced vectors.

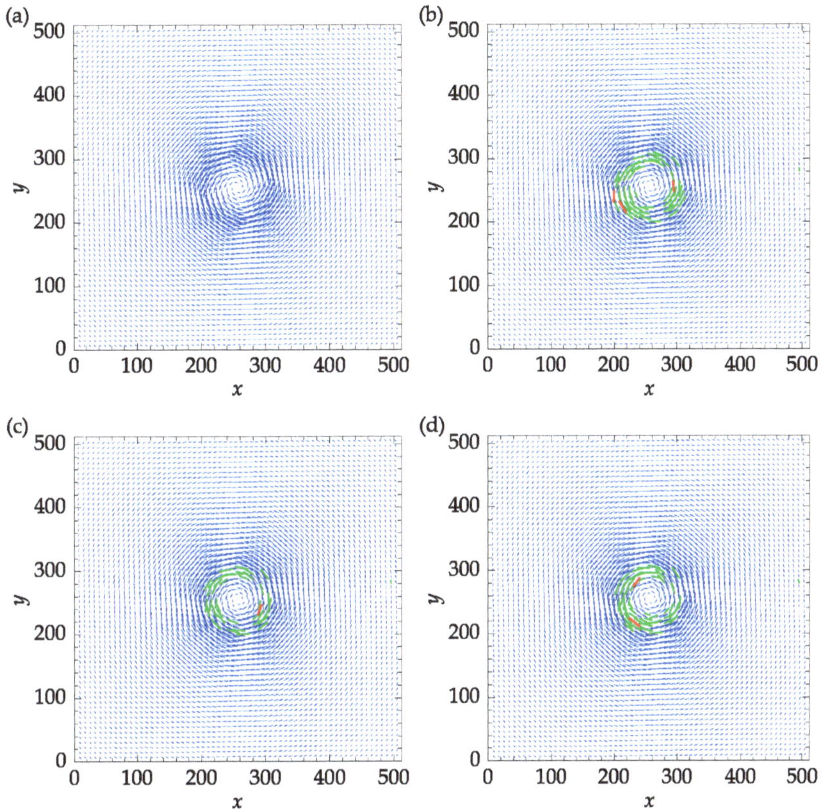

Figure 14. Effect of replacing methods on the velocity fields: (**a**) exact velocity field; (**b**) spring replacement; (**c**) mean replacement; (**d**) median replacement. Replaced vectors satisfying the validation criteria at the end of the final iteration are designated in green color. Any remaining outliers are represented in red color.

A comparison of the results of the developed software, obtained using the mean replacement methods with revalidation, and the PIVLab results is shown in Figure 16. Note that no post-processing was applied to the PIVLab results. PIVLab provides utilities for velocity-based and image-based validation during the post-processing steps. The vector field shown in Figure 16 is what is obtained solely using the validation and smoothing routines in between the PIV iterations. From Figure 16, it is clear that with the exception of one vector at $(x, y) = (352, 32)$, the remaining different vectors are focused near the center of the vortex. Using the local median filter with a threshold of two, the post-processing step in PIVLab showed that this vector is an outlier. Using the present software, however, this vector was reliably obtained using cross-correlation. More such instances were identified among the outliers near the center of the vortex. Although PIVLab allows for replacement of the remaining outliers using interpolation, the vectors replaced during the post-processing phase or the last PIV iteration are not reliable, as was confirmed by the DNS data.

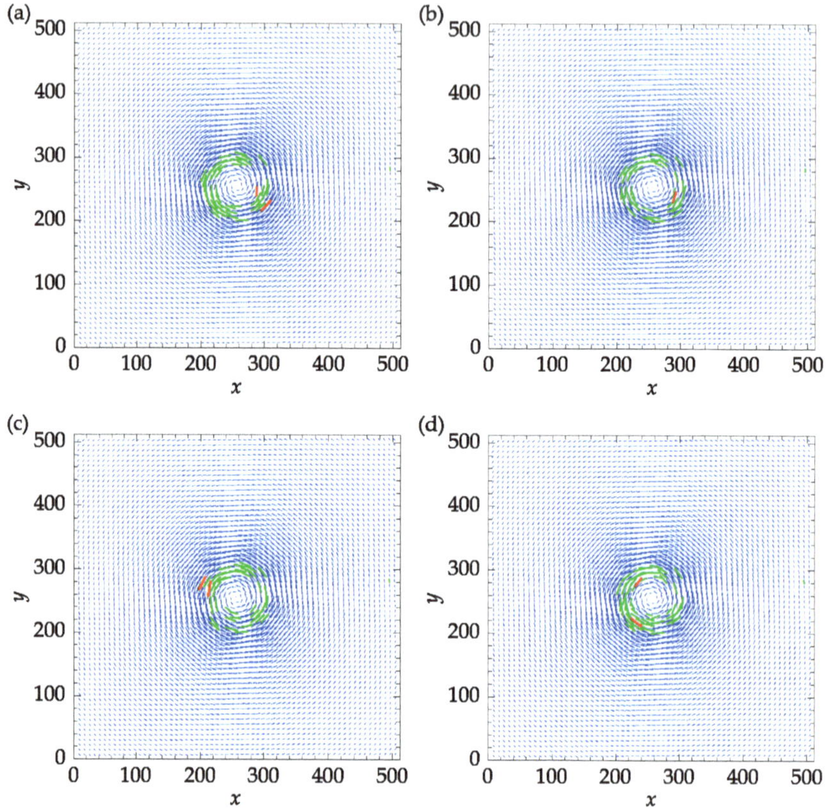

Figure 15. Effect of revalidation on the velocity fields: (**a**) mean replacement and no revalidation; (**b**) mean replacement with revalidation; (**c**) median replacement and no revalidation; (**d**) median replacement with revalidation. Replaced vectors satisfying the validation criteria at the end of the final iteration are designated in green color. Any remaining outliers are represented in red color.

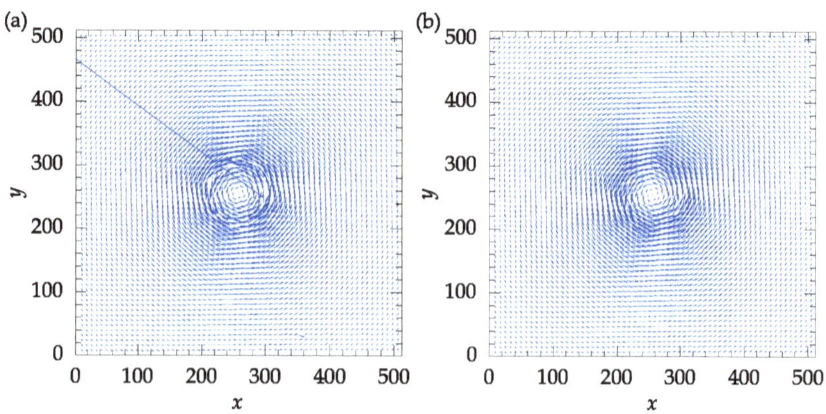

Figure 16. Comparison between the results obtained with the developed software and PIVLab: (**a**) PIVLab original results with no further post-processing; (**b**) results obtained with the present software using mean replacement with revalidation.

5. Conclusions

An open-source CPU–Based PIV data processing software was developed and validated. The software was written entirely in Python and was compatible with the Microsoft Windows and Linux operating systems. The software was functional on different hardware platforms, ranging from small embedded systems to large supercomputing clusters. Synthetic PIV images generated from the John Hopkins turbulent rectangular channel DNS data were processed to test the accuracy of the software in resolving the mean and instantaneous flow fields and computation performance.

Both the mean and instantaneous velocities closely matched the DNS data except near the walls, where the lengths scales are much smaller than the smallest PIV interrogation window, near the entrance and exit cross-sections, and in high-velocity regions. Isolated regions of low Q-criterion values were detected in the PIV results, which were not observed in the DNS data. Further investigations showed that such discrepancies in high-velocity regions may be avoided by using a larger window size for the final iteration. These analyses showed that the vectors replaced during the last iteration, even when they are not detected as outliers after the PIV process, must be marked as unreliable.

The computational performance of the software was evaluated on a workstation and the Niagara cluster. For both cases, the computation time initially decreased significantly as more processors were used. Meanwhile, the profiles almost remained flat once enough processors were utilized. The relative computation times of different subroutines of the software were measured during five runs and averaged to obtain the performance profile. The cross-correlation and deformation subroutines constituted the major portion of the total computation time. The relative computation times for validation, replacement, and smoothing subroutines, on the other hand, were significantly lower, indicating the efficiency of the algorithms.

The effect of several PIV parameters on the output velocity field for the Rankine vortex dataset was investigated. Using a double precision floating-point data did not affect the quality of the results. Using more PIV iteration or zero-padding the correlation generally tends to improve the accuracy. The errors associated with using different replacement methods were all of the same order of magnitude. Revalidation was shown to be effective in reducing the number of iterations needed for convergence of the resulting fields. Finally, the velocity field obtained with the best set of the discussed parameters showed good agreement with the PIVLab results. The discrepancies between the results of the two software were mostly confined to a few vectors near the center of the vortex.

Overall, the present study showed that the developed software is accurate and computationally efficient, and can be reliably used on the Microsoft Windows and Linux platforms to process large datasets. The software was tested on a standard computer workstation, a CPU cluster, and Google Colaboratory cloud computing service as part of this study. Hence, the software could be run on computer clusters or cloud computing services to avoid investing in hardware or needing affiliation with a university. The software offers great flexibility, enabling the users to adjust the settings according to their needs. Although the parameters were tuned to obtain more accurate results throughout this paper, it is also possible to adjust the parameters for fast PIV processing at the cost of accuracy.

Author Contributions: Conceptualization, A.S. and P.E.S.; methodology, A.S. and K.J.; software, A.S. and K.X.; validation, A.S.; formal analysis, A.S.; investigation, A.S.; resources, P.E.S.; data curation, A.S.; writing—original draft preparation, A.S.; writing—review and editing, K.X., K.J. and P.E.S.; visualization, A.S.; supervision, P.E.S.; project administration, P.E.S.; funding acquisition, P.E.S. All authors have read and agreed to the published version of the manuscript.

Funding: This research was funded by the Natural Sciences and Engineering Research Council of Canada (NSERC) grant number RGPIN-2022-03071 and the Digital Research Alliance of Canada (4752).

Data Availability Statement: The data presented in this study are openly available at (24 October 2023) https://drive.google.com/drive/folders/1IuzZlz7DjjKHptILpuzRAFRUioiAiqaX and https://drive.google.com/drive/folders/185O3NOtyl0rn5GTLUIYYEOVLuipZddjW.

Acknowledgments: The authors are grateful to Bertrand Lecordier for providing the PIV synthetic image generator, as well as his instructions on best practices. The authors would also like to acknowledge the Johns Hopkins Turbulence Database for providing access to their data repository. Computations were performed on the Niagara cluster at the SciNet High-Performance Computing consortium, which was supported by the Canada Foundation for Innovation under the auspices of Compute Canada, the Government of Ontario, the Ontario Research Fund-Research Excellence, and the University of Toronto.

Conflicts of Interest: The authors declare no conflict of interest.

Abbreviations

The following abbreviations are used in this manuscript:

PIV	Particle Image Velocimetry
CPU	Central Processing Unit
GPU	Graphical Processing Unit
WIDIM	Window Deformation Iterative Multigrid
PyPI	Python Package Index
DFT	Discrete Fourier Transform
FFTW	Fastest Fourier Transform in the West
LIL	LIst of Lists
CSR	Compressed Sparse Row
DNS	Direct Numerical Simulation
CDI	Central Difference Interrogation
STD	Standard Deviation
CI	Confidence Interval
MAE	Mean Absolute Error
RMSE	Root Mean Squared Error
NSERC	Natural Sciences and Engineering Research Council of Canada

Appendix A

The software is available for download on 24 October 2023 https://github.com/ali-sh-96/OpenPIV--Python-cpu via Github. The users may follow the instructions to install the necessary packages and download the software. The procedure for processing the dataset used in the present study, i.e., the synthetic images generated from the Johns Hopkins DNS data, is available as a working tutorial through Google Colaboratory cloud computing service. User feedback through the GitHub platform is welcomed.

Appendix B

As discussed in Section 3, the PIV process could be performed with a minimum window size of 16 pixels and 75 % instead of 8 pixels and 50 % to achieve the same vector spacing of 4 pixels. The corresponding mean and instantaneous flow fields are presented in Figures A1 and A2, which could be compared to Figures 8 and 11, respectively.

Comparing Figure A2 with Figure 11, it may be observed that the small flow structures near the walls are lost due to the larger minimum window size used to obtain the velocity field. The results in high-speed flow regions, on the other hand, are significantly improved as there are fewer discrepancies between the Q-criterion contours of the DNS data and PIV results in Figure A2. Note that the small isolated regions of negative Q values highlighted in Figure 11 are not present in Figure A2. Similarly, for the mean stream–wise velocity contours, the most evident improvement is in high-speed regions, such as the $U = 7$ iso-contour. Overall, depending on the region of interest, the users may need to adjust the smallest interrogation window size.

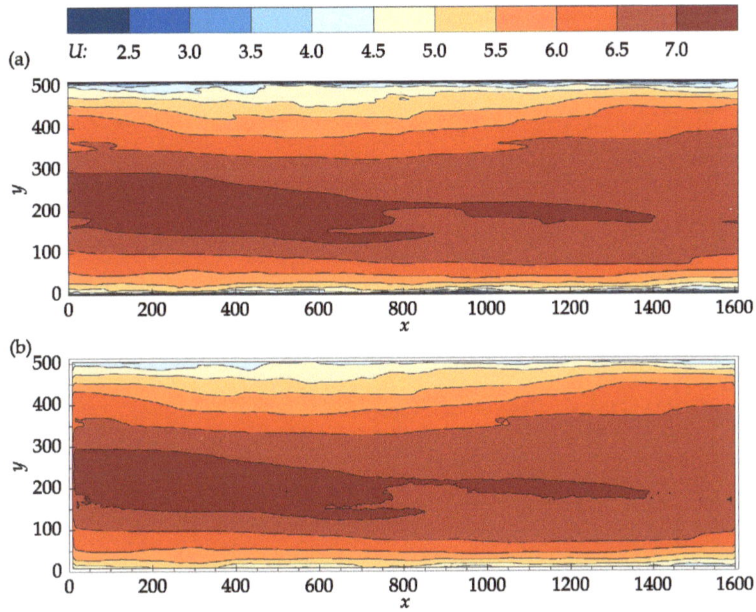

Figure A1. Contour plots of the mean stream–wise velocity: (**a**) John Hopkins DNS data; (**b**) results obtained with OpenPIV–Python using a window size of 16 pixels and 75% overlap for the final iteration.

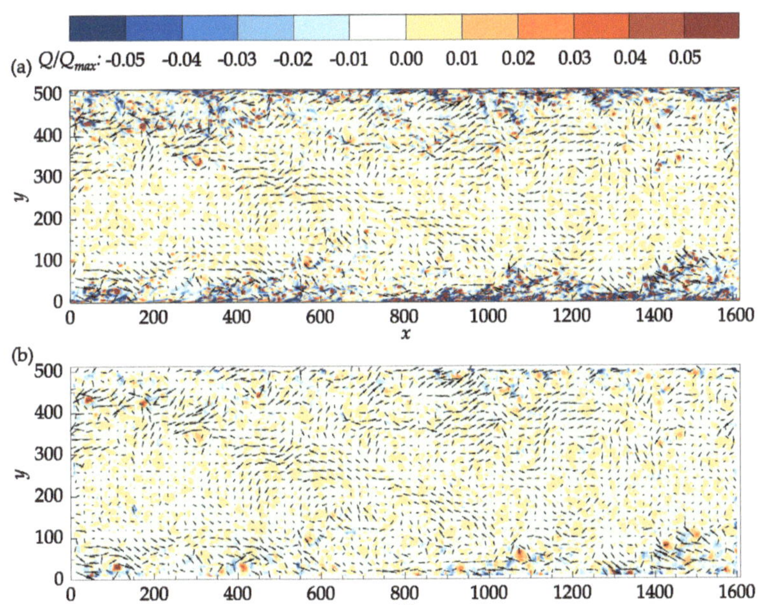

Figure A2. Vectors of the fluctuating velocities superimposed on the contour plots of the normalized *Q*-criterion: (**a**) John Hopkins DNS data; (**b**) results obtained with OpenPIV–Python using a window size of 16 pixels and 75% overlap for the final iteration.

References

1. Adrian, R.J. Twenty years of particle image velocimetry. *Exp. Fluids* **2005**, *39*, 159–169. [CrossRef]
2. Raffel, M.; Willert, C.E.; Scarano, F.; Kähler, C.J.; Wereley, S.T.; Kompenhans, J. *Particle Image Velocimetry: A Practical Guide*, 3rd ed.; Springer International: Cham, Switzerland, 2018 . [CrossRef]
3. Booth-Gauthier, E.A.; Alcoser, T.A.; Yang, G.; Dahl, K.N. Force-induced changes in subnuclear movement and rheology. *Biophys. J.* **2012**, *103*, 2423–2431. [CrossRef] [PubMed]
4. Sarno, L.; Carravetta, A.; Tai, Y.C.; Martino, R.; Papa, M.; Kuo, C.Y. Measuring the velocity fields of granular flows–Employment of a multi-pass two-dimensional particle image velocimetry (2D-PIV) approach. *Adv. Powder Technol.* **2018**, *29*, 3107–3123. [CrossRef]
5. Voorneveld, J.; Kruizinga, P.; Vos, H.J.; Gijsen, F.J.; Jebbink, E.G.; Van Der Steen, A.F.; De Jong, N.; Bosch, J.G. Native blood speckle vs ultrasound contrast agent for particle image velocimetry with ultrafast ultrasound-in vitro experiments. In Proceedings of the 2016 IEEE International Ultrasonics Symposium (IUS), Tours, France, 18–21 September 2016; pp. 1–4. [CrossRef]
6. Scarano, F.; Riethmuller, M.L. Iterative multigrid approach in PIV image processing with discrete window offset. *Exp. Fluids* **1999**, *26*, 513–523. [CrossRef]
7. Meunier, P.; Leweke, T. Analysis and treatment of errors due to high velocity gradients in particle image velocimetry. *Exp. Fluids* **2003**, *35*, 408–421. [CrossRef]
8. Dallas, C.; Wu, M.; Chou, V.; Liberzon, A.; Sullivan, P.E. Graphical Processing Unit-Accelerated Open-Source Particle Image Velocimetry Software for High Performance Computing Systems. *J. Fluids Eng.* **2019**, *141*, 111401. [CrossRef]
9. Ben-Gida, H.; Gurka, R.; Liberzon, A. OpenPIV-MATLAB—An open-source software for particle image velocimetry; test case: Birds' aerodynamics. *SoftwareX* **2020**, *12*, 100585. [CrossRef]
10. Thielicke, W.; Sonntag, R. Particle Image Velocimetry for MATLAB: Accuracy and enhanced algorithms in PIVlab. *J. Open Res. Softw.* **2021**, *9*, 12. [CrossRef]
11. PIVLab. Available online: https://www.mathworks.com/matlabcentral/fileexchange/27659-pivlab-particle-image-velocimetry-piv-tool-with-gui (accessed on 15 August 2023).
12. OpenPIV. Available online: http://www.openpiv.net/ (accessed on 15 August 2023).
13. Fluere. Available online: https://www.softpedia.com/get/Science-CAD/Fluere.shtml (accessed on 15 August 2023).
14. Fluidimage. Available online: https://pypi.org/project/fluidimage/ (accessed on 15 August 2023).
15. mpiv. Available online: https://www.mathworks.com/matlabcentral/fileexchange/2411-mpiv (accessed on 15 August 2023).
16. JPIV. Available online: https://eguvep.github.io/jpiv/ (accessed on 15 August 2023).
17. UVMAT. Available online: http://servforge.legi.grenoble-inp.fr/projects/soft-uvmat (accessed on 15 August 2023).
18. Yang, E.; Ekmekci, A.; Sullivan, P.E. Phase evolution of flow controlled by synthetic jets over NACA 0025 airfoil. *J. Vis.* **2022**, *25*, 751–765. [CrossRef]
19. Scarano, F. Iterative image deformation methods in PIV. *Meas. Sci. Technol.* **2001**, *13*, R1. [CrossRef]
20. Garcia, D. Robust smoothing of gridded data in one and higher dimensions with missing values. *Comput. Stat. Data Anal.* **2010**, *54*, 1167–1178. [CrossRef] [PubMed]
21. Gomersall, H. pyFFTW: Python Wrapper around FFTW. *Astrophys. Source Code Libr.* **2021**, ascl:2109.009. Available online: http://xxx.lanl.gov/abs/2109.009 (accessed on 15 August 2023).
22. Westerweel, J.; Scarano, F. Universal outlier detection for PIV data. *Exp. Fluids* **2005**, *39*, 1096–1100. [CrossRef]
23. Perlman, E.; Burns, R.; Li, Y.; Meneveau, C. Data Exploration of Turbulence Simulations Using a Database Cluster. In Proceedings of the 2007 ACM/IEEE Conference on Supercomputing, SC'07, New York, NY, USA, 10–16 November 2007. [CrossRef]
24. Saikrishnan, N.; Marusic, I.; Longmire, E.K. Assessment of dual plane PIV measurements in wall turbulence using DNS data. *Exp. Fluids* **2006**, *41*, 265–278. [CrossRef]

MDPI AG
Grosspeteranlage 5
4052 Basel
Switzerland
Tel.: +41 61 683 77 34

Fluids Editorial Office
E-mail: fluids@mdpi.com
www.mdpi.com/journal/fluids

www.ingramcontent.com/pod-product-compliance
Lightning Source LLC
Chambersburg PA
CBHW041410210926
51850CB00013B/3229
9783725835362